Mechatronics

Mechatronics

Electronics in products and processes

D.A. Bradley

Chair of Mechatronics Systems
University of Abertay, UK

D. Dawson

Engineering Department
Lancaster University, UK

N.C. Burd

National Semiconductor GmbH
Furstenfeldbruck, Germany

A.J. Loader

Digitron
Redditch, UK

First published in 1991 by Chapman & Hall
Reprinted in 2000 by Stanley Thornes (Publishers) Ltd

Reprinted in 2002 by:
Nelson Thornes Ltd
Delta Place
27 Bath Road
CHELTENHAM
GL53 7TH
United Kingdom

02 03 04 05 06 / 10 9 8 7 6 5 4 3 2

A catalogue record for this book is available from the British Library

ISBN 0 7487 5742 2

Page make-up by Thomson Press (India) Ltd

Printed and bound in Great Britain by Athenaeum Press

Contents

Part Two   Embedded Microprocessor Systems

Part Three   Motion Control

Part Four Systems and Design

Preface

The need for an integrated approach to the design of complex engineering systems involving electronic engineering, mechanical engineering and computing has become increasingly apparent in recent years and has led to the growth of the concept of mechatronics. However, it is a concept which is as yet not particularly well defined; a broad range of interpretations has been placed upon it. The following definition has been adopted within the EEC:

Mechatronics is the synergetic combination of precision mechanical engineering, electronic control and systems thinking in the design of products and processes.

From this definition it is clear that mechatronics is not itself a separate discipline within the overall spectrum of engineering but rather represents an integration across a number of different fields within engineering.

This text is therefore an attempt to set out the nature of mechatronics for a broad engineering audience. In order to achieve this objective the text aims to provide an indication of the range and scope of a mechatronic approach to the design of engineering systems and to identify the major areas of technology involved in such systems. It has its origins in the engineering degree course at Lancaster University and, specifically, in Professor Michael French's concept that engineering design should form a connecting theme throughout the whole of this course. As the course developed it became clear through links wth industry and through involvement in the Cambridge University based advanced course in design, manufacturing and management, for which Lancaster is the northern outpost, that there was an increasing need for engineering graduates who could function in an interdisciplinary environment. This led to the establishment in the late 1970s of a final year option course entitled 'the electromechanical interface' linking the two disciplines specifically in the area of drive technology. Then in 1985 an MEng course in mechatronics was established, the first such in the UK and from which the first graduates appeared in 1988.

The authors were all involved in the definition, setting up and teaching of the mechatronics MEng course and it was from this that the idea came about for a book bringing together some of the concepts and material from the course. That the book has eventually seen the light of day owes much to a number of people. Michael French, the Professor of Engineering Design at Lancaster University, has already been mentioned and we owe much to his ideas on the nature of engineering design. Tony Dorey, the Professor of Electronic Engineering at Lancaster University, has also provided much advice and encouragement throughout this period, while other members of staff of the Engineering Department have advised and commented on items within the text. Away from Lancaster, many thanks must go to Sid Dunn for reading the manuscript and for his invaluable comments. At Chapman and Hall, Dominic Recaldin displayed admirable perseverance. Indeed, this was particularly notable in view of the problems created by the fact that although all the authors were members of the same department when the book was started, two later left to take up other appointments in Germany and the Midlands respectively. Thanks must also go to those companies who supplied information either for inclusion in the text or as background. It must, however, be emphasized that the opinions and conclusions are throughout the authors' own and do not represent those of the companies concerned.

Finally, thanks must go to the members of the authors' families for putting up with strange hours and bursts of work as deadlines approached and for the support they provided throughout; this contribution was perhaps the most valuable of all.

D.A. Bradley
D. Dawson
N.C. Burd
A.J. Loader

Chapter 1

What is mechatronics?

The success of industries in manufacturing and selling goods in a world market increasingly depends upon an ability to integrate electronics and computing technologies into a wide range of primarily mechanical products and processes. The performance of many current products – cars, washing machines, robots or machine tools – and their manufacture depend on the capacity of industry to exploit developments in technology and to introduce them at the design stage into both products and manufacturing processes. The result is systems which are cheaper, simpler, more reliable and with a greater flexibility of operation than their predecessors. In this highly competitive situation, the old divisions between electronic and mechanical engineering are increasingly being replaced by the integrated and interdisciplinary approach to engineering design referred to as mechatronics.

In a highly competitive environment, only those new products and processes in which an effective combination of electronics and mechanical engineering has been achieved are likely to be successful. In general, the most likely cause of a failure to achieve this objective is an inhibition on the application of electronics. In most innovative products and processes the mechanical hardware is that which first seizes the imagination, but the best realization usually depends on a consideration of the necessary electronics, control engineering and computing from the earliest stages of the design process. The integration across traditional boundaries that this implies and requires lies at the heart of a mechatronic approach to engineering design and is the key to understanding the developments that are taking place.

Engineering design and product development are, as illustrated by Figs 1.1 and 1.2, complex processes involving an interaction between many skills and disciplines. Mechatronics is not a distinctly defined, and hence separate, engineering discipline but is an integrating theme within the design process. In achieving this integration it combines, as shown by Fig. 1.3, its core disciplines – electronic engineering, computing and mechanical engineering – with links into areas as diverse as manufacturing technology, management and working practices.

The foundations of a mechatronic approach to engineering design are

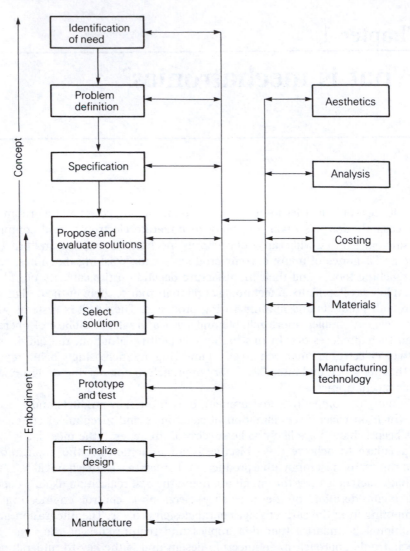

Figure 1.1   The engineering design process.

considered to lie in information and control. Indeed, it may be objected in some quarters that mechatronics is 'only control engineering' in another guise. Such objections would, however, fail to recognize the direct impact on the approach to the design of a mechanical system of the introduction and incorporation of electronics and computing technologies. Indeed, a feature of a mechatronic approach to engineering design is that the resulting mechanical systems are often simpler, involving fewer components and moving parts than their wholly mechanical counterparts. This simplification is achieved by the transfer of

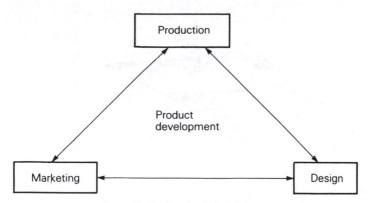

Figure 1.2 Product development.

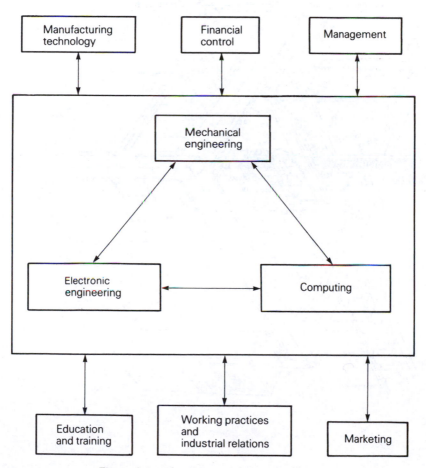

Figure 1.3 The elements of mechatronics.

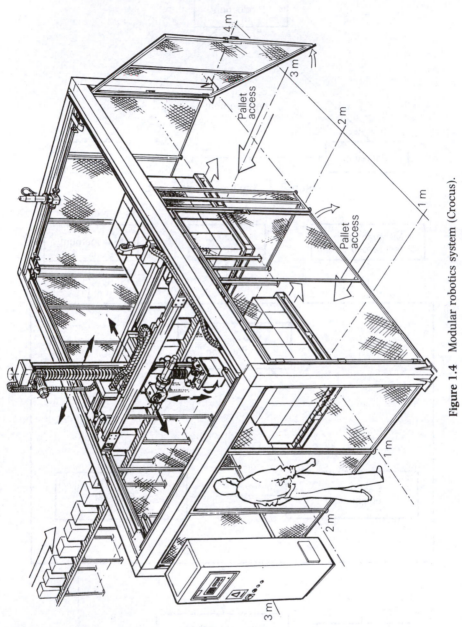

Figure 1.4 Modular robotics system (Crocus).

complex functions such as accurate positioning from the mechanical system to the electronics. An example of this is seen in the modular robotics system of Fig. 1.4. Here, encoders for position measurement together with electronic motor control have enabled the use of simple chain drives and induction motors instead of the expensive leadscrews and servomotors that would previously have been required, and which perhaps would have been too highly specified in terms of accuracy and resolution for the intended applications.

To be successful, a mechatronic approach needs to be established from the very earliest stages of the conceptual design process, where options can be kept open before the form of embodiment is determined. In this way the design engineer, and especially the mechanical design engineer, can avoid going too soon down familiar and perhaps less productive paths.

The first examples of mechatronic systems, albeit in a somewhat cumbersome and less capable and adaptable form, were to be found in the early computer numerically controlled (CNC) machine tools and in large scale automated processes such as chemical plant or rolling mills. However, in the majority of such systems the basic mechanical design was largely unaffected by the addition of electronically based control systems. Indeed, such control systems were regarded by many design engineers as a somewhat suspicious and mysterious bolt-on adjunct. Such inhibitions would now be a serious constraint on the ability of mechanical engineers in particular to exploit the opportunities made available through the incorporation of electronics.

1.1   Mechatronics in manufacturing

In the manufacturing industries there is a demand for production systems which are capable of responding rapidly to changing market conditions, accommodating a range of product types with short production runs involving relatively small numbers of items. Neither manual manufacturing processes nor mass production lines can meet these requirements. The former, though highly adaptable, suffer from low levels of productivity. The assembly and transfer lines associated with the latter lack flexibility, with changeovers involving significant time costs.

Within a wide range of manufacturing systems and processes, a mechatronic design approach has had as its primary benefit the ease with which the process can be reconfigured while, at the same time, offering enhanced product quality and consistency. The effect of a failure to adapt can be seen in the inability until quite recently of a range of UK capital equipment manufacturers, particularly of machine tools, injection moulding equipment, shoe making machinery, textile machinery and confectionery equipment, to supply systems of comparable performance to those of overseas competitors at a comparable price. Even where extensive government support was provided, a trend towards increasing import penetration was recorded.

A major reason for this failure can be identified in terms of the machine or system capability. UK produced equipment tended to be robust but lacking in the range of sensors and advanced state-of-the-art control systems offered by their competitors. The resulting equipment was slower in operation, less adaptable to changes in product specification (with long set-up times) and, most importantly, less able to deliver a product of consistent quality.

Where full attention has been given to market trends, the adoption of an integrated mechatronic approach to design has led to a revival in areas such as high speed textile equipment, metrology and measurement systems, and special purpose equipment such as that required for the automatic in-wafer testing of integrated circuits. In most cases the revival or new growth is brought about by the enhancement of process capability achieved by the integration of electronics, often in the form of an embedded microprocessor, with the basic mechanical system.

This demand for increased flexibility in the manufacturing process has led to the development of the concept of flexible manufacturing systems (FMSs) in which a number of elements such as computer numerically controlled machine tools, robots and automatically guided vehicles (AGVs) are linked together for the manufacture of a group of products. Communication between the individual elements of the system is achieved by means of local area networks (LANs). Such interconnected systems operate in many implementations as a stand-alone grouping or island of automation within the production environment.

As levels of automation increase, so does the need for communications. This requirement has led to the introduction of the manufacturing automation protocol (MAP), the technical office protocol (TOP) and the open systems interconnection (OSI) standard to provide a communications structure for the passage of information throughout the whole of the manufacturing environment, as illustrated by Fig. 1.5. These concerns are also having a major effect

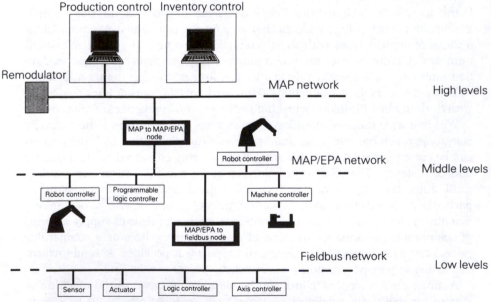

Figure 1.5 A hierarchical communications system for manufacturing control.

Table 1.1 Comparison between a traditional and a mechatronic approach to process control and manufacturing

Traditional approach	*Mechatronic approach*
Chemical process	
Centralized computer control	Distributed processing power with localized decision making capacity
All instruments, transducers wired back to central computer and control room	Reduced cable runs; machine cycles stored and executed via local control loops
Sequence control predominates	More parts of process are capable of individual control
Individual parts of the process relatively inflexible	Optimization facilitated by stages and as a whole
Control systems involve the large scale use of electropneumatic interfaces because of hazard concern	Reduced size of components via VLSI results in lower power levels; may be intrinsically safe, therefore electronic controls become acceptable
'Hard' process, e.g. bottling plant	
Sequence control predominates	Some proportional control where variable speed drives are used
Whole process controlled by relay logic	Microprocessor based programmable logic controllers (PLCs) providing distributed control incorporating a degree of intelligence
Highly inflexible as to product size and type; an 'army with spanners' needed to change conveyor guides etc.	Stepless actuators incorporating position feedback widely employed; process flexibility enhanced, downtime reduced
Two-position air cylinders or simple electropneumatics widely used, control by limit switches	Robust electrical drives with variable speed and position feedback; proportional electropneumatics in limited applications
Inspection/QA stages towards or at end of process	In-process automatic inspection
Manufacturing unit e.g. machine tool	
Dedicated CNC controller, perhaps including automatic tool changes	Networked CNC controllers with remote or central control of cycles
Essentially stand-alone mode	Linked flexible manufacturing system (FMS) with integrated parts handling and transfer, the latter based on the use of AGVs
Largely manual inspection and quality assurance procedures	In-process gauging, automated inspection, data collection and reporting
Predominantly manual-handling processes for loading and unloading	Use of general purpose robots for handling; automatic tool changing
Plant maintenance on a breakdown or preventive basis	Plant maintenance on a predictive basis, based on in-line diagnostics and condition monitoring

on the design integration of mobile systems such as vehicles. For example, where a vehicle has modular electronic controllers to handle the requirements of different systems such as a semi-active suspension and an anti-lock braking system, it is important that the modules should be able to interface to a common data bus. This is especially the case where the output of an individual sensor, say of wheel rotational speed, is used for different purposes. The more common use of such sophisticated techniques in vehicles has given impetus to the development of controlled area network (CAN) techniques – and a group of emerging standards to which all subsystem suppliers will be under pressure to conform if they are to remain competitive.

The process industries have seen the working together of mechanical, electrical and electronic engineering for many years, and the 'wet' process industry in particular has exhibited a significant degree of integration in its approach to system design, as exemplified by the high level of control engineering expertise to be found in companies such as ICI and Shell. Indeed, a chemical plant may be regarded as a mechatronic system in its own right, and in which the introduction of microprocessors and associated communications is having a significant impact through the opportunities offered for distributed control incorporating local decision making capacity. The introduction of decentralized systems has also brought about significant changes in procedures in areas of plant optimization, diagnostic based maintenance and data handling.

In both the manufacturing and process industries the development of low cost microprocessor based programmable controllers has enabled their introduction as the basic control element for many plants. In addition, there has been a progressive development in sensor technology, often employing very large scale integration (VLSI) technologies to provide 'smart' features such as self-calibration, monitoring and test.

Some indication of the nature of the changes in capability and emphasis that have occurred as a result of the adoption of a mechatronics approach is illustrated, albeit in a very limited way, in Table 1.1, which shows some of the features which distinguish the trend.

1.2   Mechatronics in products

Within products the diversity and opportunity offered by a mechatronic approach to engineering design is to date largely unrealized. End user products are substantial revenue earners and it is possible here to distinguish between existing products offering enhanced capabilities and completely new product areas which would not have existed without a mechatronic design approach having been adopted from the outset.

In the first category the following are illustrative from many examples:

Automotive engines and transmissions   Engine and driveline management systems leading to reduced emissions, improved fuel economy, protection against driver misuse by, for example, prohibiting excessive fuel flow at low speeds, anti-lock braking systems and selectable gear characteristics.

Cameras Automatic adjustment of focus, aperture and shutter speed to suit the prevailing conditions, leading to a virtual elimination of technical skills – except those of picture composition!

Power tools Modern power tools such as drills offer a range of features including speed and torque control, reversing drives and controlled acceleration.

Examples in the second category include the following:

Modular robotics Conventional industrial robots are often limited in their operation by their geometry. By providing a range of structural components and actuators together with a central controller a modular robotics system has been made available, allowing users to assemble robot structures directly suited to their needs.

Autopilots for small boats   These are able to accept an input from a windvane, fluxgate compass or radio beacon and drive a tiller or steering wheel.

Video and compact disc players Video and compact disc systems involve complex laser tracking systems to read the digitally encoded signal carried by the disc. This control is achieved by means of a microprocessor based system which also provides features such as multiple track selection, scanning and preview.

A common factor in consumer mechatronics as exemplified by the above is the continuous improvement in capability achieved against a constant or reducing real cost to the end user. The capability of a mechatronic system, based as it often is on inexpensive components or modules, also provides a means to execute bespoke solutions to special problems.

1.3   Mechatronics and engineering design

In engineering design, a mechatronics approach represents and requires the integration of a wide range of material and information aimed at providing systems which are more flexible and of higher performance than their predecessors, and which incorporate a wider range of features. Thus, for full benefit and effect, mechatronics must be a feature of both the conceptual and embodiment phases of the design process.

In manufacturing, users are demanding a much higher degree of control of both the overall process and its components. This requires a knowledge both of the capabilities of these components and of the means by which they are integrated within the complete system.

In products, the concern is primarily with the provision of enhanced performance and increased ease of use and the user is not generally concerned with the means by which these objectives are achieved. This means that in most products the microcontroller responsible for the enhanced performance is transparent to the user, whose interface effectively allows the selection of an option from a range of predetermined programs.

At present, and though many organizations are involved in the development of products and systems which are inherently mechatronic in concept and realization, the responsibility for managing the technical integration is often left to a systems department. Such departments often have no responsibility for the design process; their role is simply to ensure that any changes made by one functional design team are passed on to other related groups, a largely clerical activity. In a true mechatronic design environment the coordinating role is central to the design process and would contain much of the fundamental design effort, in both the conceptual and embodiment phases. This fact is increasingly being recognized, particularly in Japan where many organizations have a mechatronics department whose role is central to the design and development of the company's product range.

It is therefore fundamental to a mechatronics approach to engineering design that the integration of the electronics and computing technologies with the mechanical system is considered throughout the design process. For an established product or process a mechatronic approach may well be used to provide an enhancement of performance for the same basic design. In such an application it is likely that this represents only one stage in an evolutionary process and that subsequent developments will take account of the inclusion of electronics within the basic system.

1.3.1   A modular approach to mechatronics and engineering design

A mechatronic approach to engineering design is concerned with the provision of a structure within which the integration of the various technologies can be established and evaluated. In order to achieve this objective, a top down and information based strategy is suggested in which the overall system is broken down into a series of blocks or modules as in Fig. 1.6. The role of each of these modules is then as follows:

Environment module   The environment module is concerned with those external parameters such as temperature range and load factors which will influence the operation of the complete system. Within the overall design they constitute a series of parameter boundaries within which the system must exist and function. The environment modules must therefore encompass features such as standards and codes of practice.

Assembly module   The assembly module represents the physical realization of the mechanical and structural elements of the system. It is primarily concerned with parameters such as the properties of materials, structural

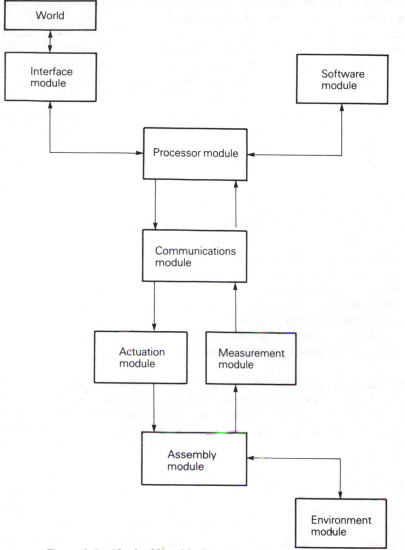

Figure 1.6   The building blocks of a mechatronic system.

behaviour, form and context. Inputs to the assembly module consist of the motions provided by the actuation modules together with the conditions defined by the environment module. Output from the assembly module is provided by the measurement module. As the assembly module is concerned with the appearance of the system it must also contain an aesthetic element.

Measurement module   This is concerned with the gathering of information about system status and behaviour. Input parameters are physical properties of the assembly module while output parameters are concerned with the nature of the information to be transmitted.

Communications module This is concerned with the transmission of information between modules within the system. Input and output conditions relate to the nature of the information to be transmitted, the distance over which it is to be transmitted and the operating environment.

Processor module This is concerned with the processing of the information provided by the measurement and interface modules. Input parameters include measured parameters and demand settings together with system parameters such as speed of operation. The outputs from the processor modules determine the operation of the actuation modules and provide information to the interface modules.

Software module This contains the operating instructions and defining algorithms for the system and controls the operation of the processor module. The nature and form of the software module is linked to that of the associated processor module.

Actuation module This represents the 'muscle' required in the system to change system conditions. Input conditions are set by the output of the processor module and outputs are defined by the type of motion required.

Interface module This is concerned with the transfer of information between levels within the system and, at the highest level, with providing the necessary man–machine interface for the transfer of user information. Inputs and outputs are concerned with the nature of the information transfer involved.

At each level in the design process the nature of the functions to be provided by an individual module can be established, and so can the nature but not necessarily the form of the informationn to be transferred. Modules defined at a higher level in the design process will themselves form a mechatronics system in their own right, in which case a further set of modules may be defined describing the next level down in the design.

As an example of this approach, consider a requirement for a mobile robot to be used in firefighting and rescue. Such a robot would be required to negotiate regions where debris was lying, to climb over such debris in order to deliver a hose to the seat of the fire, and to remove casualties from danger. The robot would also be required to communicate with and receive instructions from a control centre and to deploy a range of sensors and systems, many of which may not be capable of being specified at an early stage of the design process.

In the first stage of the design process the modules are used to define, in conceptual terms, the functions of the robot and its constituent elements. At this level, a decision may be made that a legged robot is required to provide the obstacle crossing capability defined in the environment module. Thus, in the high level representation, an actuation element may be simply defined within the actuation module as *leg*. This actuation element may now be considered as a mechatronics system in its own right, giving rise to a second set of modules which describe the functioning of the leg in more detail. Information about the operation of the leg, for example a movement instruction, is passed from the high level communications system to the leg via the lower level interface module. The actual functioning of the leg to meet the required demand is then determined by the leg's own processor module.

The design process will proceed in a similar manner through successive levels, defining further sets of modules at each stage and eventually providing detailed designs for the individual components making up the leg. The adoption of this type of approach means that the detail design of each component of the complete system can be undertaken in the knowledge of its relationship with all other components within the system. The problem of managing such a design process is a complex one and has not as yet been fully addressed.

1.4   The engineer and mechatronics

As has already been seen, within engineering design mechatronics is concerned with both the conceptual and embodiment phases of the design process. This requires that throughout the design process the individuals involved must have a sufficient knowledge of what is, or is not, available in a particular area of technology in order to assess any possible contribution to the final design and to maintain control of the project as it progresses. This latter aspect is of particular importance in respect of the inevitable changes and modifications to the initial concept and specifications that will occur as the design process proceeds to the final, detailed stages.

The skills and capabilities needed by an engineer to operate with confidence in a mechatronic environment can best be illustrated with reference to a specific system on a prototype product, namely an adaptive or smart suspension for a medium weight truck.

This suspension system is based upon air springs, one for each of the six wheels carried on three axles, with sensors to measure the current vertical position, velocity and acceleration of each wheel relative to the body of the vehicle. The data from the sensors is transmitted to the on-board microprocessor based controller at 20 ms intervals, where it is used as the input to control algorithms based on a mathematical model of the suspension system. Fast acting cross-flow valves, operating under the control of the on-board microprocessor, are then used to modulate the pressures of the individual air springs; electrically switched, dual rate dampers permit adjustment of the vehicle's ride behaviour. The resulting system is shown in Fig. 1.7.

Three modes of operation are possible:

Static   Static levelling is provided to enable the vehicle to adjust to loading dock height and to ensure a horizontal platform for live pallets.

Quasi-static   Manoeuvres such as cornering and lane changing impose a lateral acceleration which would normally cause the vehicle to roll. In the design adopted, the onset of roll is anticipated by a sensor on the steering wheel. The controller then uses this information in combination with the

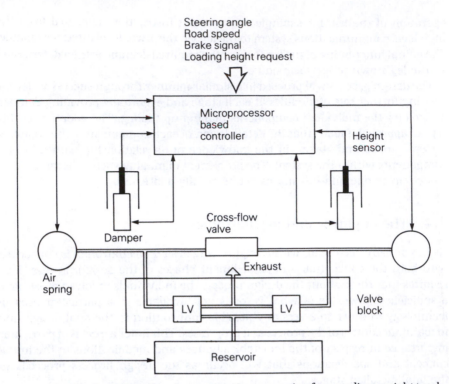

Steering angle
Road speed
Brake signal
Loading height request

Figure 1.7   Configuration of a smart semi-active suspension for a medium weight truck.

vehicle speed data to determine the operation of electropneumatic valves to transfer air to the outer air springs, simultaneously switching the dual rate dampers to the hard mode.

Dynamic   The response of the vehicle suspension to short period events such as those caused by crossing rough ground is monitored at each wheel. The damper rates are than adjusted to give the desired ride quality.

The realization of the system from concept to hardware and software, primarily by a single individual, requires the exercise or development of competence in the following areas of skill and knowledge:

1. The mathematical modelling of mass, spring and damper systems with multiple degrees of freedom; system representation and programming using numerical integration techniques;
2. The selection and application of sensors;
3. Instrumentation generally, including the use of portable data collection and test equipment;
4. The design and application of electrohydraulic and electropneumatic control systems;

5. The use of microprocessor based development systems;
6. The establishing of control algorithms for system development;
7. The more general mechanical design of test equipment and pre-production items;
8. A general knowledge of analogue circuit design and electrical installations as appropriate to vehicles, sufficient to specify and in some cases construct signal conditioning and other circuits.

Taken altogether, this is a complex and challenging problem for an engineer!

1.5   Mechatronics and technology

Mechatronics is concerned with the bringing together and integration of certain key areas of technology, particularly:

sensors and instrumentation systems
embedded microprocessor systems
drives and actuators
engineering design.

In the case of sensors, instrumentation, drives and actuators the incorporation of local dedicated processing power within a device, enabling it to act independently of the main controller, provides increased system flexibility. Such smart sensors and intelligent actuators incorporating embedded microprocessors as part of a distributed system play a significant role in the design and development of a large scale mechatronic system.

The employment of a distributed system of embedded microprocessors means that communications and software engineering have important roles to play in any mechatronic design process. In the latter area, the adoption of a top down modular approach to the design of complex mechatronic systems means that techniques such as object oriented programming can readily be introduced for software development as the information transfer is defined by the module structure.

This text is therefore structured around a discussion of these four major elements in order to indicate the form and nature of the questions and problems involved in developing a mechatronic approach to the design process. The discussion is intended not to provide a detailed description of all the possible options that exist in an individual area, but rather to identify the technologies and their mode of deployment in relation to the functioning of a mechatronic system.

Part One

Sensors and Transducers

Sensors and Transducers

Chapter 2

Measurement systems

The role of measurement is to provide information on system status which, in a mechatronic system, is used to control the operation of the system. This function is performed by the measurement modules of Fig. 1.6, which incorporate the necessary sensors and transducers together with any local signal processing.

Following the production of the high level functional specification for an individual measurement module, the designer is faced with the problem of choosing the appropriate measurement technology from a wide and growing range of such technology. For this reason, in the discussion of sensors, transducers and measurement within mechatronics, the concern is with the identification of the range and scope of measurement technologies available, of the questions that must be asked and of the decisions that must made. A detailed consideration of the full range of measurement technologies is, however, outside the scope of this presentation.

Measurement technology is not a new science but can be traced back to origins in trade, where the need to quantify, on a basis which was both repeatable and representative, the nature of goods on offer led to the introduction of standards for measures such as volume and weight. With the growth in science and technology as exemplified by the Renaissance, more and specific measuring instruments evolved to support the growing interest in and investigation of the physical world.

With the advent of the industrial revolution, instrumentation and measurement science began to be applied to manufacturing. The early instruments were analogue in nature and provided information about the basic physical parameters concerned with the operation of the process. Feedback was achieved by the operator who read the instruments and then made the necessary adjustments to maintain the process within the required bounds.

The next step forward came with the introduction of the steam engine and automatic feedback control mechanisms such as the Watt governor for speed control. Following this lead, instrumentation gradually became more integrated with the process, providing feedback information to a controller which when regulated the system behaviour with a minimum of manual intervention.

By the late 1960s the advantages of discrete control had been recognized and production processes were often controlled by relay based systems. Such relay controllers, though complex in configuration and layout, were relatively unsophisticated and incapable of handling complex information. The performance requirements of their associated sensors and transducers was correspondingly restricted to the supply of simple data at a level capable of being utilized by the controllers.

These conditions were largely maintained with the introduction of computer monitored, as opposed to controlled, systems in which the computer was used primarily in a supervisory role as a central machine receiving information sequentially from a number of radially connected outstations. With the advent of minicomputers such as the DEC PDP series, computer controlled systems increased in both number and scope, though they were limited in application by both processing power and cost. More recently, sophisticated microprocessor based programmable logic controllers together with distributed and embedded microprocessor systems have resulted in the complex interconnected designs that are now found in both products and manufacturing processes.

The growth in microelectronics and computing technologies from the mid 1970s that made available low cost processing power has resulted in a parallel growth in the demand for information about system conditions. In many cases this increased demand could not be met by existing sensors and transducers and resulted in the development of a range of advanced sensors and transducers, often incorporating local processing power, either integral with the sensing element on the same silicon slice or in the form of a single chip microprocessor.

2.1   Sensors, transducers and measurement

In any measurement system, sensors and transducers are used to provide information about system conditions. Unfortunately, the usage of the terms 'sensor' and 'transducer' is complicated by the variety of different meanings adopted internationally, including in some instances an apparent interchangeability. For the purpose of this book the following definitions are adopted:

Sensor   That part of the measurement system that responds to the particular physical parameter to be measured.

Transducer   That component of the system that transfers information in the form of energy from one part of the system to another, including in some cases changing the form of energy containing the information.

In addition to a sensor/transducer, a measurement may include a pre-processing stage and a post-processing stage, as shown in Fig. 2.1. The pre-processing stage serves to characterize the information in the incoming signal before presentation to the sensing element. This then detects and responds to the physical stimulus at its input and provides the input to the transducer. The output signal from

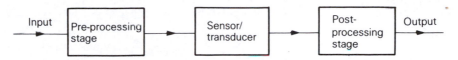

Figure 2.1   Pre- and post-processing of measurement data.

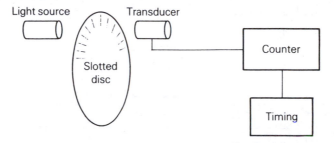

Figure 2.2   Speed measurement using a slotted disc with optical sensors and transducers.

the transducer is then operated on by the post-processing stage to produce the final output. The relationship between these stages is illustrated by the optical encoder used for speed measurement shown in Fig. 2.2. Here, the light from the source is pre-processed by the slotted disc to produce a series of pulses. These light pulses are received by the transducer which converts the signal into a series of electrical pulses. The post-processing stage then counts the number of pulses occurring in a defined time interval to determine the speed of rotation of the shaft carrying the encoder.

As has been stated, the physical basis for the transmission of information by a transducer is that of energy transfer. This concept has been developed by Middelhoek (see Bibliography) to identify six signal energy domains for the transfer of information. These domains are as follows:

Radiant   This covers the full spectrum of electromagnetic radiation. Major parameters are frequency, phase, intensity and polarization.

Mechanical   This covers parameters such as distance, velocity, size and force.

Thermal   This covers temperature effects in materials, including parameters such as thermal capacity, latent heat and phase change properties.

Electrical   This covers electrical parameters such as current, voltage, resistance and capacitance.

Magnetic   This covers magnetic field parameters such as field strength and flux density.

Chemical   This covers the internal structure and behaviour of matter and includes parameters such as concentrations of material, crystal structure and aggregation state.

Of these, the electrical output domain is the most significant in a mechatronics

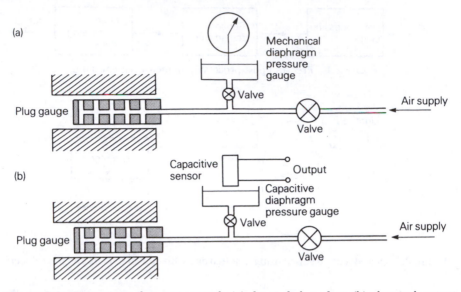

Figure 2.3 Pneumatic plug gauges with (a) direct dial readout (b) electrical output using a capacitive transducer.

context because of its information handling capability, and transducer outputs in other domains may well be further processed to transfer the information into the electrical domain. Figure 2.3 illustrates this condition in relation to a pneumatic plug gauge used for measuring the internal dimensions of a cylinder. In the original system of Fig. 2.3a, the back pressure developed is read by a diaphragm pressure gauge in which the displacement of the diaphragm operates a mechanical linkage to move the pointer on the dial. By replacing the mechanical linkage with a capacitive displacement transducer as in Fig. 2.3b, an electrical output signal is provided which can then be transmitted onwards via a communications link.

In terms of their utilization of energy, transducers may be considered as *direct* or *passive*, requiring no energy source other than that provided by the input signal in order to function, and *indirect* or *active*, when an additional energy source must be provided. Examples of direct transducers are photoelectric devices and thermocouples, where the input energy is in each case directly converted to electrical energy. Indirect transducers include strain gauges and Hall effect devices, in which the incident energy is used to modulate an electrical signal. The additional, or modulating, energy input required by an indirect transducer must itself be derived from one of the six energy groups already referred to. This has resulted in the three-dimensional representation of Fig. 2.4, on which all transducers, whether direct or indirect, can be located by reference to their input energy domain, their output energy domain and any modulating energy source.

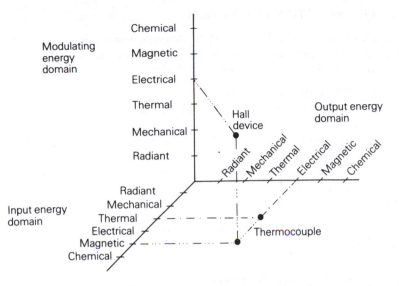

Figure 2.4   Three-dimensional representation of sensor and transducer operating energy domains.

2.2 Classification

It is possible to further classify transducers in a variety of ways, for example in terms of their performance, their function or their output signal. Each of these, and any other, classification can be used, as illustrated by Fig. 2.5, to identify transducers for a particular application.

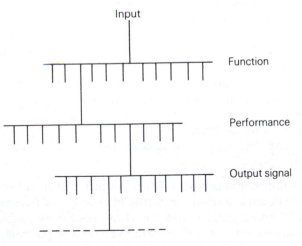

Figure 2.5   Transducer selection chain.

2.2.1   CLASSIFICATION BY FUNCTION

In a mechatronic system the majority of measurements will be concerned with providing information about parameters in the mechanical domain. Typical measurands within this domain are as follows:

Displacement   Linear and angular.
Velocity   Linear and angular; flow rate; (speed).
Acceleration   Vibration; (shock).
Dimensional   Position; size; area; thickness; volume; roughness; strain.
Mass   Weight; load; density.
Force   Absolute, relative, static, dynamic and differential pressures; torque; power; (stress).
Other   Hardness; viscosity.

Similar functional tables can be drawn up for each of the other domains identified earlier.

2.2.2   CLASSIFICATION BY PERFORMANCE

Performance can be assessed in terms of parameters such as accuracy, repeatability, linearity, sensitivity and range. The selection of an appropriate device then requires that the available devices must be examined against each of the relevant performance parameters in terms of the application under consideration to produce a subset of suitable devices.

2.2.3   CLASSIFICATION BY OUTPUT

Output signals fall into one of a number of groupings as follows:

Analogue output   Provides a continuous output signal, some property of which is directly related to the value of the measurand.
Digital output   Generates a digital representation of the measurand in either serial or parallel form. Information may be produced either at regular time intervals or on demand.
Frequency output   A signal is produced, the frequency of which is a function of the measurand. Output may be a continuous or a pulsed waveform which can be readily converted to a digital format by the use of counters and timers.
Coded output   Various forms of coded signal can be produced including amplitude and frequency modulation, pulse width and pulse position modulation.

Provided overall performance parameters are met, the choice of transducer is not necessarily dependent upon the nature of its output, as the use of analogue to digital (A/D) and digital to analogue (D/A) converters enables a signal in either the analogue or the digital domain to be used to generate an equivalent signal in the other domain.

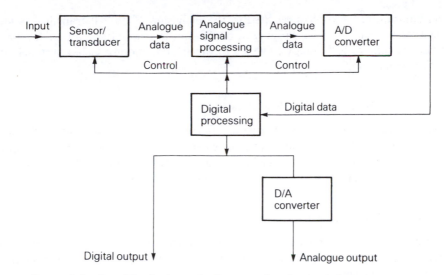

Figure 2.6   Simplifiyed schematic diagram of an integrated smart sensor.

2.3   Developments in transducer technology

Developments in transducer technology include the introduction of solid state silicon based transducers, optical systems incorporating fibre optics, piezoelectric devices and ultrasonic transducers, together with enhanced signal processing techniques. In particular, the introduction of local signal processing, often on the same chip as the sensor and transducer, has led to the development of the concept of the smart sensor, shown in schematic form in Fig. 2.6.

2.3.1   SOLID STATE TRANSDUCERS

The development of thin and thick film technologies has resulted in the production of a range of largely silicon based transducers for a wide selection of physical properties. In addition, micro-machining has enabled the creation of complex mechanical structures at the chip level, particularly the beams and diaphragms required for strain gauges, accelerometers and pressure gauges. However, even more complex structures such as gears, levers and motors have been shown to be capable of production on chip.

The continued development of solid state transducers based on silicon and other semiconductor materials is likely to have advantages in terms of reliability, signal conditioning, reduced signal-to-noise ratios and cost.

2.3.2   OPTICAL TRANSDUCERS

Fundamental to optical sensing technology are developments in lasers, photodetectors, fibre optics, optical materials and signal processing, the combination of which has resulted in the provision of optical transducers for measurement applications ranging from distance to strain. In addition, developments in camera technology have led to the introduction of vision systems for measurement purposes.

2.3.3   PIEZOELECTRIC TRANSDUCERS

Piezoelectric devices have been used for many years as the basis for miniature accelerometers and ultrasonic ranging systems. The development of surface acoustic wave (SAW) technology has led to the introduction of transducers for applications such as gas detection and pressure sensing

2.3.4   ULTRASONIC TRANSDUCERS

Ultrasound is increasingly being used as a means of providing non-invasive measurement in environments ranging from process control to medical imaging. Such non-invasive measurements can:

1. Reduce the level of hazard when working with explosive, radioactive, poisonous, corrosive or flammable materials;
2. Avoid the contamination of materials such as drugs or foodstuffs;
3. Simplify maintenance procedures by enabling direct access to the sensors and transducers.

Measurements using ultrasound rely on the Doppler effect or the reflection, absorption or scattering of a pulse of ultrasound to generate the required information, often after significant signal processing.

2.4   Signal processing and information management

It is in the area of signal processing that computing technology has had the greatest impact upon the development of sensors and transducers. The availability of cheap local processing power has enabled compensation to be introduced for effects, such as a non-linear temperature coefficient, which would otherwise have rendered some technologies unusable, and has allowed for the conversion of the measured information into a form in which it can be transmitted over a communication link for further processing. Local processing power has also enabled the introduction of self-test and self-checking routines and the devolution of a decision making capability to the sensor/transducer.

At a higher level, the use of sophisticated data and signal processing techniques has enabled the development of the concepts of sensor fusion and data fusion in which information from a number of sources is combined to build up a picture of a complex system. Such an approach, when integrated with expert systems and artificial intelligence, offers increasing scope for development.

Figure 2.7   Airbus A320 cockpit (BAe).

Information management and the presentation of data based on advanced signal and information processing techniques have an important role in the construction of a measurement system by ensuring that at any instant all necessary information is made available as required and that appropriate man–machine interfaces are available to present the data. An example of this is seen in the development of cockpit displays for civil and military aircraft, of which the A320 cockpit of Fig. 2.7 is an example. Here, the large number of individual gauges that were characteristic of cockpits have largely been replaced by visual display units on which the pilot receives the current flight and systems data. Should more detailed information be needed, either on demand or automatically in the case of a fault, these same displays are used for the presentation of the required data.

2.5   The design of a measurement system

Mechatronics implies an information based approach to engineering design, with measurement systems forming one element within the design concept. The primary function of the high level design process is therefore to establish the information required. Once this has been achieved, the design of the individual measurement modules can proceed to identify appropriate sensors, transducers and signal processing requirements to provide the required information. In this way, areas of deficiency will be identified at an early stage in the design of the measurement module, forcing either a reconsideration within the high level design process of the information requirements or the development of new and improved means of measurement.

The production of the specification for and the high level design of a measurement system requires the designer to consider a wide range of factors covering:

1. The information required and the identification of the physical parameters of the system that must be measured in order to provide this information;
2. The nature, quality and performance of the measurement in terms of parameters such as linearity, accuracy and resolution;
3. A determination of the most *inaccurate* measurement that would be acceptable given the function of the measurement;
4. The effect on system performance of any drift in the measurement circuit, particularly on variation of the zero;
5. The environmental conditions under which the sensors and transducers are expected to operate;
6. The cost targets to be met;
7. The nature and form of the information transfer required;
8. The reliability of the system;
9. The form of the interface to adjacent modules in the system.

A major concern in the design of a measurement system lies in the provision of either too much or too little information, both of which are conditions which arise from a misunderstanding of the role of measurement.

The first condition, that of providing too much information, arises from the continuously increasing ability of measurement systems to rapidly provide high quality information over a wide range of functions and environments. There is, therefore, a tendency towards the gathering of all available information, at high levels of accuracy and resolution. As a result, it is possible that the relevant information may be lost within a large mass of irrelevant data, putting a heavy demand on the data processing stages of the system.

The second condition, that of making available too little information, generally arises as a result of insufficient consideration being given to the function and purpose of measurement within the context of the overall system. The failure to make available the necessary information imposes an obvious limitation on overall performance.

Once the high level specification has been produced, the detailed design of the measurement system can then proceed. This detailed design process will again involve a modular approach requiring:

1. The identification of means by which the required parameters can be measured taking into account the performance requirements;
2. The selection of appropriate sensors and transducers;
3. The specification and provision of any local processing power that may be required;

4. The specification and production of any special purpose electrical or mechanical hardware associated with the operation of the sensors and transducers;

5. The specification of associated software and the control of the production of such software.

Throughout the design of the measurement system it is important that an awareness is maintained of the need for integration with the overall system design, including liaison with groups engaged on the design of other elements of the system.

As an example, consider the design of a measurement system to provide the vertical position, velocity and acceleration of the wheels relative to the body of the vehicle for a smart suspension system. This information, together with vehicle speed and steering angle, will form the input data to the microprocessor acting as suspension controller to determine the output settings of the system actuators.

The high level specification has determined that the maximum travel of the suspension is 250 mm and that position is to be measured to within 1 mm. The road input to the suspension contains frequencies from 1 to 25 Hz, with an amplitude variation defined by the curve of Fig. 2.8. On the basis of this information, a sampling rate of 20 ms is chosen in line with the sampling theorem to avoid loss of information.

The design of the measurement modules can now proceed. Options for the

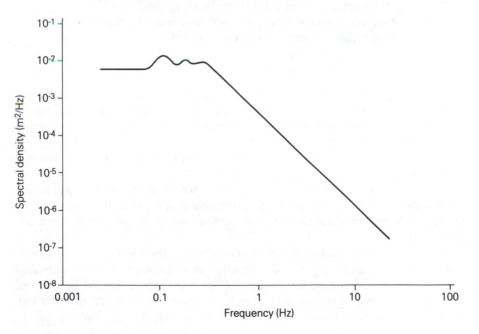

Figure 2.8   Road input to a suspension system.

measurement of the required parameters include:

1. The direct measurement of position only, with velocity and acceleration obtained by differentiation;
2. The direct measurement of acceleration only, with velocity and position obtained by integration;
3. The direct measurement of all three parameters;
4. The direct measurement of two out of the three parameters, with the third obtained by integration or differentiation of one or other of the measured parameters.

The measurement of position alone together with the use of differentiation presents problems of stability and drift. In addition, any high frequency components that may be present in the signal from the transducer will be emphasized by differentiation. The measurement of acceleration alone together with the use of integration has some problems of drift and stability, but acts to reduce the levels of high frequency noise. The measurement of all three parameters is unnecessary as the parameters are directly related. The measurement of position and acceleration together with the use of integration to obtain velocity provides a means of checking the measurement and simplifies the signal processing at the transducers.

The option chosen was to measure position and acceleration and to obtain velocity by integrating the output from the accelerometer.

Position can be measured by a number of sensors, including:

linear potentiometers (wirewound or conducting film)
driven rotary potentiometers (wirewound or conducting film)
linear variable displacement transformers (LVDTs)
linear variable inductive transducers
linear magnetoresistive transducers.

Acceleration can be measured using a piezoelectric accelerometer mounted directly on the axle. Position transducers may be mounted at various points on the suspension, and the geometry of the system must be taken into account to determine the output performance obtained.

The option chosen was to use a linear variable inductive transducer mounted on the axle to reduce the required stroke to 100 mm.

Environmental conditions are very severe, with the transducer subject to dirt and water as well as corrosive salts. This will require that the transducer is mounted in an enclosure which will provide adequate protection from the environment.

There will also be a range of extraneous signals in the form of body vibrations which will introduce noise into the system and will have to be accommodated by the signal processing system. Because of the environment this will be provided by a processor mounted remotely from the suspension; a communication link must be provided, in this case in the form of a screened twisted pair.

Chapter 3

Resistive, capacitive, inductive and resonant transducers

Transducers utilizing resistive, capacitive, inductive and resonant effects form the basis of many current measurement systems and will continue to do so in the foreseeable future, supplemented and supported as necessary by other technologies.

3.1 Resistive transducers

3.1.1 POTENTIOMETERS

The simplest form of resistive transducer is the potentiometer, available in both rotary and linear forms. In either case, a change in the output voltage is produced by varying the effective potential divider ratio of the potentiometer.

The majority of potentiometers are constructed from a large number of closely packed turns of a resistance wire with a sliding contact. The resolution of a wirewound potentiometer is therefore a function of the number of turns per unit length of track. For multiturn potentiometers, resolutions of the order of $\pm 0.01\%$ can be achieved with a linearity around 0.25%. Where higher linearity is required then techniques such as cam correction can be introduced.

Where improved resolution is required, potentiometers using a ceramic resistance element (cermet) or conductive plastic film are available. The latter provide high resolution at low noise levels and can be trimmed very precisely using lasers or computer numerically controlled (CNC) machining to generate both linear and non-linear response characteristics.

Errors include changes in resistance with temperature; variations of the order of 20 ppm $°C^{-1}$ are to be expected with wirewound potentiometers. Care must also be taken not to load the output as this will introduce an error into the measurement. Referring to Fig. 3.1a:

$$V_o = V_i R_2 / (R_1 + R_2) \tag{3.1}$$

If a load is now introduced as in Fig. 3.1b:

$$V_o = V_i R_o / (R_1 + R_o) \tag{3.2}$$

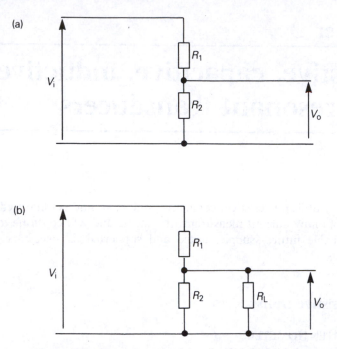

Figure 3.1   Potentiometer: (a) unloaded (b) loaded.

where V_o and V_i are the output and input voltages
and

$$R_o = R_L R_2 / (R_L + R_2)$$

3.1.2   STRAIN GAUGES

If a material is stressed by the application of an axial load as in Fig. 3.2 then
it will experience a change in length in the direction of the applied load.
The ratio (change in length)/(original length) is a dimensionless parameter
referred to as the strain in the material, and is related to the stress by Young's
modulus E such that

$$E = \text{stress/strain} \tag{3.3}$$

A typical resistance strain gauge consists of a pattern of resistive foil arranged
as shown in Fig. 3.3 and bonded to a backing material which enables it to be
handled and also transmits applied load to the gauge. If the load is applied along
the axis of the gauge then this gives rise to a strain, which causes a change in
the active length of the gauge and hence in its resistance.

The change in resistance, and hence the strain, is usually measured using a
bridge circuit such as that of Fig. 3.4. The out-of-balance voltage that results

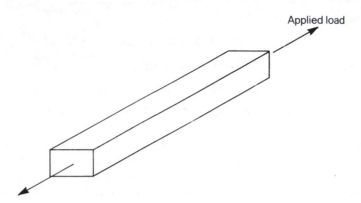

Applied load

Figure 3.2   Specimen with axial load.

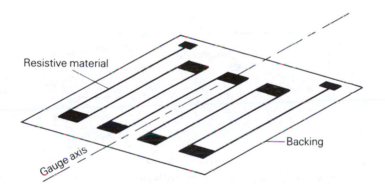

Resistive material

Gauge axis

Backing

Figure 3.3   Strain gauge.

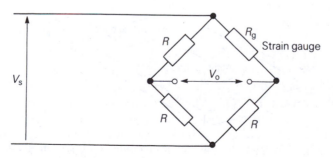

V_s

R

R_g

Strain gauge

V_o

R

R

Figure 3.4   Strain gauge bridge.

when the gauge is strained is

$$V_o = V_s \Delta R / (4R + 2\Delta R) \tag{3.4}$$

For the small changes in resistance that result,

$$V_o \approx V_s \Delta R / 4R = (GV_s/4)(\Delta L/L) \tag{3.5}$$

where G is the gauge factor of the gauge and typically lies in the range 2 to 4 for a foil gauge.

Changes in temperature also cause variations in the resistance of a strain gauge and are a possible source of error. For this reason, an unstrained compensating gauge having the same temperature characteristic as the active gauge is used in the bridge circuit, as in Fig. 3.5, to correct for temperature effects. To be effective, the compensating gauge must be mounted as close as possible to the active gauge. This may require that it is located on the member under measurement, in which case its axis is placed at right angles to the strain axis as in Fig. 3.6.

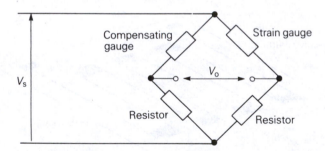

Figure 3.5   Strain gauge bridge with compensating gauge.

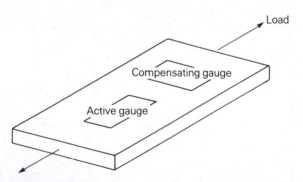

Figure 3.6   Active and compensating gauges mounted on a specimen with their axes aligned at 90°.

Where the strain axis is not known, strain gauge rosettes such as those illustrated in Fig. 3.7 are used. The principal strain axis can then be calculated from the information they provide. Other gauge types shown in Fig. 3.8 are used for the measurement of shaft torque and of the loading in circular diaphragms.

Strain gauges can be used for the measurement of other mechanical parameters such as pressure or force by converting their effect into a strain which can then

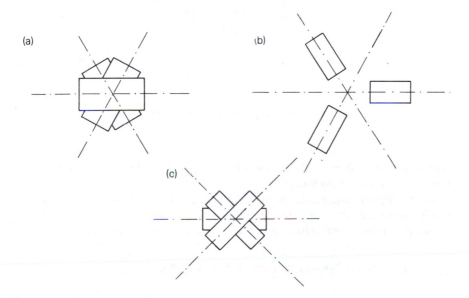

Figure 3.7   Strain gauge rosettes: (a) superimposed 120° rosette (b) delta (120°) rosette (c) superimposed 90° rosette.

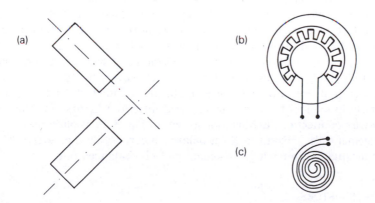

Figure 3.8   Other forms of strain gauge: (a) 90° strain gauges for shaft torque measurement (b) gauge for measuring radial strain (c) gauge for measuring tangential strain.

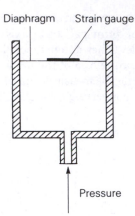

Figure 3.9   Pressure gauge using a strain gauged diaphragm.

be detected and measured. For example, Fig. 3.9 shows a pressure gauge in which the deflection of the diaphragm, and hence the applied pressure, is measured by the strain gauge.

Where higher gauge factors than those capable of being provided by a foil strain gauge are required, semiconductor strain gauges can be used. These rely on the piezoresistive effect in silicon, and gauge factors up to 200 can be achieved.

3.1.3   RESISTIVE TEMPERATURE TRANSDUCERS

(a)   RESISTANCE THERMOMETERS

The basic resistive temperature transducer is the resistance thermometer, consisting of a length of platinum wire formed into a small bobbin and protected by a sleeve. Sensitivity is low and they do not respond rapidly to changes in temperature because of the thermal inertia of the wire coil. As a result, they are normally used for the measurement of steady temperatures.

In a typical application, a resistance thermometer would form one arm of an equal ratio bridge circuit as in Fig. 3.10. However, because the changes in resistance that are anticipated are small, any contact or lead resistance will adversely influence the measurement and introduce errors. For this reason a three-wire connection such as that shown in Fig. 3.11 is often used. With this arrangement, the current in R_3 at balance is zero while the current in R_1 also flows through R_2, enabling voltage drops to be balanced out.

(b)   THERMISTORS

The thermistor is a semiconductor resistor of known thermal characteristics and is normally used to measure temperatures in the range $-30°C$ to $200°C$.

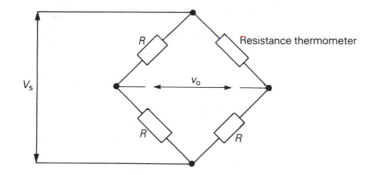

Figure 3.10   Resistance thermometer bridge circuit.

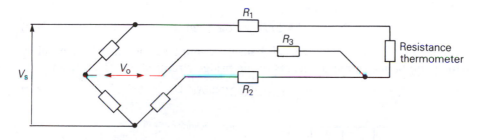

Figure 3.11   Resistance thermometer, three-wire bridge.

Their small size means that they have a low thermal inertia, enabling them to be used to monitor varying as opposed to static temperatures. Care must be taken when using a thermistor to avoid errors as a result of self-heating effects due to the passage of current through the device.

The main disadvantage of the thermistor is its non-linear characteristic, which has tended to limit its application. However, with the advent of microprocessor based measurement systems this inherent non-linearity can be accommodated in the software, enabling the use of thermistors over a wider range of applications

3.2   Capacitive transducers

Ignoring fringing effects, the capacitance of the parallel plate capacitor of Fig. 3.12 is given by

$$C = \varepsilon_0 \varepsilon_r A / d \qquad (3.6)$$

where ε_0 is the permittivity of free space ($8.854 \times 10^{-12}\,\mathrm{F\,m^{-1}}$), ε_r is the relative permittivity of the dielectric material, A is the effective cross-sectional area of the plates and d is the separation of the plates.

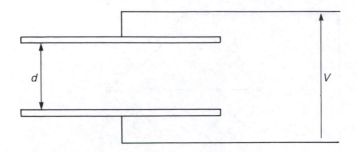

Figure 3.12   Parallel plate capacitor.

Similarly, the capacitance of the pair of concentric cylinders of Fig. 3.13 is given by

$$C = 2\pi L\varepsilon_0\varepsilon_r/\log_e(b/a) \tag{3.7}$$

Changes in $\varepsilon_r, A, L, a, b$ or d will all cause a change in the capacitance value, which can be measured by means of the bridge circuit of Fig. 3.14 when

$$v_o = v_s\Delta C/[2(2C + \Delta C)] \tag{3.8}$$

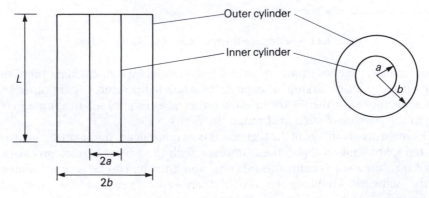

Figure 3.13   Cylindrical capacitor.

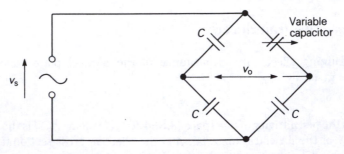

Figure 3.14   Capacitive bridge.

Alternatively, the capacitor can be incorporated as the tuning element of an oscillator. Any change in the value of capacitance then appears as a change in the oscillator frequency. This change in frequency can be detected as the difference in the amplitudes of the output voltages of a pair of resonant circuits tuned to slightly different centre frequencies, as represented by Fig. 3.15. This

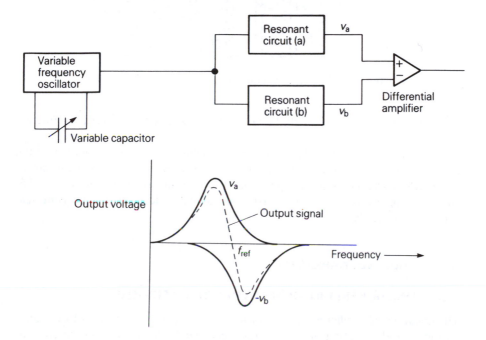

Figure 3.15   Resonant variable frequency detector.

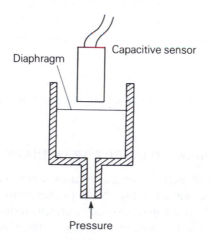

Figure 3.16   Diaphragm pressure gauge with capacitive transducer.

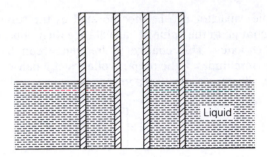

Figure 3.17 Liquid level gauge using a cylindrical capacitor.

arrangement produces a nearly linear variation about the reference frequency f_{ref}.

Applications of capacitive transducers include the measurement of small displacements; for example, a capacitive transducer could replace the strain gauged diaphragm pressure gauge of Fig. 3.9 with the arrangement of Fig. 3.16. Another application is the measurement of liquid level using a cylindrical capacitor, as in Fig. 3.17.

3.3   Inductive transducers

3.3.1   LINEAR VARIABLE DIFFERENTIAL TRANSFORMER

The linear variable differential transformer (LVDT) consists of a pair of secondary windings and a single primary winding arranged relative to a ferrite core, as in Fig. 3.18a. Movement of the core changes the coupling of the primary winding with the individual secondary windings, resulting in a change in the amplitudes of the induced secondary voltages. By connecting the output of the secondary coils in opposition, the magnitude of the output voltage v_o will vary as shown in Fig. 3.18b, since

$$v_0 = v_1 - v_2 \tag{3.9}$$

Typically, an LVDT will operate at a frequency of a few kilohertz to provide an output of several mV.

3.3.2   LINEAR VARIABLE INDUCTIVE TRANSDUCER

The linear variable inductive transducer uses a combination of a coil and a conductive spoiler, as shown in Fig. 3.19. In operation, the coil is excited at a frequency of 100 kHz or greater, and eddy currents induced in the spoiler result in a change in the effective inductance of the coil which is a function of the position of the spoiler. As no magnetic materials are used in the construction

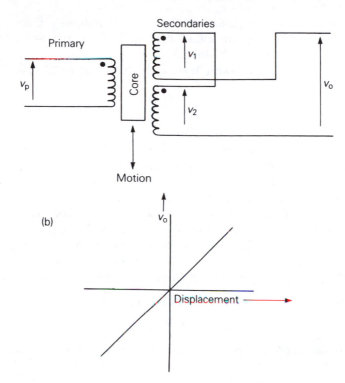

Figure 3.18   Linear variable differential transformer (LVDT): (a) configuration (b) output characteristic.

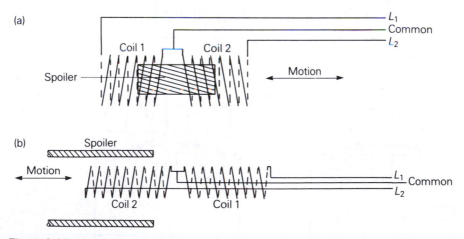

Figure 3.19   Linear variable inductive transducer: (a) internal spoiler configuration (b) external spoiler configuration.

of the transducer, its vulnerability to any stray magnetic fields is reduced compared with other inductive transducers, and effects such as hysteresis and magnetic non-linearities are eliminated.

Linear variable inductive transducers with strokes up to 220 mm have been produced.

3.3.3   THE INDUCTOSYN

The linear inductosyn consists of a fixed track, which may be several metres long, and a slider, as in Fig. 3.20a. The two parts are separated by a small air gap, typically between 0.1 and 0.15 mm, and are magnetically coupled. The slider carries a pair of windings displaced relative to each other by one-quarter of the cycle pitch (90° electrical), as shown in Fig. 3.20b. When the track winding is energized with a sinusoidal signal, typically of the order of a few kilohertz, voltages are induced in the slider windings. These voltages are 90° out of phase and will vary in amplitude with slider position as the coupling between the windings varies. The slider voltages can be expressed as

$$v_1 = kV \sin(\omega t) \sin(2\pi x/p) \tag{3.10}$$

$$v_2 = kV \sin(\omega t) \cos(2\pi x/p) \tag{3.11}$$

where x is the displacement from the null position, k is a constant related

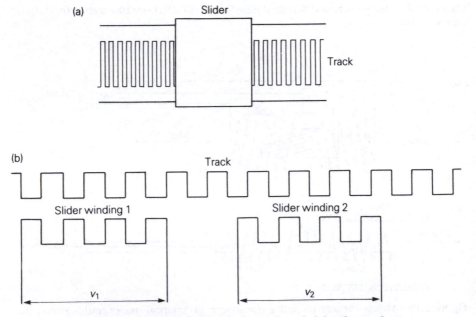

Figure 3.20   Linear inductosyn: (b) track and slider (b) windings.

to the coupling between the slider windings and the track winding and p is the winding pitch.

From equations 3.10 and 3.11 it can be seen that the output from the inductosyn varies sinusoidally over a single pitch of the windings before repeating. Within this distance, typically a few millimetres, resolutions of the order of 2 to 3 μm are achievable.

The inductosyn is an incremental encoder with a short repetition period. For applications involving large displacements, such as machine tools, it must be combined with other coarser measurement systems to obtain an absolute position.

Where angular measurement is required, a rotary inductosyn can be used. This has its tracks arranged radially on a pair of discs, and resolutions of the order of 0.05 seconds of arc can be achieved.

3.3.4   INDUCTIVE VELOCITY TRANSDUCERS

Velocity can be measured using the arrangement of Fig. 3.21 in which a permanent magnet is moved relative to a coil. The voltage induced in the coil is a function of the velocity of the magnet and its strength together with the dimensions of the coil.

Where rotational motion is involved, either DC or AC tachometers can be used to produce an output voltage which is directly proportional to angular velocity.

3.4   Thermoelectric transducers

When the junction between two dissimilar metals is heated or cooled relative to a second or reference junction, an EMF is produced which is a function of the metals used and the temperature difference between the two junctions. This

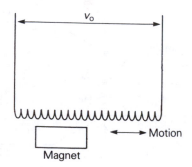

Figure 3.21   Direct measurement of linear velocity using a magnet and coil.

is known as the Seebeck effect. The generated EMF can be expressed by an equation of the form

$$e_0 = c_1 T + c_2 T^2 + c_3 T^3 + \cdots \tag{3.12}$$

where c_1, c_2, c_3, etc. are constants.

For certain specific combinations of metals the higher order coefficients in equation 3.12 become small, and an approximately linear relationship of the form

$$e_0 = c_1 T \tag{3.13}$$

can be assumed.

The performance of a thermocouple is normally specified in relation to a reference temperature of 0°C. In many practical applications the reference junction is located in a controlled environment whose temperature is maintained at some non-zero reference temperature. In this case, correction can be applied using the relationship

$$e(T_m, T_0) = e(T_m, T_r) + e(T_r, T_0) \tag{3.14}$$

where T_m is the temperature of the measuring junction, T_r is the temperature of the reference junction, T_0 is 0°C, and $e(T_m, T_0), e(T_m, T_r)$ and $e(T_r, T_0)$ are the EMFs that would be generated by junctions at the pairs of temperatures indicated.

In practice, the thermocouple will normally be connected to the detector by extension leads of a different material, as in Fig. 3.22. In this configuration there are three junctions in the system, generating EMFs e_1, e_2 and e_3 respectively, giving a total EMF at the detector of

$$e_0 = e_1 + e_2 + e_3 \tag{3.15}$$

This condition is an example of the thermoelectric law of intermediate metals. Referring to Fig. 3.23, this states that the EMF resulting at the junction between two materials a and b at a temperature T is equal to the sum of the EMFs that would result from maintaining junctions between materials a and c and b and

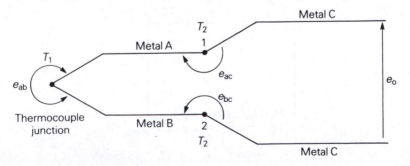

Figure 3.22 Thermocouple with extension leads.

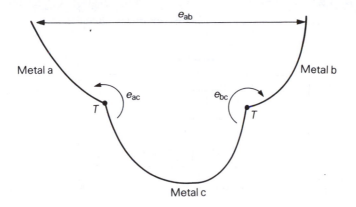

Figure 3.23   Thermoelectric law of intermediate metals.

c at the same temperature T. This can be expressed as

$$e_{ab} = e_{ac} + e_{bc} \tag{3.16}$$

or alternatively as

$$e_{ab} = e_{ac} - e_{cb} \tag{3.17}$$

Referring to Fig. 3.22, this means that if junctions 1 and 2 are maintained at the same temperature the effect will be that of a reference junction between metals a and b at that temperature.

Thermocouples are simple and straightforward in application. However, they have low sensitivities and their output signal level is small and subject to noise effects. They are also subject to errors due to mechanical loading, particularly if the hot junction is strained.

3.5   Resonant transducers

In a resonant transducer a mechanical element is excited to vibrate at a frequency which is a function of the parameter to be measured. The advantages of resonant transducers are stability, freedom from interference and low power requirements. In addition, the frequency output is easily transmitted and is simple to inter-face to digital systems by means of counter/timers. The disadvantages are a temperature dependent performance and a generally non-linear response to the measurand.

3.5.1   VIBRATING WIRE TRANSDUCERS

A vibrating wire transducer for the measurement of force or strain is shown in simplified form in Fig. 3.24. The wire, usually of tungsten or indium, is excited

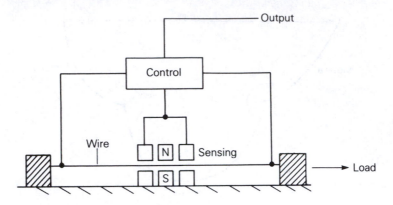

Figure 3.24   Vibrating wire strain gauge.

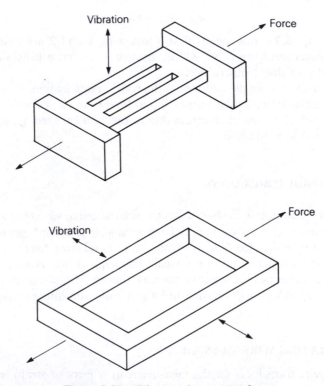

Figure 3.25   Vibrating beam transducers.

at its resonant frequency by the driver circuit, using feedback to maintain the resonant condition. The resonant frequency f_0 is given by

$$f_0 = \frac{1}{2l}\sqrt{\frac{T}{m}} \qquad (3.18)$$

where l is the length of the wire, m is the mass per unit length and T is the applied load.

3.5.2 VIBRATING BEAM TRANSDUCERS

Vibrating beam transducers such as those of Fig. 3.25 are available for force measurement. These transducers provide a very high Q factor and can be very small; devices are available manufactured from quartz or micromachined in silicon with signal processing incorporated on the same chip.

By arranging the configuration of the beam, transducers can be constructed for a variety of measurement functions, including flow, viscosity and density.

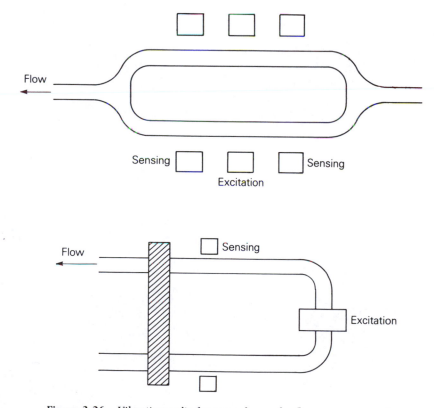

Figure 3.26   Vibrating cylinder transducers for flow measurement.

3.5.3   VIBRATING CYLINDER TRANSDUCERS

Vibrating cylinder transducers operate in a similar manner to other resonant transducers in that their frequency of oscillation is influenced by the parameter they are designed to measure. They are, however, harder to excite than other resonating transducers and produce a complex output which requires significant processing to extract information. Nevertheless, vibrating cylinder transducers such as those shown in Fig. 3.26 are available for the measurement of flow and density.

Chapter 4

Optical measurement systems

Optical transducers are available for a wide variety of measurement applications. Operating from the infrared to the ultraviolet in the electromagnetic spectrum of Fig. 4.1, they are in many instances non-contacting and are often capable of operating in hostile environments ranging from extreme temperatures to high levels of electrical noise.

The principal elements of an optical measurement system are as follows, as shown in Fig. 4.2:

1. A radiant energy source;
2. A transmission system to direct the light energy from the source as required;
3. A transducer to modify the characteristics of the light energy in some way;
4. A detector to monitor the changes introduced by the transducer.

Though these basic elements have been available and used for many years in instruments such as polarimeters for stress determination, it is developments in the following areas that have led to the range and scope of optical measurement systems now available:

lasers
fibre optics
optical materials
photodetectors
signal processing.

4.1 Radiant Energy sources

4.1.1 INCANDESCENT LAMPS

Incandescent lamps have an electrically heated filament within an evacuated glass or quartz envelope and are widely used for general illumination and projection systems. They are a low cost source and radiate over a range

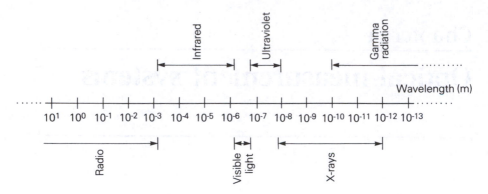

Figure 4.1   The electromagnetic spectrum.

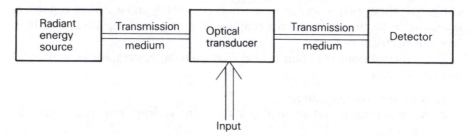

Figure 4.2   A basic optical measurement system.

covering the visible spectrum and extending into the infrared. They have a comparatively short lifetime, and thermal inertia limits the rate at which intensity can be modulated.

4.1.2   DISCHARGE LAMPS

Discharge lamps produce light as a result of collisions between free electrons and gas in a sealed envelope. The spectrum of the light emitted is a function of the type of gas used, the gas pressure and the system geometry. High pressure discharge lamps usually produce a high intensity, broad band light output containing a number of broad spectral lines, while low pressure lamps generate a number of well separated spectral lines which can be isolated and used independently.

4.1.3   LIGHT EMITTING DIODES

Light emitting diodes (LEDs) provide a range of high efficiency, low cost, low intensity, near monochromatic light sources with relatively narrow spectral bandwidths. The range of LEDs available covers the full visible spectrum and extends into the infrared. Output intensity can be varied linearly at frequencies of the order of 1 GHz over an approximately 30 to 40 dB range.

4.1.4 LASERS

A laser provides a highly collimated light source of narrow spectral bandwidth and high intensity by the stimulated emission of coherent photons in the lasing material, which may be a gas, a liquid, an insulating crystal or a semiconductor.

The performance of a laser is expressed in terms of a number of parameters, the chief of which are as follows:

Radiant power This is a measure of the energy the laser can deliver. It ranges from a few milliwatts for a typical instrumentation laser to several kilowatts in the case of CO_2 lasers used for machining.

Coherence This is an expression of the degree of directionality, parallelism and monochromaticity of the light produced by the laser. Temporal coherence is related to the coherence length of the laser, which is itself a function of the monochromaticity of the light. Spatial coherence is related to the divergence of the beam. This ranges from about 0.5 mrad for gas lasers to 20 mrad for semiconductor lasers.

Table 4.1 lists some of the lasers commonly available.

4.1.5 ILLUMINATION

The way in which illumination is used can have a significant effect on the nature of the information that is made available. The two principal techniques used are backlighting and forelighting, as illustrated by Fig. 4.3. Each technique may be summarized as follows:

Backlighting Backlighting results in an image with clearly defined boundaries, which simplifies the separation of objects from their background and reduces the volume of data to be handled. However, only the outline of the object and through features such as holes are distinguishable.

Forelighting Forelighting enables surface features to be distinguished, providing additional information about the shape of an object in the field of the illumination. Extraction of information is complicated by the presence of shadows and highlights.

4.2 Photodetectors

4.2.1 THERMAL PHOTODETECTORS

Thermal photodetectors operate by monitoring the generation of heat in the detector by the incident radiation. A typical device such as a bolometer or a

Table 4.1 Laser characteristics

Type	Wavelength	Power	Notes
Gas			
Argon	Various in range 454 to 529 nm	5 mW to 20 W continuous	High powered laser of variable output wavelength; most intense at 488 and 514 nm
He—Ne	633 nm	0.1 mW to 50 mW continuous	Low cost, general purpose laser with a wide range of applications in measurement
CO_2	10.6 μm	20 W to 5 kW continuous	High power laser used for machining
CO_2 (TEA)	10.6 μm	30 mJ to 150 J pulsed	Pulses of 50 to 100 ns duration at frequencies to 1 kHz
Solid state			
Ruby	694 nm	30 mJ to 100 J pulsed	
Nd-YAG	1.064 μm	10 mJ to 150 J pulsed	10 ns to 1 ms pulses at frequencies to 50 kHz
Nd-YAG	1.064 μm	1 mW to 10 mW continuous	Diode pumped
Semiconductor			
GaAlAs	750 to 905 nm	1 mW to 40 mW continuous Up to 100 W pulsed	Solid state laser with a broader output spectrum and higher beam divergency than other lasers; can be pulsed repetitively at frequencies to the order of 20 kHz
InGaAsP	1.1 to 1.6 nm	1 mW to 10 mW continuous	
Dye			
Flash tube	Variable, 340 to 960 nm	To 50 W average pulsed	200 ns to 4 μs up to 50 Hz
Ion	Variable, 400 to 1000 nm	To 2 W continuous	
Excimer			
Argon fluoride	193 nm	50 W pulsed	5 to 25 ns pulses up to 1 kHz
Krypton fluoride	248 nm	100 W pulsed	5 to 50 ns pulses up to 500 Hz
Xenon fluoride	351 nm	30 W pulsed	1 to 30 ns pulses up to 500 Hz

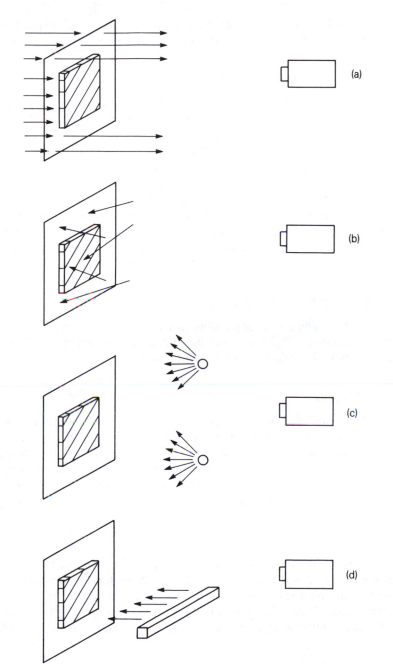

Figure 4.3   Forms of illumination: (a) backlighting (b) diffuse lighting (c) flash or strobe lighting (d) linear illumination.

thermopile will respond to wavelengths in the range 0.3 to 30 μm but has a low sensitivity and a slow response time.

4.2.2   QUANTUM PHOTODETECTORS

Quantum photodetectors rely for their operation on the generation of an electron–hole pair within the detector by an incident photon. In the case of a silicon based detector, the release of this electron–hole pair requires the provision of enough energy to bridge the bandgap of 1.1 eV at the junction. This in turn implies a wavelength for the incident radiation of 1.1 μm or less. A limiting feature of quantum photodetectors is that their dark currents and hence their noise levels are highly temperature dependent.

Silicon based quantum photodetectors utilize either the photovoltaic effect, in which a voltage is generated across a junction between p-type and n-type material, or the photoconductive effect, where the incoming radiant energy causes a change in the conductivity of the material.

(a)   PHOTOVOLTAIC DETECTORS

When incident radiation of appropriate wavelength falls on the depletion layer of the p-n junction of Fig. 4.4, a potential difference proportional to the intensity of illumination is developed across the junction.

(b)   PHOTODIODES

The configuration of a photodiode is shown by Fig. 4.5. When the reverse biased p-n junction is subject to incident radiation on or near the depletion layer, electron–hole pairs are created which are swept up by the bias voltage. The resulting reverse current through the diode is a measure of the intensity of the incident radiation. Photodiodes are available with responses covering a range from infrared through to ultraviolet.

(c)   PHOTOTRANSISTOR

The phototransistor of Fig. 4.6 may be considered as a photodiode in parallel with the base-collector junction of a bipolar transistor. The resulting reverse current of the photodiode becomes the base current of the transistor and is amplified accordingly. The phototransistor provides increased sensitivity by introducing an internal signal processing function in the form of amplification.

(d)   PHOTOCONDUCTORS

The silicon photoconductor of Fig. 4.7 consists of a p-type base on to which n-type material has been diffused. When this n-type layer is illuminated there

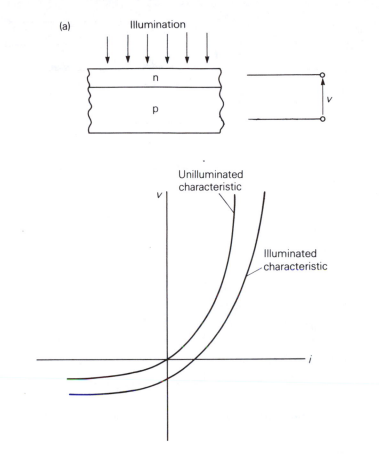

Figure 4.4 The photovoltaic effect: (a) illuminated junction (b) photovoltaic characteristics.

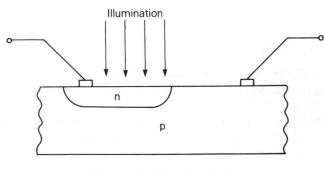

Figure 4.5 The photodiode.

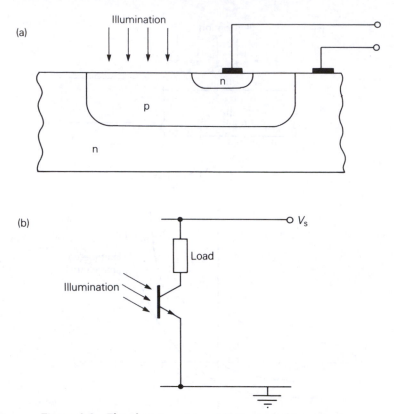

Figure 4.6 The phototransistor: (a) structure (b) application.

is an effective change in conductivity which can be detected by a bridge circuit.

By introducing specific materials on to a layer of silicon dioxide (SiO_2), photoconductors can be produced which respond to specific wavelengths. Typical materials used are indium antimonide (InSb), which responds to a wavelength of 7 μm, and cadmium sulphide (CdS), which is sensitive at 0.7 μm.

(e) PHOTOMULTIPLIERS

The photomultiplier consists of a vacuum tube containing a photosensitive cathode (photocathode), a series of intermediate electrodes (dynodes) and an anode, as shown in Fig. 4.8. When incident radiation is received at the photocathode, electrons are released which are accelerated by the applied voltage on the dynodes. When these electrons reach a dynode a secondary emission of electrons occurs, resulting in a multiplication of the current. The anode current is a function of the number and structure of the dynodes, the material used for their construction and the bias voltage applied.

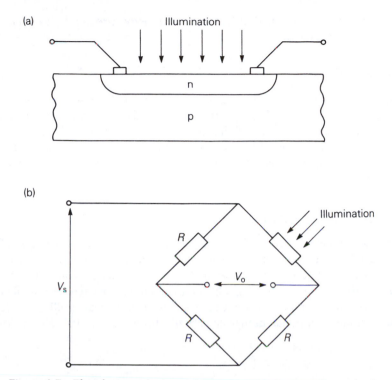

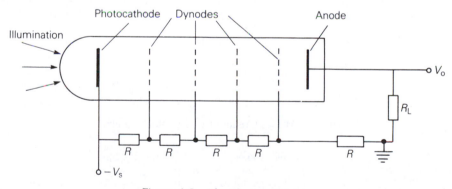

(a)

(b)

Figure 4.7   The photoconductor: (a) structure (b) photoconductor bridge.

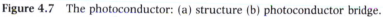

Figure 4.8   The photomultiplier.

4.2.3   ARRAY DETECTORS

Photodiodes in particular can be arranged to form either linear or two-dimensional arrays. When an array is exposed to a light source, charge accumulates on each photodiode at a rate which is proportional to the intensity of the illumination, thus building up an image on the array.

The information contained in the array after exposure can be extracted in a number of ways as follows.

(a)   MOSFET READOUT

An integral MOSFET device is used to connect each element of the photodiode array to a sense line as in Fig. 4.9, the output end of which is connected to an amplifier. The output voltage of the amplifier then corresponds to the charge on the photodiode connected to the sense line. In large arrays the sense line will have a significant capacitance and the output signal level is low, reducing sensitivity.

(b)   CHARGE COUPLED DEVICES

In a charge coupled device (CCD), electrodes are introduced on to a layer of SiO_2 as in Fig. 4.10. The charge packets generated at each of the photodiode elements are then transported from electrode to electrode by the application of a voltage signal. At the output of the CCD channel the charge packets are applied to a low capacitance amplifier to produce a serial output at a significantly higher output voltage, and hence higher sensitivity, than with a MOSFET readout.

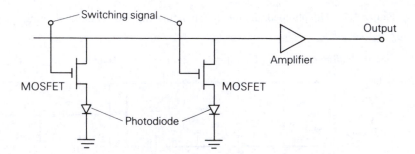

Figure 4.9   MOSFET readout of a photodiode array.

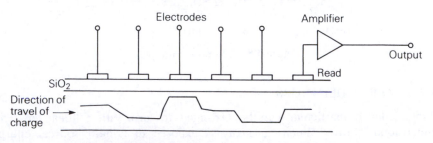

Figure 4.10   Charge coupled device.

(c)   CHARGE INJECTION DEVICE

A charge injection device (CID) extracts information from the photodiode array by injecting a current to cancel the accumulated charge at each element of the array. This approach permits the digital scanning of the array and offers the opportunity of direct access to individual image elements (pixels).

(d)   DYNAMIC RANDOM ACCESS MEMORY

If the elements of a conventional dynamic random access memory (DRAM) chip are exposed to light then a charge pattern will build up on these elements which corresponds to the illumination. This charge will be read by the DRAM circuitry as a series of 0s and 1s to generate a binary image.

4.3   Vision systems

The basic configuration of a vision system is shown in Fig. 4.11. The camera is used to form an image which is converted for storage in the frame store. The data within the frame store is then available to be accessed by a computer for processing and analysis.

Cameras operating in the visible spectrum are primarily of the vacuum tube or solid state type using television (TV) raster scan (two-dimensional) or single line scan (one-dimensional) techniques.

Vacuum tube technology is well established (Videcon, Plumbicon, Staticon and Newicon) and has as its advantages that it provides high resolution at low cost and with spectral characteristics that range from the infrared to the ultraviolet. However, the cameras are fragile, with a relatively short life of around 10 000 hours; have characteristics which change with age; exhibit significant geometric distortion (barrelling and pincushioning) and high noise levels; and are vulnerable to external magnetic fields. They are also prone to an image lag, with a residual image of a moving object carried over into the next image. Operation is normally in an interlaced mode, with odd numbered lines in the image field read first followed by the even numbered lines. This procedure is adopted to reduce flicker, but it increases the complexity of the signal processing.

Solid state cameras by comparison are compact and robust, with high definition at low noise levels, little geometrical distortion, a low image lag and a low

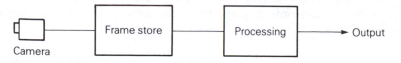

Figure 4.11   A basic vision system.

power consumption. Their disadvantages are limited spectral response and sensitivity and a high component complexity. The information from the camera is used to form a grey scale image in which the intensity level at each picture element (pixel) of the image is represented by a binary value. Typically, 6 or 8 bits will be used to encode the grey scale. If colour is required then this can be achieved by the use of appropriate filtering but at the expense of increased data storage requirements and processing times. Picture resolution is determined by the number of pixels used. For most industrial purposes a 256×256 or 512×512 array, requiring 64 Kbytes and 256 Kbytes of storage for an 8 bit grey scale respectively, will be adequate. Other applications such as medical or satellite imaging will require much higher resolutions, necessitating array sizes of the order of $10^5 \times 10^5$ pixels.

Line scan cameras use a linear array of photodetectors and provide a low cost means of producing simple images. A typical application for a line scan camera is the measurement of the dimensions and orientation of objects passing through the image field, as for example on a conveyor.

Thermal imaging systems utilize the pyroelectric effect observed in certain semiconductor materials. When such a material is exposed to infrared radiation, the resulting distribution of surface charge in the material is proportional to the amount of heat energy absorbed. This surface charge is then read by a scanning electron beam to generate the image. The spectral response of the thermal imager is controlled by the response of the window to the incident radiation with, for example, a germanium window responding to wavelengths in the region from 8 to $15\,\mu m$.

The diffusion of the absorbed heat energy within the pyroelectric material is the primary limiting factor on the performance of thermal imaging systems. To permit recovery and enable the production of a continuous image of a stationary object, the input signal is chopped using a mechanical scanner to block the incoming radiation.

For a commercial system operating at ambient temperature, images containing 400 or more lines at bandwidths in excess of 5 MHz are achievable, resulting in a temperature resolution of the order of 0.2°C. For greater performance, cooling is required, often to liquid nitrogen temperatures as in the case of the thermal imaging common module (TICM) system. Such systems are, however, extremely expensive.

4.3.1   IMAGE PROCESSING

Once the image has been captured by the camera and placed in the frame store it must be processed in order to extract the information contained. This involves a number of stages:

Pre-processing   Pre-processing is used to correct for any distortion introduced by the camera and to smooth and enhance the image prior to analysis.

Table 4.2 Some examples of commonly used calculation based data in image analysis

Parameter	Comment
Area	Independent of position and orientation of object
Perimeter	Shape factor = area/(perimeter)^2
Minimum enclosing rectangle	Orientation dependent
Centre of area	Independent of position and orientation of object
Minimum radius vector	Measured from centre of area to nearest edge point
Maximum radius vector	Measured from centre of area to furthest edge point
Moment of area	About defined axes

Feature extraction   Following pre-processing, feature extraction techniques such as binary imaging and edge detection can be used to isolate objects within the field of the image.

Analysis   Objects within the image field can be analysed using a variety of techniques, of which the following are some of the most common:

1. Calculation: a variety of calculation techniques can be applied to objects within the image field; table 4.2 lists some of the more frequently used forms of analysis;
2. Template matching: this is a process of pattern recognition in which objects in the image field are compared with a known image for identification;
3. *n*-tuple classification: this is a mathematical technique which extracts information about unknown objects by reference to a set of training objects.

4.4   Laser scanning

In a camera based vision system, the camera is used to generate an image from an illuminated field. In a 'flying spot' laser scanning system an image is produced by scanning a narrow beam of monochromatic light across the object field and collecting the scattered light by a wide area detector. The use of a laser enables ambient illumination to be excluded by filtration, allowing operation in daylight or other lit environments.

Laser scanning systems are capable of achieving high resolutions ranging from 5000 to 30 000 resolved spots per scan line, the latter requiring complex scanning, detector and amplifier arrangements. Scanning speeds can be very high, ranging up to 10^8 pixels per second.

Scanning of the laser beam across the image field is normally achieved by an arrangement of prisms or mirrors. The simplest configuration, providing low scanning speeds and relatively low resolution, uses an oscillating plane mirror as in Fig. 4.12. For higher speeds and resolutions a rotating polygonal mirror, as shown in Fig. 4.13, is used. To achieve a raster scan, two actuators are required, typically an oscillating mirror and a rotating drum.

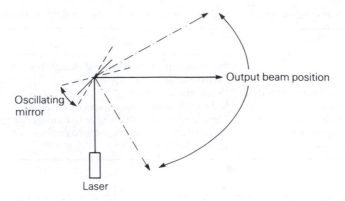

Figure 4.12   Scanning laser using an oscillating mirror.

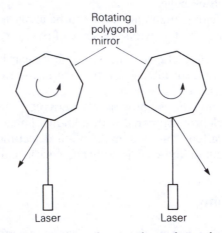

Figure 4.13   Scanning laser with a polygonal mirror.

The resulting scan line is focused along an arc centred at the scanner, requiring special lenses to correct for a flat image field. These same lenses can also be used to convert a constant angular velocity scan into one with a constant linear velocity.

4.5   Fibre optic transducers

Optical fibres possess a number of significant advantages which make them suitable for use as the basis for a wide range of transducers. Because of their mode of operation and the near impossibility of electromagnetic radiation gaining access to the fibre, as a propagating medium they are not vulnerable to many forms of interference, thus reducing noise problems. The fibre is inert and the absence of an electrical signal means that optical fibre transducers are well suited to operation in hazardous environments. For mechanical measurements, the low inertia of the fibre enables transducers to respond over a wide frequency range.

Optical fibre transducers operate by modulating transmitted energy in some way. The available forms of modulation are as follows:

intensity
phase
polarization angle
wavelength
spectral content and distribution.

In some fibre optic transducers the optical fibre plays a major role in modulating the energy from the source; these are referred to as *intrinsic* systems. In others the optical fibre plays no part in achieving the modulation but simply acts as a transmission medium to direct the energy to and from the modulating environment; these are *extrinsic* systems.

4.5.1   INTENSITY MODULATION

The modulation of intensity is the simplest form of modulation to introduce. In order to compensate for the tendency of the transmission properties of optical fibre to vary with ageing, a reference is usually incorporated as part of the measurement system. The techniques for achieving intensity modulation are as follows.

(a)   SHUTTERS

If a shutter is introduced between a pair of fibres as in Fig. 4.14, then the shutter aperture can be measured by the change in the intensity level of the received light.

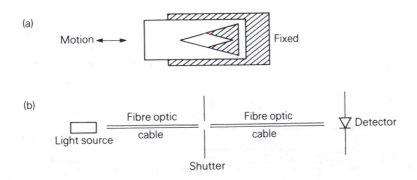
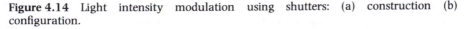

Figure 4.14 Light intensity modulation using shutters: (a) construction (b) configuration.

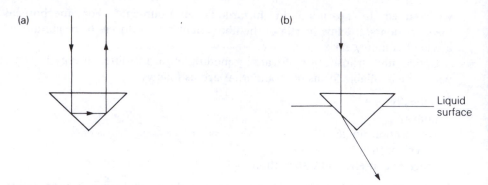

Figure 4.15   Optical liquid level detector: (a) total internal reflection in absence of liquid (b) light loss due to change in refractive index at surface of prism in the presence of liquid.

(b)   LIQUID LEVEL DETECTOR

For the liquid level detector of Fig. 4.15, in the absence of liquid the light is initially totally internally reflected by the prism and returns to the detector. When the liquid reaches the prism the refractive index at the surface of the prism changes and the light escapes, causing the disappearance of the signal at the detector.

(c)   MICROBENDING

When an optical fibre is subjected to bending as in Fig. 4.16, the change in the intensity of the transmitted light is a function of the degree of bending. Applications include the measurement of force, pressure, strain, displacement or temperature.

(d)   REFLECTION

If light from a fibre optic cable is directed on to a reflective surface as in Fig. 4.17, the intensity of the returned light will be a function of the position of the reflecting surface. If the transmitting and receiving cables are combined into a

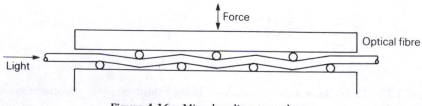

Figure 4.16   Microbending transducer.

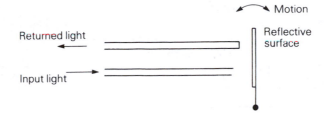

Figure 4.17   Position measurement using reflected light.

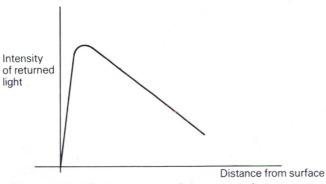

Figure 4.18   The Fotonic transducer output characteristic.

single bundle with their individual fibres arranged at random, the response will have the form shown in Fig. 4.18.

(e)   LATERAL DISPLACEMENT

If the ends of a pair of fibres are subject to a lateral displacement, the degree of displacement is measured by the change in the intensity of the transmitted light. This principle is used in the fibre optic hydrophone of Fig. 4.19, in which the acoustic signal causes a vertical displacement of the receiving fibre, modulating the intensity of the light at the receiver.

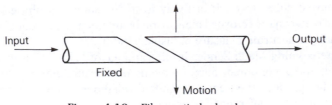

Figure 4.19   Fibre optic hydrophone.

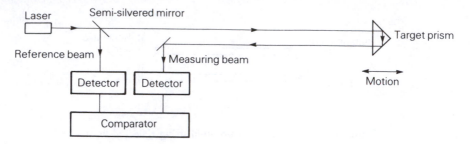

Figure 4.20   A basic interferometer.

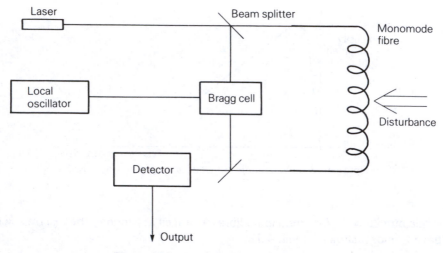

Figure 4.21   Mach-Zender interferometer.

4.5.2   PHASE MODULATION

The detection of a phase change in an optical system provides a means of measuring very small variations in a number of physical parameters. Figure 4.20 shows the basic interferometer arrangement for distance measurement. Here, light from a laser source, to provide the necessary coherence, is passed to a beam splitter to generate the measuring and reference beams. As the target prism is moved either towards or away from the source, a phase shift will be introduced between the returned measuring beam and the reference beam which will provide an accurate measure of displacement.

The heterodyning Mach-Zender interferometer of Fig. 4.21 uses a coherent light source and a monomode fibre. An acousto-optic modulator such as a Bragg cell is used to produce the frequency shifted reference beam. The phase shift is then detected at the difference frequency.

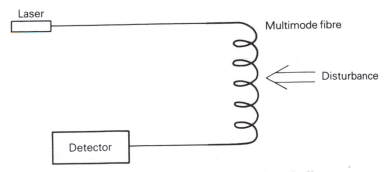

Figure 4.22   Interferometer using a multimode fibre.

Alternatively, as in Fig. 4.22, a multimode fibre can be used when a phase modulation occurs between the different modes within the fibre. This is a very simple system in concept, but has problems of linearization and stability in practice.

Phase modulation can also be introduced by variations of temperature, pressure or force, and strain, causing the optical path length in the fibre to change. Measurements are usually carried out for oscillatory parameters as the measurement of slowly varying parameters is subject to the effects of drift in both the electronics and the optical system.

One application in which the effects of drift have been compensated is the fibre optic gyroscope, shown in schematic form in Fig. 4.23. Light from the source is passed in opposite directions around the coil. The phase shift that Δ results when the coil is rotated about its central axis is then given by

$$\Delta\phi = 2\pi^2 d^2 N\omega/\lambda c \tag{4.1}$$

where N is the number of turns on the coil, d is the coil diameter, ω is the rotation in radians per second, λ is the wavelength of light used and c is the velocity of light.

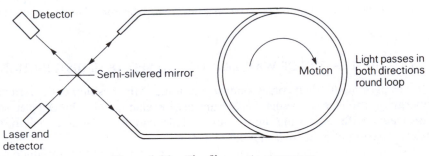

Figure 4.23   The fibre optic gyroscope.

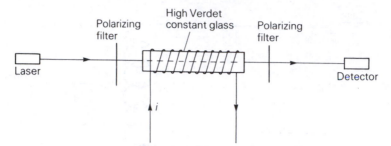

Figure 4.24   Current sensor using the Faraday magneto-optical effect.

4.5.3   MODULATION OF THE ANGLE OF POLARIZATION

The main application of modulation of the angle of polarization is in the measurement of the strength of a magnetic field by means of the Faraday magneto-optic effect.

If polarized light is passed along a magnetic field of strength H, the plane of polarization is rotated by an amount given by

$$\phi = V \int H \, dL = VNI \tag{4.2}$$

where ϕ is the angle of rotation, V is the Verdet constant for the medium, H is the strength of the magnetic field, L is the path length, I is the coil current and N is the number of turns on the coil.

A system for current measurement using the Faraday effect is shown in Fig. 4.24. Light from the laser source is passed through a polarizing filter and then through a high Verdet constant glass rod in the magnetic field of the current to be measured. The transmitted light passes through a second polarizing screen, aligned at 45° to the first screen, and then to the detector. With no current flowing a steady signal will be received at the detector. In the presence of a current the plane of polarization will be rotated clockwise or anti-clockwise depending on the direction of the current, while the angle of rotation will be a function of the current magnitude.

In addition to the Faraday magneto-optic effect, the polarization angle can be rotated by the application of an electric field using the Pockel and Kerr effects.

4.5.4   MODULATION OF WAVELENGTH AND SPECTRAL DISTRIBUTION

Wavelength modulation corresponds to a change of the position of the returned energy on the electromagnetic spectrum as, for example, in the Doppler shift associated with a moving light source. This condition is observed in the backscattering of light by a moving object.

Use can also be made of the fact that certain phosphors will reradiate at a different wavelength to that of the stimulating radiation. Other phosphors radiate with a spectral mix which is a function of their temperature, an effect which has been used for temperature measurement.

Wavelength modulation, and hence a varying spectral distribution, are found in the emissions from a body as its temperature changes. An optical fibre pyrometer uses the fibre as a means of transmitting the gathered radiation to the measuring instrument.

4.6   Non-fibre optical transducers

4.6.1   OPTICAL ENCODERS

Optical encoders are widely used in machine tools and robots to provide either an absolute or an incremental measurement of both linear and angular position. They are one of the longest established and simplest forms of optical transducers.

(a)   INCREMENTAL ENCODERS

Figure 4.25a shows a typical incremental angular encoder using transmitted light. Light from the source is pre-processed by passing through the slots in the disc and is received by a pair of photodiodes offset by one slot. As the disc rotates, light falls on each diode in turn to produce the output pulse trains of Fig. 4.25b. By counting the number of pulses produced, the distance moved can be obtained. The direction of rotation is found from the order in which the on/off transitions of the respective photodiodes occur. A refinement is the inclusion of a reference slot in the disc together with a third diode to provide a single reference in each complete revolution. Instead of a slotted disc, a disc

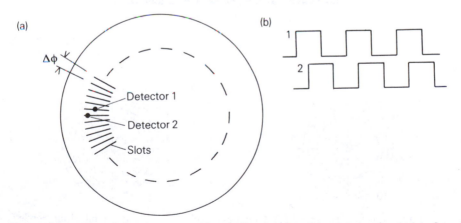

Figure 4.25   Slotted disc incremental encoder: (a) disc (b) output pulse trains from detectors.

carrying reflecting strips could be used, in which case both transmitter and receiver are mounted on the same side of the disc.

The angular resolution of the encoder is a function of the number of slots n_s on the disc and the active length D_w of the photodetector in the plane of the disc. Referring to Fig. 4.25:

$$\Delta\phi = 360/n_s \qquad (4.3)$$
$$D_w = r \sin(\Delta\phi/2)$$

Angular resolution can be increased by gearing the drive to the encoder so that it completes more than one revolution for each revolution of the shaft. Such as approach is expensive and introduces the possibility of error due to factors such as backlash in the gears.

Linear incremental encoders are commonly based on the use of Moiré fringes. By moving two optical gratings relative to each other a regular pattern of fringes is obtained at the detector, as in Fig. 4.26. Directional information is obtained by inclining the slots on one grating with respect to the other and using a number of detectors, usually four, to generate two sinusoidal signals 90° out of phase with each other. Positional information is obtained from the magnitude of the sine waves and the direction of travel from the zero crossing sequence.

Optical gratings can be made relatively cheaply and are capable of providing

Figure 4.26   Moiré fringes.

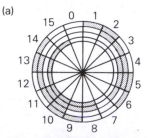
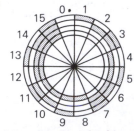

Figure 4.27   Absolute encoders: (a) 4 bit Gray coded encoder (b) 4 bit binary coded encoder.

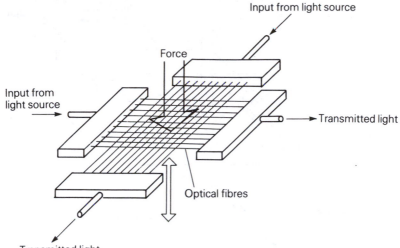

Figure 4.28   Tactile force transducer based on the microbending of optical fibres.

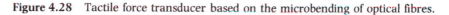

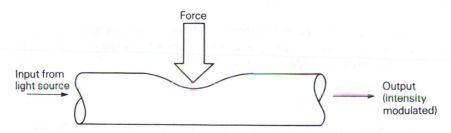

Figure 4.29   Tactile force transducer using a deformable plastic material to modulate the intensity of transmitted light.

resolutions of the order of micrometres, with accuracies of $1\ \mu$m over a 1 m length.

(b)   ABSOLUTE ENCODERS

Absolute encoders, as their name implies, provide positional information without any need for the external reference required with an incremental encoder. They are used where knowledge of position is required at all times, for example following start-up where they avoid the need for resetting. However, when the encoder is required to operate over more than one complete revolution, a record of the number of completed revolutions is required.

A typical absolute encoder will use the Gray code, the 8 bit version of which

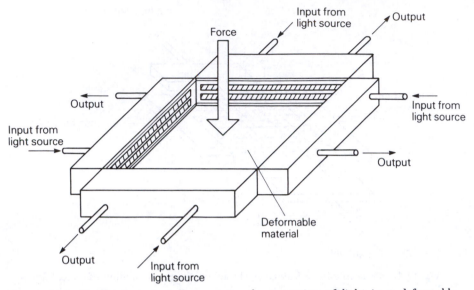

Figure 4.30 Tactile force transducer using the scattering of light in a deformable material.

is shown in Fig. 4.27a, in preference to the binary coded form of Fig. 4.27b, as only one bit changes state at any transition. This reduces the possibility of error due to the misreading of bits at a transition.

4.6.2 TACTILE SENSING

Tactile or touch sensing is used in robotics to detect when an object has been gripped and also the gripping force. This can be achieved optically using the

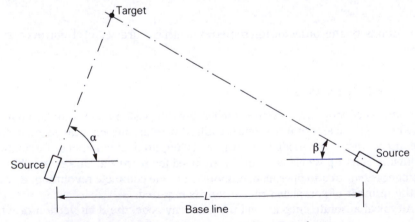

Figure 4.31 Distance measurement by triangulation.

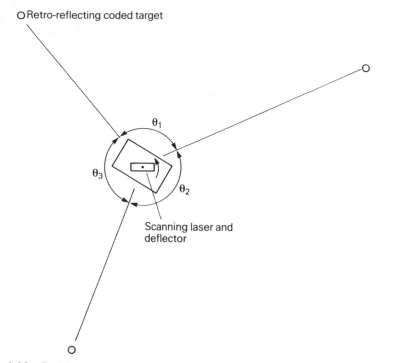

Retro-reflecting coded target

Scanning laser and deflector

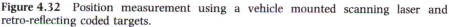

Figure 4.32 Position measurement using a vehicle mounted scanning laser and retro-reflecting coded targets.

array of optical fibres shown in Fig. 4.28, in which the microbending of the fibres that occurs when an object is gripped modulates the intensity of the transmitted light.

Use can also be made of the changes that take place in the intensity of transmission of light through a deformable plastic material on the application of a force. In one example, shown in Fig. 4.29, deformation of the material alters its absorption characteristic to introduce the intensity modulation. In an alternative version, the deformation causes a scattering of the light within the material which can be monitored by detectors along the edges, as in Fig. 4.30.

4.6.3   TRIANGULATION

Optically based triangulation systems using lasers and/or cameras can be used for the measurement of position at ranges of a few hundred metres, with typical configurations as shown in Fig. 4.31.

In the case of a fixed base laser tracking system, the presence of the laser beam on the target is detected either by photodetectors on the target or by mounting a retro-reflecting prism on the target. The camera based equivalent of this configuration uses a flashing beacon (to enable it to be distinguished from other light sources) mounted on the target, which is then tracked by the cameras.

An alternative triangulation system, shown in Fig. 4.32, uses a rotating laser mounted on a mobile base to scan retro-reflecting bar coded markers at fixed locations and hence obtain the position of the vehicle. This technique is currently employed as a navigation system for free-ranging automatic guided vehicles.

Chapter 5

Solid state sensors and transducers

Developments in manufacturing and production technologies have permitted the introduction of a variety of solid state sensors and transducers. Many of these use silicon as the active material by virtue of the sensitivity that it exhibits to a wide range of physical phenomena, together with the ease with which it can be purified and formed into numerous circuit configurations. In addition, the mechanical properties of silicon enable the production of complex three-dimensional structures on the chip.

The sensitivity of silicon is also a disadvantage; it creates problems in isolating the required information at the output of the sensor or transducer because of unwanted signals caused by the response to physical phenomena other than that to be measured. Analogue and digital signal processing and signal analysis, combined with appropriate design of the sensing elements, are required to provide the necessary compensation.

5.1 Magnetic measurements

5.1.1 HALL EFFECT

The Hall effect device of Fig. 5.1 is probably the most commonly used magnetic sensor. When the current carrying semiconductor is subjected to a magnetic field at right angles to the plane of the semiconductor, the effect of the interaction between the moving charge carriers in the material and the applied field, the Lorentz force, will be to produce a voltage between the edges of the semiconductor at right angles to the current and field which is proportional to the current and the applied magnetic field. The polarity of this Hall voltage is determined by the semiconductor material used (p-type or n-type), the direction of the current and the direction of the magnetic field.

The output from the basic silicon Hall plate is generally too small to be of direct use and must therefore be amplified. This can be achieved by integrating the Hall effect device and the necessary circuitry on to the same chip. Integrated single chip devices incorporating amplification, stabilization and temperature

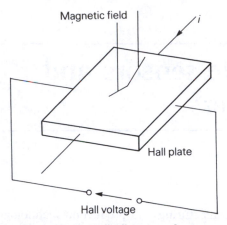

Figure 5.1   Hall effect transducer.

compensation circuitry are currently in use for a range of applications including rotor position sensing in permanent magnet (brushless) motors, displacement transducers and compasses.

The MOSFET structure can be adapted to exhibit the Hall effect by increasing the area of the gate channel and introducing additional Hall contacts as in Fig. 5.2. The resulting Hall voltage is proportional to the device current and the magnetic field, and sensitivities of the order of $100\,\text{mV}\ \text{T}^{-1}$ are achievable. Hall effect devices based on junction FETs (JFETs) have been reported in which the variation in the drain current is proportional to the applied magnetic field.

Hall effect devices are vulnerable to offset effects, mainly caused by misalignment of the Hall contacts and piezoresistive effects. Offset compensation

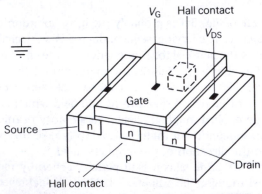

Figure 5.2   An n-channel Hall effect MOSFET.

can be achieved by making measurements with the Hall and bias contacts interchanged and combining the results.

5.1.2 MAGNETORESISTOR

Magnetoresistors utilize the change in resistivity that occurs in certain ferromagnetic materials in the presence of an external magnetic field. Known as the Gauss effect, this is a solid state phenomenon which can be realized using thin film technology. A simple magnetoresistor has a square law resistance/field relationship which must be linearized for most applications. This may be achieved by the use of a stable bias field from an external magnet or coil, or by combining two or four magnetoresitive elements into a bridge circuit on the same chip.

Magnetoresistive transducers are capable of detecting lower fields than Hall effect devices, and this has led to their use in reading heads for magnetic storage mediums. Other applications include rotor position sensing for brushless DC machines and, in combination with a low strength permanent magnet, for position measurement.

Magnetoresistors are susceptible to temperature variation and appropriate compensation must be applied in use. This can be achieved by the incorporation of a temperature compensating diode into the detector circuit, by combining four magnetoresistors formed on to the same chip into a bridge circuit, or digitally.

5.1.3 MAGNETODIODE

The presence of a magnetic field will cause a change in the characteristics of a diode due to the effect of the magnetic field in concentrating the injection carriers towards or away from regions with differing recombination rates. Magnetodiodes with sensitivities greater than that of Hall effect devices but with reduced linearity can be produced using silicon on sapphire (SOS) technology.

5.1.4 MAGNETOTRANSISTOR

The structure of the dual collector magnetotransistor is shown in Fig. 5.3. In the absence of an external magnetic field, current is shared equally between the collectors. If an external magnetic field is applied in the plane of the chip this will result in an unequal distribution of current between the collectors. The difference between the collector currents is a linear function of the applied magnetic field.

The MAGFET is analogous to the magnetotransistor and comprises a MOSFET with two drain contacts. Performance is very similar to that of the magnetotransistor.

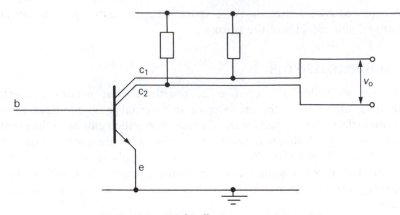

Figure 5.3   Dual collector magnetotransistor.

5.2   Temperature measurements

5.2.1   THERMISTOR

The semiconducting oxides of materials such as iron, cobalt, chromium, manganese, nickel and titanium exhibit a large, exponentially varying resistance with negative temperature coefficient, as expressed by the following equation:

$$R = R_o \exp[\beta(1/T - 1/T_o)] \qquad (5.1)$$

The resulting characteristic is highly non-linear, and appropriate compensation is normally applied when these materials are used for temperature measurement. An application of the thermistor which makes use of the negative temperature coeffficient is the compensation of positive temperature coefficient effects in other devices as, for example, in the control of the gain of an amplifier.

Thermistors are low cost devices and their small size means that they have a relatively low thermal inertia, enabling them to respond to changes in temperature. They are usually used in bridge circuits, and care must be taken to ensure that they are not subject to self-heating caused by the passage of the measuring current as this will distort the measurement.

5.2.2   THERMODIODES AND THERMOTRANSISTORS

(a)   THERMODIODES

The semiconductor diode equation can be written as

$$V_j = kT \log(I/I_s)/q \qquad (5.2)$$

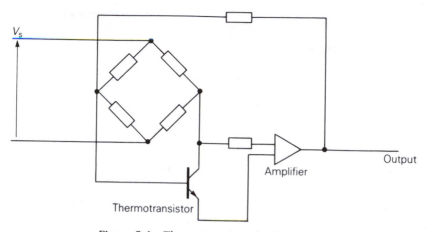

Figure 5.4   Thermotransistor bridge circuit.

where I_s is the reverse leakage current, q is the electronic charge (1.602×10^{-19} C), V_j is the voltage applied across the junction, k is Boltzmann's constant (1.38×10^{-23} J K^{-1}) and T is the temperature.

The voltage drop across a diode carrying a constant current therefore provides a measure of temperature. Devices with sensitivities of the order of 10 mV K^{-1} are available.

The operating range of thermodiodes is limited by the variation of the reverse leakage current I_s with temperature. In the case of silicon, I_s doubles approximately every 7°C from a value of around 25 nA at 25°C.

(b)   THERMOTRANSISTORS

If the collector current of a transistor is held constant then the base-emitter voltage V_{be} will exhibit a near linear variation with temperature expressed by

$$V_{be} = 1.27 - K_t T_a \qquad (5.3)$$

where K_t is a constant and T_a is the absolute temperature (K). The value of 1.27 for V_{be} at absolute zero is independent of chip geometry, biasing and manufacturing tolerances. This means that differences between individual devices of nominally the same characteristics can be compensated by adjusting their respective bias currents to obtain the same value of V_{be} at the calibration temperature. Figure 5.4 shows a single collector thermotransistor incorporated as part of a bridge circuit for temperature measurement.

Thermotransistors for the measurement of absolute temperature are available on a single integrated circuit incorporating all the necessary biasing circuitry and some output signal processing.

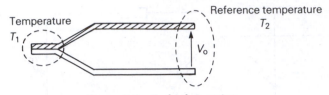

Figure 5.5   Seebeck junction.

5.2.3   SEEBECK EFFECT DEVICES

When the junction between a pair of semiconductor materials is heated or cooled relative to a reference, as in Fig. 5.5, a voltage will be produced at the reference terminals which is a function of the temperature difference T and the Seebeck coefficient for the system, such that

$$V_o = \alpha_s T \tag{5.4}$$

where α_s is the Seebeck coefficient ($V\,K^{-1}$) and $T = T_1 - T_2$.

Silicon thermopiles based on the Seebeck effect have been developed for the measurement of the true RMS value of electric currents, for flow measurement and as radiation detectors.

5.2.4   SOLID STATE PYROMETERS

If a thin slice of a material such as lead zirconium titanate is heated to just below its Curie temperature in the presence of an electric field, the crystals in the slice will become aligned with the applied field and this polarization will be maintained on cooling the slice. This is the pyroelectric effect, used as the basis for a range of solid state pyrometers.

The resulting polarization is then a function of the material used and the wavelength of the incident radiation. Variations in the temperature of the material result in a change in the captive charge at the surface of the slice, which is detected as a change in the potential between a pair of electrodes or can be read using a scanning electron beam. In operation, the pyroelectric element is usually only exposed to the incoming infrared radiation for a short period. In the case of a thermal imaging system using the pyroelectric effect, this is achieved by mechanically chopping the incoming signal at frequencies up to 25 Hz, enabling the continuous monitoring of static images.

A solid state pyrometer can be considered as a temperature sensitive capacitance shunted by a large non-linear resistance with a characteristic such that

$$v = T\,(\lambda A/C) \tag{5.5}$$

where λ is the pyroelectric coefficient of the material, A is the cross-sectional area of the slice, C is the capacitance between the electrodes and T is the temperature.

Because of the fall-off in generated voltage with increasing frequency of the received radiation, solid state pyrometers are normally designed to operate at specific wavelengths within the infrared region of the spectrum. A typical package would include an n-channel FET as an impedance matching element between the pyroelectric element and the preamplifier.

5.3   Mechanical measurements

5.3.1   STRAIN

The piezoresistive effect in semiconductor materials has been utilized to develop silicon strain gauges with gauge factors of the order of 200. Though extremely sensitive, such gauges are non-linear and highly temperature dependent, with changes in temperature appearing as a strain.

The use of micromachining to generate on-chip structures in silicon has enabled the production of pressure transducers with diaphragm thicknesses of 10 to 20 μm on a 200 μm deep chip as in Fig. 5.6, and the cantilever beam accelerometers such as that of Fig. 5.7. In both of these examples the necessary interconnections for the strain gauge elements are incorporated on the chip together with some signal processing.

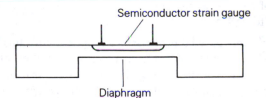

Figure 5.6   Semiconductor diaphragm pressure transducer.

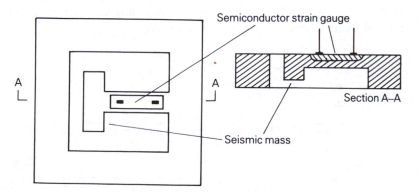

Figure 5.7   Semiconductor cantilever beam accelerometer.

5.3.2   FORCE

The piezoresistive effect is also observed when a p-n junction in silicon is subjected to a mechanical force. The application of this force causes the current/voltage characteristic at the junction to change, enabling measurement of the applied force.

5.4   Chemical measurements

5.4.1   HUMIDITY

Silicon based humidity transducers make use of the Peltier effect, in which the passage of current across the junction between two materials results in the heating or the cooling of the junction. By combining a cooling Peltier junction with a silicon layer of the correct geometry there will be a significant change in the measured capacitance as moisture begins to condense at the surface of the transducer.

An alternative form of humidity transducer uses a porous film sandwiched between a porous electrode and a silicon dioxide (SiO_2) base. As water is absorbed by the film there will be a change in the capacitance measured between the electrodes. Using this technique, a 1 mm² sensing element has been produced which can be incorporated into the gate of a MOSFET mounted on the same chip.

5.4.2   GAS DETECTORS

A range of gas detectors based on the use of catalytic metals has been developed. The selectivity and sensitivity of the transducer is a function of the metal used and the operating temperature. Using this technique, transducers capable of responding to a range of gases including hydrogen, oxygen, carbon monoxide, ammonia and oxides of nitrogen have been produced with sensitivities to 1 ppm in air.

A typical gas detector incorporates the catalytic metal into the gate circuit of a MOSFET as in Fig. 5.8. Practical devices may also incorporate a heating element and temperature sensing on the same chip.

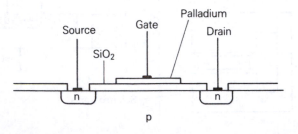

Figure 5.8   MOSFET gas sensor.

Chapter 6

Piezoelectric and ultrasonic sensors and transducers

6.1 Piezoelectric devices

Crystals of materials such as quartz and Rochelle salt, the ceramics lead titanate zirconate and sodium potassium niobate, and polymers such as polyvinylidene, when deformed by the application of an external force, generate a charge between specific planes in the material. Conversely, if voltage is applied to such a crystal a mechanical motion will result. This piezoelectric effect is used as the basis of transducers, with applications ranging from microphones to humidity measurement.

6.1.1 ACCELEROMETERS

Piezoelectric materials are well suited for use as the basis of a range of force transducers for both static and dynamic measurements and form the active element in miniature accelerometers, typical configurations for which are shown in Fig. 6.1a. In each case the effect of acceleration is to cause the seismic mass to exert a force on the piezoelectric material causing a charge to be produced. This charge is monitored by a charge amplifier as in Fig. 6.1b when, assuming an ideal amplifier,

$$e_o = - S_q a/C_f \tag{6.1}$$

where S_q is the accelerometer charge sensitivity and a is the acceleration. In a practical charge amplifier, a feedback resistor would be included as in Figure 6.1c to prevent the build-up of charge on the feedback capacitor C_f.

For accurate measurement an accelerometer should load the structure as little as possible. This is particularly important on light structures when the mass of the accelerometer becomes significant. The effect of introducing the accelerometer can be approximated using the following equations:

$$a_1 = a_2(M_s + m_a)/M_s \tag{6.2}$$

$$f_1 = f_2[(M_s + m_a)/M_s]^{1/2} \tag{6.3}$$

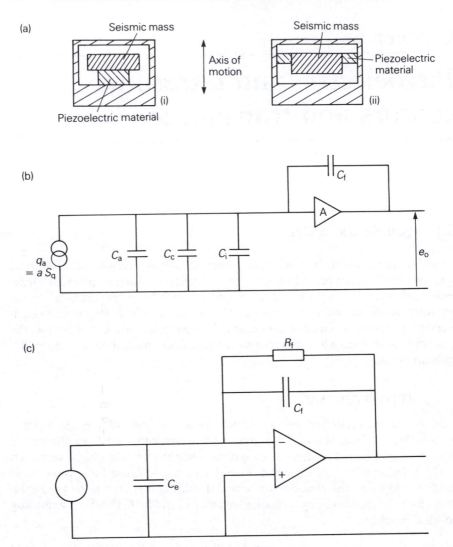

Figure 6.1 Piezoelectric accelerometers. (a) Accelerometer configuration: (i) piezo-electric material in compression (ii) piezoelectric material in shear. (b) Charge amplifier used with a piezoelectric accelerometer: C_a accelerometer capacitance, C_c cable capacitance, C_i input capacitance of amplifier, (c) Practical charge amplifier.

where a_1 and a_2 are the accelerations without and with the accelerometer mounted; f_1 and f_2 are the resonant frequencies without and with the accelerometer mounted; M_s is the effective (dynamic) mass of the structure; and m_a is the mass of the accelerometer. As a general rule the mass of the accelerometer should be no greater than one-tenth of the effective mass of the structure on which it is mounted.

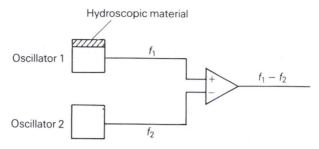

Figure 6.2   Humidity sensor.

6.1.2 HUMIDITY MEASUREMENT

If a piezoelectric crystal oscillator is coated with a hygroscopic material then its natural frequency of oscillation will change as the coating absorbs moisture. By using two matched crystals as in Fig. 6.2, one of which carries such a hygroscopic coating and the second of which is held under stable conditions, the difference frequency between the two oscillators provides a measure of humidity.

6.1.3 SURFACE ACOUSTIC WAVE DEVICES

A surface acoustic wave (SAW) is an elastic wave which is propagated along the surface of a piezoelectric material. For a homogeneous material the amplitude and phase velocity of the SAW is determined by the piezoelectric, elastic, conductive and dielectric properties of the material. For multilayered devices, the physical properties and thickness of each layer determine the frequency of the SAW.

SAW devices are capable of operating in the gigahertz region and can be formed into oscillators, filters, delay lines and resonators. If one of these physical properties can be modulated there will be a change in the nature of the SAW which can be used for measurement purposes.

SAW transducers are available for the measurement of voltage, pressure, force and temperature. These use self-oscillating SAW structures whose frequency of oscillation is changed by the presence of the appropriate external influence. The main area of application to date has been in gas detectors. These use a dual delay line oscillator such as that of Fig. 6.3, one-half of which is coated in an interface material which interacts with the gas molecules to produce a velocity change in the acoustic wave which is proportional to the concentration of the gas. The resulting phase shift between the two halves of the oscillator gives a measure of the gas concentration. The use of the dual delay line structure provides a first order compensation for factors such as temperature, humidity and ageing.

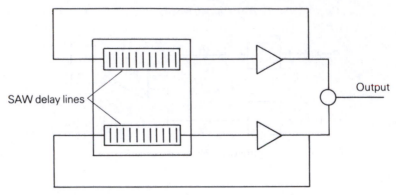

Figure 6.3   Dual delay line surface acoustic wave (SAW) structure.

6.1.4   LIGHT MODULATION

Piezoelectric crystals exhibit an electro-optical effect in which the change in their dielectric properties on the application of a voltage causes a corresponding change in their optical properties. Light modulators based on this principle are used in association with lasers for communication systems.

6.1.5   PIEZOELECTRIC ACTUATORS

Piezoelectric ceramics are increasingly being used as actuators to provide precise linear displacements of the order of 1 μm with accuracies approaching 10 nm. Such actuators are finding increasing application in optical systems, scanning electron microscopes, manufacturing and micromachining. On a larger scale, resonant piezoelectric devices have been used to produce miniature fans for use in electronics installations, while piezoelectric motors have been developed for applications such as autofocus camera lens drives.

6.2   Ultrasonic systems

6.2.1   SOURCES

(a)   PIEZOELECTRIC

A crystal of piezoelectric material will oscillate mechanically when subjected to an alternating voltage across appropriate faces of the crystal. Maximum intensity of oscillation will be achieved when the frequency of the exciting signal matches the resonant frequency of the crystal. This is determined by the thickness of the crystal between the exciting electrodes; resonance occurs when the thickness of the crystal is equal to half the mechanical wavelength in the material. A number of crystals can be combined in a mosaic to provide a larger radiating surface and to increase the energy of the source. By using ceramic materials, which are themselves polycrystalline, sources can be shaped to focus their energy into a small volume to produce a localized region of high intensity ultrasound.

For very high frequencies of ultrasound, sources based on a very thin layer of piezoelectric material have been developed. To produce these sources, semiconductor technology is used either to place a thin insulating layer within a semiconducting piezoelectric material or to deposit a thin layer of such material on to a suitable substrate.

(b) MAGNETOSTRICTIVE

In a magnetostrictive material the application of a magnetic field causes a change in the physical dimension in the direction of the field, enabling its use as an ultrasonic source. A typical magnetostrictive source will consist of a central core and winding and a focusing cone, as shown in Fig. 6.4. Direct current is applied to produce an initial extension of the core, and the AC signal is then superimposed upon this DC level. In operation at resonance, a node will be established at some point along the length of the transducer and this is used as the mounting point to avoid any loss of energy to the mounting. Overall efficiency will, however, be less than that of a piezoelectric source because of losses in the magnetostrictive material of the core.

(c) MECHANICAL

Sirens operate by mechanically interrupting a flow of gas through a series of holes, and can provide several hundred watts of ultrasonic energy at frequencies of around 30 kHz with an efficiency of the order of 70%. Whistles utilize the kinetic energy in a gas stream and can produce several watts of power at frequencies of the order 40 to 50 kHz; the efficiency is, however, lower than that of a siren.

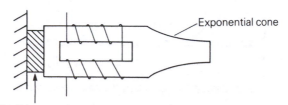

Figure 6.4   Magnetostrictive ultrasonic source.

(d)   ELECTROMAGNETIC AND ELECTROSTATIC

Configurations similar to conventional audiofrequency loudspeakers can be used to generate ultrasonic frequencies and may have advantages in certain environments, for example where high temperatures are anticipated.

6.2.2   COUPLING OF THE SOURCE

In many applications the ultrasonic transmitter is required to be closely coupled to some part of a structure or system. In practice, for example due to surface irregularities, it may not be possible to achieve the required level of coupling by mounting the transmitter directly on to the surface, in which case a coupling medium is used. Water is the most common form of coupling medium; special materials are available for use where high or low temperatures are involved.

6.2.3   RECEIVERS

The transducers used for the generation of ultrasonic frequencies usually work in the reverse sense and can be used as receivers for ultrasonic frequencies. In pulsed systems this means that it is possible for the same transducer to act as both the transmitter and the receiver of the ultrasonic pulse.

6.2.4   ULTRASONIC FLOW MEASUREMENT

Ultrasonics provides a means for the non-invasive measurement of the flow of liquids in pipes. A variety of methods are available operating on very different principles, and the appropriate arrangement must be chosen for a particular application.

In order to operate, an ultrasonic flowmeter requires the establishment of a stable flow profile in the pipe at the measurement point. This in turn means that an appropriate length of straight pipe must be introduced before the measurement point to allow the required flow profile to become properly established.

(a)   DOPPLER SHIFT FLOWMETER

The Doppler shift flowmeter of Fig. 6.5 utilizes the presence within the fluid of scattering elements such as solid particles, bubbles or eddies. These produce a

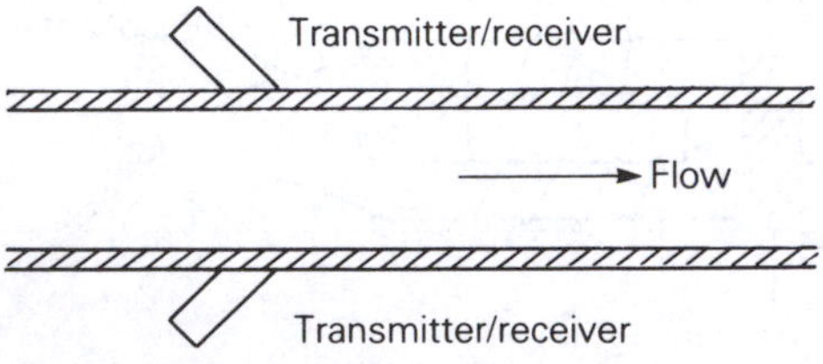

Figure 6.5   Doppler flowmeter.

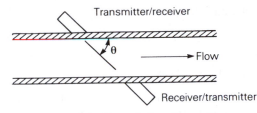

Figure 6.6   Transit time flowmeter.

shift in the return frequency which is a function of their speed of motion, hence enabling the speed of the fluid to be measured.

Measurement performance is a function of the flow profile; the number, size and distribution of the scattering particles; the pipe wall thickness and the quality of the acoustic coupling; and the knowledge of the speed of sound in the fluid. By mounting the transmitter and receiver flush with the inside of the pipe, greater accuracy can be achieved.

(b)   TRANSIT TIME FLOWMETER

The transit time or ring-around flowmeter is shown in Fig. 6.6. The flow of fluid in the pipe causes a variation in the transit times of the downstream and upstream signals. The difference between these two times is a function of the flow rate.

The difference in the transit times can be measured directly as a change in phase or as a frequency shift. The latter has the advantage that it eliminates the need for knowledge of the velocity of sound in the fluid medium, since

$$f_u = 1/t_u = (v_s - v_f\cos\theta)/d \qquad (6.4)$$

$$f_d = 1/t_d = (v_s + v_f\cos\theta)/d \qquad (6.5)$$

where t_u and t_d are the upstream and downstream transit times respectively, v_s is the velocity of sound in the fluid, v_f is the velocity of the fluid, and f_u and f_d are the effective frequencies of the signals travelling upstream and downstream

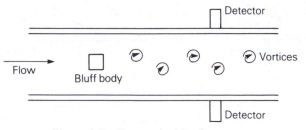

Figure 6.7   Vortex sheddin flowmeter.

respectively. The difference frequency is

$$f = f_d - f_u = (2v_f \cos \theta)/d \tag{6.6}$$

(c)   VORTEX SHEDDING FLOWMETER

The vortex shedding flowmeter of Fig. 6.7 uses a bluff body placed in the flow to generate a series of vortices in the flow which can then be detected by ultrasonic or other means. As the rate at which the vortices are produced is a function of the flow velocity, this can be determined by counting the number of vortices passing a particular point in a given time.

(d)   CROSS-CORRELATION FLOWMETER

The cross-correlation flowmeter shown in Fig. 6.8 makes use of the presence of random disturbances in the body of the fluid to modulate the transmitted signal across the pipe. By correlating the received signal at one part of the pipe with that from a point displaced in the direction of flow, the transit time of the disturbance between the two measuring points and hence the fluid velocity can be obtained.

6.2.5   ULTRASONIC DISTANCE MEASUREMENT

Ultrasonic rangefinders measure the time delay between the transmission of a pulse and the receipt of the reflected pulse. For a velocity of sound in the medium of v_s, the distance to the target object is given by

$$d = v_s t \tag{6.7}$$

where t is the time delay.

As the velocity of sound in a material differs from material to material, and also with the temperature of the material, appropriate compensation must be included in the measurement to minimize the error introduced.

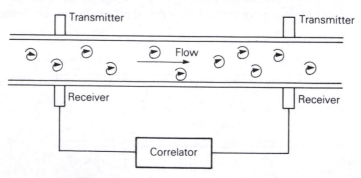

Figure 6.8   Cross-correlation flowmeter.

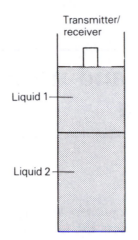

Figure 6.9   Ultrasonic liquid level gauge.

Rangefinders operating on this basis have been used for obstacle detection, liquid level measurement and the positioning and control of robot manipulators.

6.2.6   ULTRASONIC MEASUREMENT USING VARIATION IN TRANSMISSION VELOCITY

The velocity of sound in a material differs from material to material and can therefore be used as a means of measurement. Consider the situation of Fig. 6.9, in which two immiscible liquids are present in the same tank. The transit time for a pulse of ultrasound from the transmitter in the base of the tank to the receiver on the surface will be a function of the relative depths of the two liquids.

6.2.7   ULTRASONIC IMAGING

Ultrasonic scanning has been in use for a number of years in medicine as an alternative to X-rays for the production of internal images of the body. Current techniques include the production of real-time two-dimensional images and three-dimensional tomography. Processing of the information in the images is carried out in a similar fashion to that for vision systems.

Sonar imaging systems based on side-scan sonar and within-pulse sector scanning sonar are used to build up profiles of a seabed or other surface. The nature of sonar means that it can compete effectively with vision in certain environments, such as muddy or turbid water.

Chapter 7

Interference and noise in measurement

All measurements are subject to error. Indeed, the very act of introducing a transducer into a system distorts the system and forms a possible source of error. It is important in any measurement system to reduce the sources of error to a level consistent with the quality of the information required and to quantify the maximum error that can result.

Errors fall into two main categories: systematic errors arising from the configuration of the system; and random errors due to noise and interference.

7.1   Interference

Interference is a random effect resulting from the influence or external parameters on the system, the major sources of which are time varying electrostatic and magnetic fields.

7.1.1   COMMON MODE REJECTION RATIO

Referring to Fig. 7.1, the input voltages v_1 and v_2 can be expressed in terms of their sum and difference as

$$v_1 = (v_1 + v_2)/2 + (v_1 - v_2)/2 \qquad (7.1)$$

$$v_2 = (v_1 + v_2)/2 - (v_1 - v_2)/2 \qquad (7.2)$$

The differential mode voltage v_{dm} and the common mode voltage v_{cm} are then expressed by

$$v_{dm} = (v_1 - v_2)/2 \qquad (7.3)$$

$$v_{cm} = (v_1 + v_2)/2 \qquad (7.4)$$

These voltages will give rise to output voltages v_{odm} and v_{ocm} such that

$$v_{odm} = G_{dd}v_{dm} + G_{cd}v_{cm}$$

$$v_{ocm} = G_{dc}v_{dm} + G_{cc}v_{cm} \qquad (7.5)$$

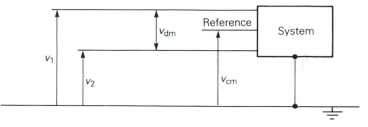

Figure 7.1   Common and differential mode voltages.

where G_{dd} is the differential mode to differential mode gain, G_{cd} is the common mode to differential mode gain, G_{dc} is the differential to common mode gain and G_{cc} is the common mode to common mode gain.

The ability of a circuit to function in the presence of a common mode voltage is expressed by the common mode rejection ratio (CMRR), defined as

$$CMRR = 20 \log (G_{dd}/G_{cd})$$

This ratio typically lines in the range 60 dB (poor) to 120 dB (good). Improvement in the CMRR can be achieved by the isolation of inputs with respect to ground, the balancing of amplifier inputs, the use of guard circuits and active screening.

7.1.2   GROUND OR EARTH LOOPS

Ground or earth is ideally an equipotential reference which is used to provide a common return path in a system. In practice this equipotential condition is not always maintained, as is illustrated by the fact that significant potential differences can exist between the various parts of a ship or aircraft. The presence of transients such as those caused by switching or lightning can also result in the production of a potential difference, in this case of several thousand volts, if proper precautions are not taken.

The existence of such high voltages not only threatens the operation of equipment but presents a major risk to its operators. For this reason, safety earths may well need to be deployed to limit any rise in potential.

Consider the situation illustrated by Fig. 7.2. Here, a system is earthed at two points with an impedance between them. The presence of stray currents in this earth path will result in the production of a common mode voltage which, when $R_1 \neq R_2$, will result in unequal line currents and hence unequal lead voltage drops. The resulting differential mode voltage produces an error in the measured voltage referred to as the series mode voltage error. A similar problem arises when equipments are mounted in a metal rack to which they are electrically connected.

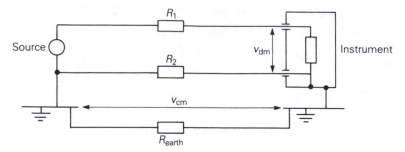

Figure 7.2   Earth loop effects, with common mode and differential mode voltage.

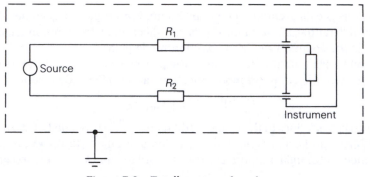

Figure 7.3   Totally screened enclosure.

7.1.3   ELECTROSTATIC INTERFERENCE: SCREENING AND GUARDING

Electrostatic interference occurs as a result of the presence of stray or parasitic capacitance which provides a path for the interfering signal. The function of screens and guards is to reduce or eliminate the effect of this stray capacitance.

Consider the screened enclosure of Fig. 7.3. The presence of stray capacitance between the screen and the internal components often requires that the internal ground reference of the screened circuit is connected to the screen at some point to eliminate possible feedback paths, as illustrated by Fig. 7.4a and 7.4b. However, care must be taken to ensure that all connections to the screen are made at a single common point in order to prevent the production of common mode voltages by circulating currents in the screen.

Where one of the input terminals to the system is earthed then only partial screening can be achieved, as in Fig. 7.5. However, circulating currents can flow through the screen and earth leads, and a loop is created into which additional voltages and currents can be induced by any time varying magnetic fields that are present.

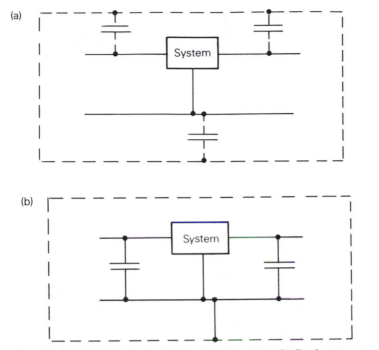

Figure 7.4 Control of capacitive feedback: (a) capacitive feedback via screen (b) elimination of capacitive feedback by connection of screen to system.

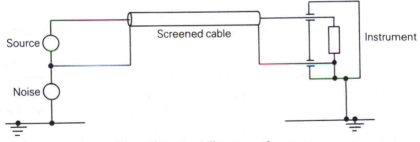

Figure 7.5 Partially screened system.

If a differential input stage is provided then the ground loop can be eliminated to provide improved screening and common mode rejection by the arrangement of Fig. 7.6, in which a twin core screened cable is used to connect the signal to the instrument.

A further improvement in performance can be achieved through the introduction of a guard circuit such as that of Fig. 7.7. This consists of either an internal screening box or an arrangement of tracks on a printed circuit board which is

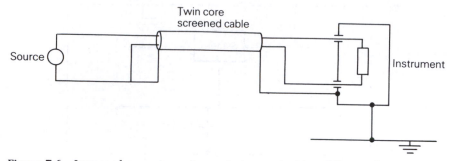

Figure 7.6   Improved screening using as instrument with a differential input stage.

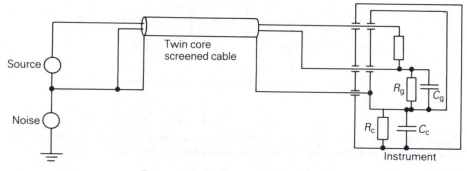

Figure 7.7   Instrument with guard.

then connected to the screen. The guard circuit is driven at the common mode voltage, preventing the flow of current between the other terminals and the guard.

While screening can protect a system from external sources of interference, the presence of induced screen currents is itself a source of interference as a result of capacitive coupling between the screen and the internal conductors. Active screens reduce this effect by using a double screened cable and buffer amplifier as in Fig. 7.8. The inputs of this amplifier are connected to the signal and outer screen with the output connected to the inner screen. This has the effect of forcing the voltage of the inner screen to become equal to that of the signal and renders ineffective any capacitance between the signal conductor and the inner screen. Any current flowing in the capacitance between the inner and outer screens is provided by the buffer amplifier.

Where earthing is a problem, and for reasons of operator and equipment safety it is not possible to use a single earth point, then some means of breaking any possible earth loop must be provided as in Fig. 7.9. This can be achieved by the use of transformers or isolation amplifiers in the data transmission line and by the use of fibre optic or radio telemetry links.

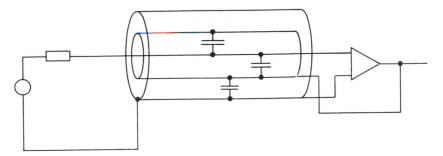

Figure 7.8   Active screen arrangement.

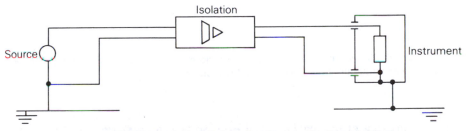

Figure 7.9   Breaking the earth loop by the introduction of isolation.

7.1.4   ELECTROMAGNETIC INTERFERENCE

If the system is operated in the presence of a time varying magnetic field produced by an alternating current then a voltage will be induced into the system which will appear as noise at the terminals. The nature of the introduced interference is such that it is much more difficult to eliminate than that caused by electrostatic effects, particularly if the system currents are themselves the source of the interfering magnetic fields.

Protection from external magnetic fields is achieved by careful design of the circuit layout to place sensitive components as far as possible from sources of interference; by avoiding any conductors running parallel to a magnetic field; by crossing regions of known magnetic field with conductors at right angles to the direction of the field; by reducing the lengths of conductors; and by minimizing the area of any loops in the circuit. The use of twisted pairs of wires for interconnections assists in this; such pairs produce a cancellation of magnetic field when the conductors are carrying equal and opposite currents.

Further protection can be achieved by placing the system in an enclosure made of a high permeability material. The choice of this material will depend on the frequency of the interference and the resulting penetration of the field into the material.

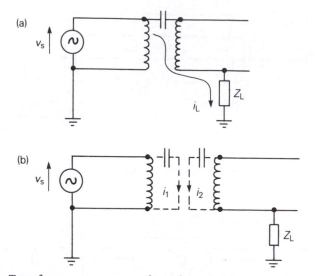

Figure 7.10   Transformers as a source of interference: (a) capacitive coupling between transformer windings (b) double screened transformer.

7.1.5   POWER SUPPLIES AS A SOURCE OF INTERFERENCE

Power supply transformers are a major source of interference, particularly in the form of the mains hum caused by capacitive currents in the secondary of a transformer producing a common mode voltage. Where an isolation transformer is included in an instrument the capacitive coupling between the primary and secondary shown in Fig. 7.10a results in parasitic currents to earth in the secondary which are a further source of common mode voltages. These currents may be controlled by the introduction of a double screen into the transformer as in Fig. 7.10b.

7.2   Noise

Noise is considered to be a signal varying randomly in both frequency and amplitude. It may be generated within circuit components by a combination of thermal effects and the structure of the material. Alternatively, it may result from external factors such as the presence of electrostatic and electromagnetic fields.

The ability of a system to function and to transmit information in the presence of noise is defined by its signal-to-noise ratio. In any particular application, the combination of measuring technique and associated signal processing used will significantly influence this ratio and define the levels of signal that can be

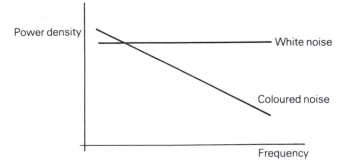

Figure 7.11   Noise characteristics.

detected and used. Digital techniques, particularly for communication, have had a significant impact in reducing the influence of noise downstream of the transducer. However, the sensing element remains vulnerable to noise effects, especially where the intensity of the signal being measured is low.

7.2.1   WHITE AND COLOURED NOISE

White noise is characterized by a constant power spectral density over a wide range of frequencies. Where the noise signal has a Gaussian amplitude distribution, as for example in the case of thermal or shot noise, then this is referred to as Gaussian white noise.

Band limited noise, in which the power spectral density rolls off at a constant rate, is referred to as coloured noise.

Spectral density characteristics representing both types of noise are shown in Fig. 7.11.

7.2.2   SOURCES OF NOISE

(a)   THERMAL NOISE

The random movement of electrons within a conductor causes a variation in the voltage at the terminals of the conductor. The power contained in this noise signal can be expressed in terms of the absolute temperature T, Boltzmann's constant k $(1.38 \times 10^{-23}\,\mathrm{J\,K^{-1}})$ and the bandwidth B as

$$\text{noise power} = P_n = 4kTB \quad \text{watts} \tag{7.6}$$

Referring to Fig. 7.12, this power can also be expressed in terms of an equivalent noise current or equivalent noise voltage and a noise resistor as follows:

$$P_n = e_n^2/R_n \tag{7.7}$$

$$P_n = i_n^2/R_n \tag{7.8}$$

Figure 7.12   Noise circuits.

(b)   SHOT NOISE

In a semiconductor device there is a random motion of charge carriers across potential barriers which results in a noise component in the output signal of the device. This may be expressed in terms of a noise current as

$$i_n^2 = 2qI_{DC}B \qquad (7.9)$$

where q is the charge on an electron (1.6×10^{-19}C), I_{DC} is the direct current across the potential barrier and B is the signal bandwidth.

The current I_{DC} can be found from the diode equation as

$$I_{DC} = I_s[\exp(qV_j/kT) - 1] \qquad (7.10)$$

where I_s is the reverse leakage voltage and V_j is the voltage across the junction.

(c)   STELLAR OR COSMIC NOISE

Stellar or cosmic noise represents the background noise level of the universe and arises from a variety of sources. The total level of noise is referred to as the sky noise temperature and is generally low in relation to other forms of noise, except in the direction of galactic radio sources. The sky noise temperature is lowest in the microwave region, where there is window of low noise levels which is used for communications purposes.

(d)   FLICKER

Flicker or low frequency noise is associated with the passage of charge carriers through a discontinuous medium and relates to the movement of the charge carriers across the discontinuities. To a first approximation, flicker is inversely proportional to frequency and tends to become dominant below about 100 Hz. Flicker may be described by the flicker current, given as

$$i_f^2 = KI_{DC}^a B(1/f)^\beta \qquad (7.11)$$

where K is a device dependent constant, I_{DC} is the direct current through the medium, B is the bandwidth and f is the frequency. The constant a lies in the range of 0.5 to 2 and is a function of the medium, while the constant β varies from 1 for pink noise to 2.7 for cosmic noise.

7.2.3   NOISE FACTOR

The noise factor NF is used to express the generation of noise within a system by comparing the signal-to-noise ratios at the input and the output of the system. it is expressed by

$$NF = 10 \log (SNR_{in}/SNR_{out}) \qquad (7.12)$$

7.2.4   SIGNAL-TO-NOISE RATIO

The signal-to-noise ratio is defined in terms of the ratio of the power in the input signal and the power in the noise signal referenced across a common value of resistance, typically 1 ohm. It may then be expressed by

$$SNR = 10 \log (v_{sig}^2/e_n^2) \qquad (7.13)$$

where e_n is the noise voltage.

Chapter 8

Signal processing

While it is possible that the output signal from a transducer could be used directly, this is not the case in the majority of applications. At the simplest level this may be because the transducer output is of the wrong amplitude and requires the introduction of some gain or attenuation to match its output to the next stage of the system. In a complex environment, the raw signal from the transducer may contain additional components which must be removed in order to obtain the required information. These unwanted components may be introduced by the presence at the input to the transducer of random signals of the same type as those to be monitored, as a result of noise, as a function of the characteristic of the sensing element itself (for example, thermal effects in silicon), or as a consequence of a combination of these and other factors.

The role of signal processing is to isolate the required information contained in the output signal from the transducer and present it in a suitable form to the next stage of the system. While significant signal processing is performed using analogue techniques, the use of computers has enabled the introduction of complex analytical techniques for the extraction of information.

Increasingly signal processing is being provided at the transducer, and in the case of many silicon based transducers is being incorporated on to the same chip as the sensing element. The development of such 'smart sensors', with the ability to compensate for specific effects such as temperature and containing a self-checking function as well as a local decision making capability, is a major development in measurement.

8.1 Operational amplifiers

The basic building block for analogue signal processing systems is the operational amplifier of Fig. 8.1. This is a difference amplifier, the output of which is an amplified version of the difference between the voltage signals at its inverting (−) and non-inverting (+) inputs. Referring to Fig. 8.1:

$$v_o = A(v_1 - v_2) \qquad (8.1)$$

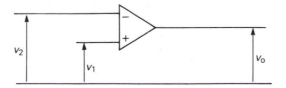

Figure 8.1   The operational amplifier.

As an operational amplifier has a high intrinsic gain and a very high input impedance it may be considered, for many applications, as an ideal amplifier operating with zero input current and voltage ($i_a = 0$ and $v_a = 0$). Referring to Fig. 8.2, in which the amplifier is used with negative feedback:

$$i_1 = (v_i - v_a)/Z_1 \tag{8.2}$$

$$i_2 = (v_o - v_a)/Z_2 \tag{8.3}$$

Since $i_a = 0$ and $v_a = 0$, point a is a virtual earth and

$$i_1 = -i_2 \tag{8.4}$$

Hence

$$v_o = -(Z_1/Z_2)v_i \tag{8.5}$$

The effective input impedance in this configuration is Z_1; the output impedance is effectively zero, in which case the amplifier is acting as an ideal voltage source.

If the two impedances Z_1 and Z_2 are both resistive then amplification together with a 180° phase change is obtained. By connecting the output voltage v_o to a potentiometer as in Fig. 8.3 the gain may be varied, since

$$v_o/v_i = -(R_3 + R_4)R_2/(R_1R_3) \tag{8.6}$$

This arrangement has the advantage that a relatively low value of R_2 can be used, minimizing time constant effects due to stray capacitance.

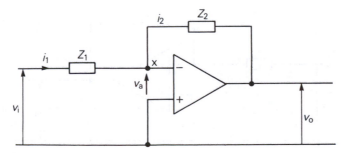

Figure 8.2   Operational amplifier with negative feedback.

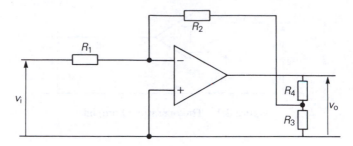

Figure 8.3   Operational amplifier with potentiometer feedback.

8.1.1   INTEGRATOR

If instead of a resistor the impedance Z_2 is a capacitor, as in Fig. 8.4, the relationship between v_o and v_i becomes

$$v_o = - (1/R_1 C) \int v_i \, dt \qquad (8.7)$$

since

$$i_2 = dq/dt = d(Cv_o)dt \qquad (8.8)$$

8.1.2   BUFFER AMPLIFIER

Many transducers have a limited output current capability and must be operated into a high impedance. An operational amplifier connected as a non-inverting amplifier in the configuration of Fig. 8.5 can be used for this purpose, in which case

$$v_x = v_o R_1/(R_1 + R_2) \qquad (8.9)$$

Since $v_a = 0$ for an ideal amplifier,

$$v_x = v_i \qquad (8.10)$$

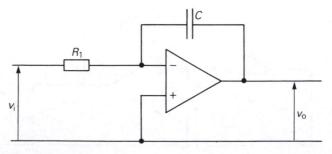

Figure 8.4   An operational amplifier used as an integrator.

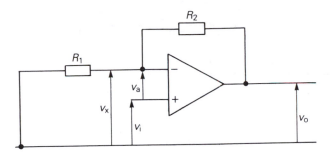

Figure 8.5   Non-inverting buffer amplifier.

and thus

$$v_o = v_i(1 + R_2/R_1) \tag{8.11}$$

8.1.3   CURRENT TO VOLTAGE CONVERTER

Many transducers provide an output in the form of a current, which in many applications must be converted into a voltage signal. This can be achieved by connecting the transducer to the inverting input of the operational amplifier as in Fig. 8.6. Assuming that the amplifier requires zero current, then

$$i_s + i_r = 0 \tag{8.12}$$

and

$$v_o = -i_s R \tag{8.13}$$

8.1.4   VOLTAGE TO CURRENT CONVERTER

Conversely, it may be required to convert a voltage signal into a current signal. If the target load is connected in the feedback loop of the operational amplifier as in Fig. 8.7, the load current is

$$i_L = -v_i/R_1 \tag{8.14}$$

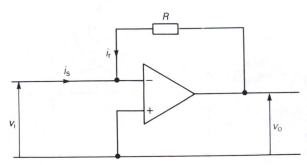

Figure 8.6   The operational amplifier as a current to voltage converter.

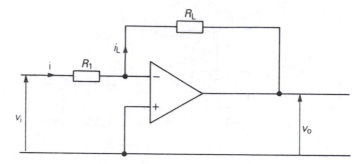

Figure 8.7   The operational amplifier as a voltage to current converter.

Care must be taken to select an amplifier capable of delivering the required load current when operated in this mode.

8.1.5   LOGARITHMIC AMPLIFIER

A logarithmic amplifier connects either a diode or a transistor in the feedback loop of the operational amplifier. For the configuration of Fig. 8.8 the relationship between i_2 and v_o will be

$$i_2 = I_{bes} \exp(qv_{be}/kT) \tag{8.15}$$

where I_{bes} is the saturation current for the base-emitter junction, q is the electronic charge (1.602×10^{-19} C), k is Boltzmann's constant (1.38×10^{-23} J K^{-1}) and T is the temperature (K).

The relationship between v_o and v_i is then

$$v_o = -c_1 \log_e (c_2 v_i) \tag{8.16}$$

where $c_1 = kT/q$ and $c_2 = 1/(R_1 I_{bes})$.

Logarithmic amplifiers using an npn transistor in the feedback loop will handle

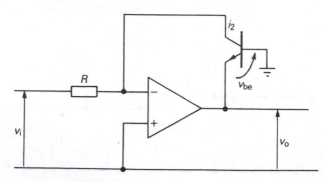

Figure 8.8   A logarithmic amplifier.

positive input signals only. Negative input signals can be accommodated by replacing the npn transistor by a pnp transistor.

8.1.6   CHARGE AMPLIFIER

The output from a piezoelectric transducer takes the form of a variation in the charge between specific faces of the crystal. A charge amplifier is used to convert this charge variation into a voltage signal. A typical configuration for a charge amplifier based on an operational amplifier is shown in Fig. 8.9, in which the resistance R_f is included to prevent a build-up of charge on the feedback capacitor C_f. Assuming an ideal amplifier, the sensitivity and time constant are independent of the capacitance of the crystal and of the connecting cable.

8.1.7   DIFFERENTIAL AMPLIFIER

In many applications a difference voltage must be amplified. This can be achieved using the circuit of Fig. 8.10. The output voltage of this arrangement is

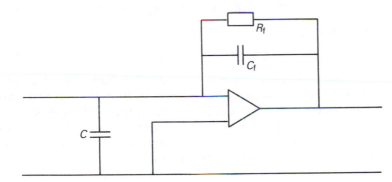

Figure 8.9   Charge amplifier.

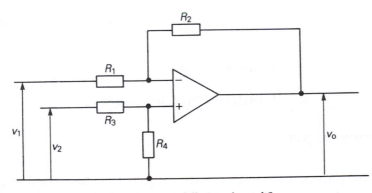

Figure 8.10   A differential amplifier.

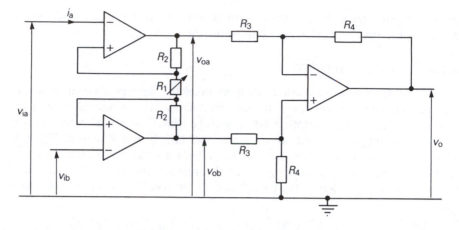

Figure 8.11   Instrumentation amplifier.

expressed in terms of the input voltages as

$$v_o = (1 + R_2/R_1)[R_4/(R_3 + R_4)]v_{ib} - (R_2/R_1)v_{ia} \qquad (8.17)$$

Setting $R_3 = R_1$ and $R_4 = R_2$,

$$v_o = (R_2/R_1)(v_{ib} - v_{ia}) \qquad (8.18)$$

In practice, the value of gain that can be obtained is limited by the need to keep the input impedance, and hence R_1, at a sufficiently high value to maintain a high common mode rejection ratio. A high input impedance form of differential amplifier, referred to as an instrumentation amplifier, uses three operational amplifiers arranged as in Fig. 8.11 to overcome this problem and is available as a device in its own right. For the arrangement shown,

$$i_a = (v_{ia} - v_{ib})/R_1 = (v_{oa} - v_{ob})/(R_1 + 2R_2) \qquad (8.19)$$

Rearranging,

$$v_{oa} - v_{ob} = (R_1 + 2R_2)(v_{ia} - v_{ib})/R_1 \qquad (8.20)$$

Referring to equation 8.18,

$$v_o = (R_4/R_3)(v_{oa} - v_{ob})$$

Hence

$$v_o = (R_4/R_3)(1 + 2R_2/R_1)(v_{ia} - v_{ib}) \qquad (8.21)$$

8.1.8   COMPARATOR

Where a comparison between two signals is required, an operational amplifier can be used in the manner of Fig. 8.12. Because of the high gain of the amplifier

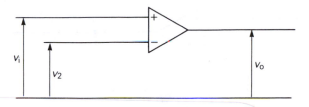

Figure 8.12   Comparator.

only a very small difference signal, of the order of 0.1 μV, is required for the output to swing from the positive to the negative rail voltage and vice versa. In practice, rather than an operational amplifier, a comparator integrated circuit would be used which is specially designed to give a rapid response with minimum error.

8.1.9   SCHMITT TRIGGER AMPLIFIER

If an operational amplifier is connected as in Fig. 8.13a, the output voltage will vary in the manner shown in Fig. 8.13b. The values of v_i at which the

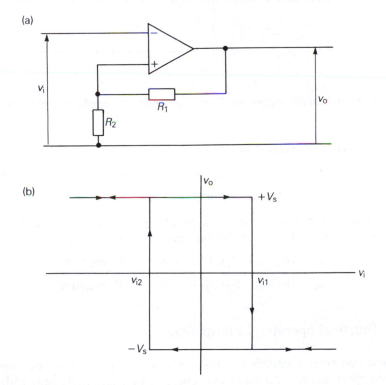

Figure 8.13   Schmitt trigger amplifier: (a) amplifier configuration (b) output.

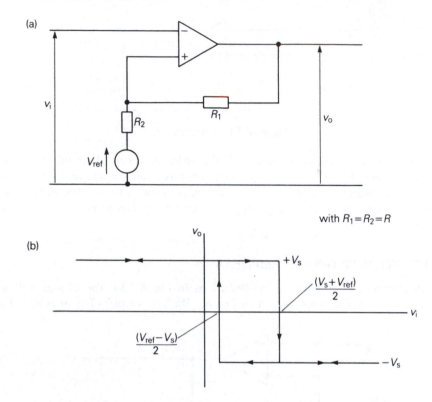

Figure 8.14 A Schmitt trigger amplifier with offset: (a) amplifier configuration (b) output.

transitions occur are defined by

$$v_{i1} = V_s R_2/(R_1 + R_2) \qquad V_s \text{ positive} \tag{8.22}$$

$$v_{i2} = V_s R_2/(R_1 + R_2) \qquad V_s \text{ negative} \tag{8.23}$$

By including a reference voltage V_{ref} as in Fig. 8.14a, the output is shifted as shown by Fig. 8.14b. The transition voltages then become

$$v_{i1} = (R_1 V_{\text{ref}} + R_2 V_s)/(R_1 + R_2) \qquad V_s \text{ positive} \tag{8.24}$$

$$v_{i2} = (R_1 V_{\text{ref}} - R_2 V_s)/(R_1 + R_2) \qquad V_s \text{ negative} \tag{8.25}$$

8.2   Practical operational amplifiers

Practical operational amplifiers are vulnerable to errors caused by variation in input offset voltage and input bias current, by changes in gain with time and by bandwidth and slew rate limitations, each of which must be accounted for in the choice of amplifier and the design of the associated circuitry. Of particular importance are factors which affect the long term stability of the amplifier. By using low drift, highly stable devices such as chopper stabilized amplifiers and auto-zeroing amplifiers these effects can be minimized, but regular checking and recalibration are still required if the long term performance of the system is to be maintained.

8.2.1   AMPLIFIER ERRORS

(a)   INPUT OFFSET VOLTAGE

If the operational amplifier is not perfectly symmetrical then a non-zero output voltage will be produced when both inputs are connected together and to earth. Negative feedback reduces the effect of the offset voltage, but full compensation is only obtained by means of external circuitry to cancel the offset.

(b)   INPUT BIAS CURRENTS

Input bias currents are particularly associated with operational amplifiers with bipolar transistor input stages. The presence of these currents results in the production of bias voltages in the external resistances, which then appear as an offset in the output voltage. The effect of input bias currents can be minimized by arranging that their associated bias voltages are the same for the inverting and non-inverting inputs when cancellation takes place. This can be achieved by including a bias resistor in the connection to the non-inverting input as in Fig. 8.15. The value of this resistor is determined by the relationship

$$R_{bias} = R_1 R_2 / (R_1 + R_2) \tag{8.26}$$

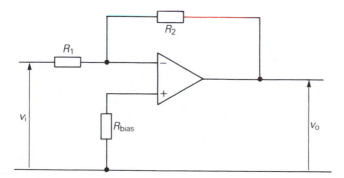

Figure 8.15   Operational amplfiier with bias resistor.

(c) DRIFT

Drift is a problem encountered in all DC coupled amplifiers. It is impossible to distinguish the effect of drift in the output voltage, brought about by long term effects such as ageing and variations in gain supply voltage, from the effect on the output signal of slowly varying input signals. Drift is expressed in terms of $\mu V\,{}^\circ C^{-1}$ for temperature effects, $\mu V\,V^{-1}$ for supply voltage variations, and μV per month or μV per year for ageing.

(d) FREQUENCY RESPONSE

The frequency response of a typical operational amplifier is shown in Fig. 8.16, and has 3 dB bandwidth of around 10 Hz with a unity gain bandwidth of several kilohertz. With the use of negative feedback, the effective gain A_{eff} is reduced to

$$A_{eff} = A/(1 + A\beta) \qquad (8.27)$$

where A is the DC gain and β is the feedback fraction. The bandwidth is correspondingly increased to

$$B_{eff} = B(1 + A\beta) \qquad (8.28)$$

and yields a constant gain-bandwidth product.

(e) SLEW RATE

The maximum frequency that an operational amplifier can accommodate may be limited not by its frequency response but by the maximum rate of change of input signal to which it can respond. This is referred to as the slew rate, and defines the maximum rate of change of output voltage that can occur for a step

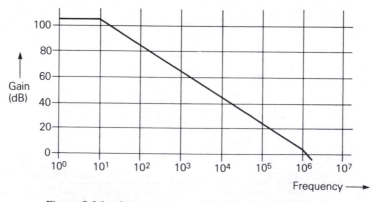

Figure 8.16 Operational amplifier frequency response.

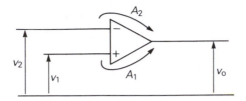

Figure 8.17   Operational amplifier with differential gains.

change at the input. Slew rate has dimensions of $V s^{-1}$ but is more usually expressed in $V \mu s^{-1}$, and is typically of the order of $0.5 V \mu s^{-1}$.

The slew rate is related to the full power bandwidth of the operational amplifier, which defines the highest frequency, full voltage sine wave that can be applied at the input without slew rate limitations applying.

(f)   VARIATION IN GAIN

When an amplifier is used in the differential mode, any variation between the inverting and non-inverting gains (A_1 and A_2 respectively) will introduce errors into the output voltage signal. This may be illustrated by reference to Fig. 8.17, the output voltage for which is

$$v_0 = A_1 v_1 - A_2 v_2 \tag{8.29}$$

Rearranging,

$$v_0 = G_{dm}(v_1 - v_2)/2 + G_{cm}(v_1 + v_2)/2 \tag{8.30}$$

where $G_{dm} = (A_1 + A_2)$ and $G_{cm} = (A_2 - A_2)$. Of these, G_{dm} is the differential mode voltage gain and G_{cm} is the unwanted common mode gain. In many applications the common mode voltages can be as great or greater than the signal voltages, and any increase of G_{cm} reduces the common mode rejection ratio (CMRR) since

$$\text{CMRR} = G_{dm}/G_{cm} \tag{8.31}$$

8.2.2   CHOPPER STABILIZED AMPLIFIERS

A chopper stabilized amplifier operates by separating the input signal into its high and low frequency components, each of which are then amplified separately before being recombined at the output.

Referring to Fig. 8.18, the high frequency component is capacitively coupled to the inverting input of the output amplifier. The low frequency component, including any drift signal, is used to amplitude modulate the output of the chopper oscillator. The result is an alternating signal whose amplitude is determined by the amplitude of the low frequency component. This modulated

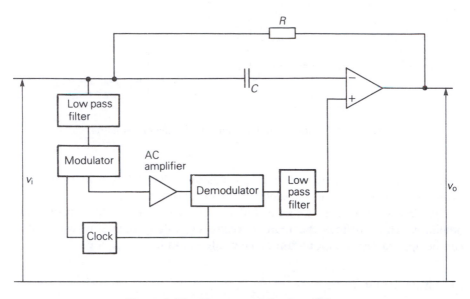

Figure 8.18 Chopper stabilized amplifier.

signal is used as the input of an AC amplifier, the output of which is demodulated and fed to the non-inverting input of the output amplifier for recombination with the high frequency component of the original signal.

The result is a reduction in the effects of drift and DC offset in the output amplifier by a factor equal to the gain of the AC amplifier, which itself introduces no offset. Levels of drift as low as $2 \mu V$ per year are quoted for chopper stabilized amplifiers.

8.2.3   AUTO-ZEROING AMPLIFIER

The auto-zeroing amplifier is a development of the chopper stabilized amplifier intended specifically for low frequency applications in the range from DC to around 10 Hz. The upper frequency limit is determined by the need for this to be much lower than the switching rate used by the amplifier.

In operation, the amplifier switches at high frequency between the modes shown in Fig. 8.19a and b, with the auto-zeroing input providing the zero reference. In mode I, amplifier 1 is connected as a unity gain amplifier so as to charge capacitor C_2 to a voltage equal to the current levels of noise and offset voltages. When the amplifier is switched into mode II, the voltage on C_2 is effectively connected in series with the input signal to eliminate the noise and offset components. At the same time capacitor C_1 is being charged to the input noise and offset levels of amplifier 2. The overall effect is to produce an amplifier whose drift is of the order of a few tenths of a microvolt per year.

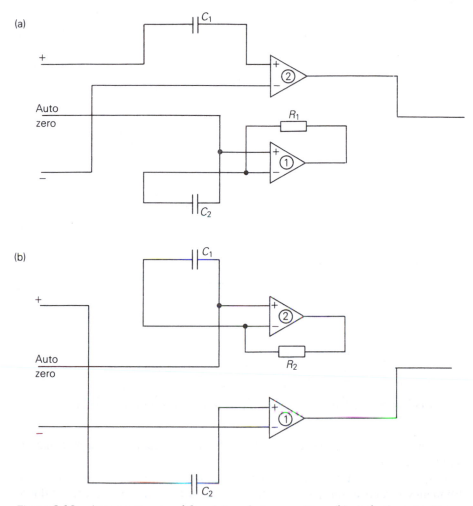

Figure 8.19   Auto-zeroing amplifier: (a) mode I connections (b) mode II connections.

8.3   Signal isolation

There are many measurement applications, for example in motor controllers, where the transducer is at some potential with respect to earth. In such a situation the transducer is required to be electrically isolated from the remainder of the circuit.

8.3.1   ISOLATION AMPLIFIER

Referring to the block diagram of Fig. 8.20a, the isolation amplifier is seen to consist of an input stage and an output stage with transformer coupling on

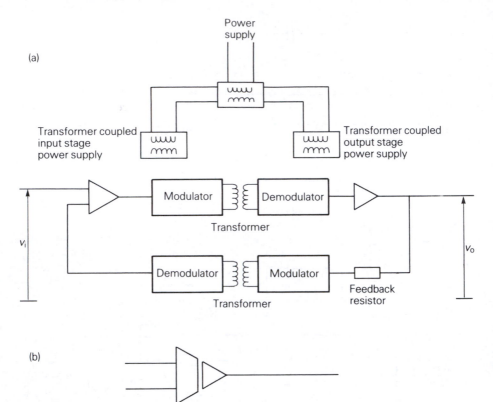

Figure 8.20   Isolation amplifier: (a) circuit arrangement (b) circuit symbol.

both the forward and feedback paths. The power supply to both stages is similarly isolated.

Amplifiers of this type can provide isolation up to a few thousand volts. Their frequency response and gain are, however, less than those of a conventional operational amplifier.

8.3.2   OPTO-ISOLATION

Isolation can also be provided by means of matched photodiodes and photo-detectors. Operation will normally be in the linear region of the operating characteristic, and care must be taken to maintain the input signal within the functional limits.

8.3.3   TRANSFORMER ISOLATION

Transformer isolation can be used for frequencies to several tens of kilohertz. The transformer is designed to ensure a linear response with minimum phase shifts over the whole of the frequency range.

8.4   Phase sensitive detector

A phase sensitive detector is effectively a two-position switch which is used to alternately connect the non-inverted and inverted forms of the input signal to the output, as in Fig. 8.21. The frequency at which the switching occurs is determined by a reference signal, and the effect is to produce a multiplication of the reference and input signals such that

$$v_o = v_i v_r \qquad (8.32)$$

If v_i and v_r are both sinusoids then

$$v_o = V_i V_r \sin(\omega_i t + \phi_i) \sin(\omega_r t)$$
$$= V_i V_r \{\cos[(\omega_i - \omega_r)t + \phi_i] - \cos[(\omega_i + \omega_r)t + \phi_i]\}/2 \qquad (8.33)$$

If the output signal v_o is fed to a low pass filter centred around 0 Hz then, from equation 8.33, this filter will produce an output only when v_i and v_r have the same frequency and hence v_o contains a DC component. Also from equation 8.33 it can be seen that the magnitude of the filter output is determined by the phase relationship of the input and reference signals, varying from a maximum when ω_i is 0° or 180° to zero at 90° and 270°. For this reason, a phase shift capability is normally included to enable the output signal to be maximized.

8.4.1   PHASE LOCKED LOOP

A phase sensitive detector can be used in association with a voltage controlled oscillator as in Fig. 8.22 to form a phase locked loop, which can be used to

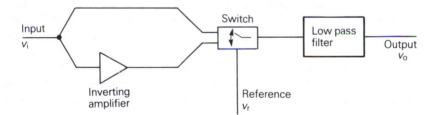

Figure 8.21   Phase sensitive detector.

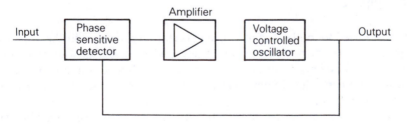

Figure 8.22   Phase locked loop.

track a frequency signal without the need for any external reference. Referring to the figure, the output from the filter stage of the phase sensitive detector is amplified and used as the control signal for the voltage controlled oscillator, the output of which provides the reference signal for the phase sensitive detector.

8.5   Multiplexing

Multiplexing is used to enable a single data channel to be shared between a number of data sources.

8.5.1   TIME DIVISION MULTIPLEXING

In time division multiplexing (TDM) each of the data sources is connected to the data channel in turn, allowing its data to be transmitted. The major problem with TDM is that the data from each of the sources becomes available to the processor at a different time, removing any comparative information unless a simultaneous sample and hold operation has been performed at each of the sources. For this reason TDM is most often used for systems containing large numbers of relatively slowly varying data sources such as pressure, temperature and static strain measurements. The nature of TDM, consisting of a series of discrete values, is particularly suited to use with digital signals.

8.5.2   FREQUENCY DIVISION MULTIPLEXING

Frequency division multiplexing (FDM) uses each data source to modulate a particular subcarrier frequency. These subcarriers are then combined in a mixer and used to modulate a high frequency carrier which is used for transmission of the information to the receiver. At the receiver, after demodulation from the high frequency carriers, the subcarriers are separated by means of a filter and the information contained is extracted. Both amplitude and frequency modulation systems are used and the technique is particularly suited to use with analogue signals.

8.6   Filters

Filters are used to inhibit the presence of certain frequencies in the final signal. In practice it is not possible to completely remove these unwanted frequency components but only to attenuate them by a definable amount. Figure 8.23 shows a number of ideal filter characteristics together with outputs from their practical realizations.

Filter design is concerned with achieving, as nearly as possible, the desired ideal characteristics for a particular application. Unfortunately, in order to

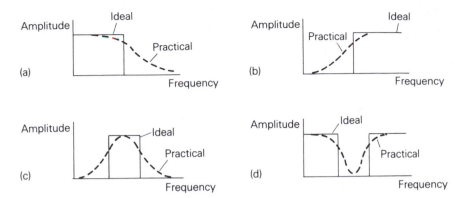

Figure 8.23   Filter characteristics: (a) low pass (b) high pass (c) band pass (d) band stop.

approach these ideal characteristics using analogue circuitry, increasingly complex and expensive circuit configurations are required. This has led to the development of digital filters which, within their frequency range, reproduce the full range of analogue filter characteristics together with further complex forms which would not be realizable using analogue techniques. The frequency range in which digital filters can be used is, however, restricted by the need to sample the incoming data and the rate at which this can be achieved. For frequencies above this limit, analogue filters must be used to achieve adequate performance.

In practice any filter, whether analogue or digital, used for a particular application is a compromise between performance, complexity and cost.

8.6.1   ANALOGUE FILTERS

Analogue filters may be realized using either active or passive circuits. Of these the passive filters are simpler, requiring no external power supplies, but they have a lower performance than active filters. A particular limitation on the performance of passive filters is that the components used in their construction are themselves non-ideal; for example, inductors have a resistive component.

Filters are characterized by their 3 dB bandwidth, effective noise bandwidth, selectivity and response time. The 3 dB bandwidth of a filter is taken as the width of the filter characteristic at a level 3 dB below the level of the pass band, as in Fig. 8.24, and is the most commonly used form of bandwidth. The effective noise bandwidth is used to relate the performance of a practical filter to that of the equivalent ideal filter, and is expressed as the bandwidth of the ideal filter which with the same amplitude in the pass band transmits the same power as the actual filter for a signal with a constant power spectral density with frequency, i.e. a white noise source.

The selectivity of a filter is a measure of its ability to separate out components

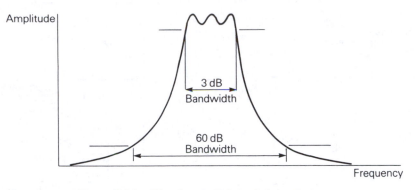

Figure 8.24   Filter bandwidth and shape factors.

of different frequency from the signal; it is expressed by a shape factor, defined as the ratio of the width of the filter characteristic at an attenuation of 60 dB from its maximum to the 3 dB bandwidth. Where the dynamic range is less than 60 dB, a 40 dB shape factor may be used, based on the width of the characteristic at 40 dB attenuation.

Filters do not respond instantaneously to changes in the input signal level. The actual time required for the output signal to reach its steady state level, the response time, is of the order of 1/(bandwidth).

Figure 8.25 shows the realization of a number of simple passive filters. In each of these filters the attenuation of the signal is quite gradual, and an improved performance with a sharper cut-off can be obtained by connecting

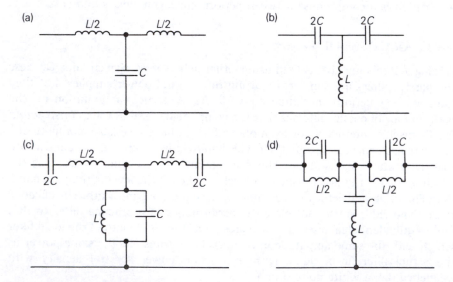

Figure 8.25   Simple passive filters: (a) low pass (b) high pass (c) band pass (d) band stop.

a number of similar stages to produce a ladder network of the form shown in Fig. 8.26. By this means a range of complex filter characteristics can be produced, including:

Butterworth Butterworth filters are characterized by a monotonically decreasing amplitude function for $\omega \geqslant 0$. Also referred to as maximally flat filters.

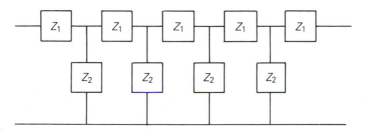

Figure 8.26   Ladder network.

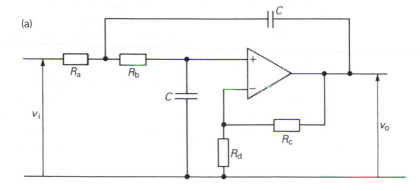

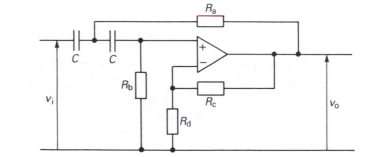

Figure 8.27   Simple active filters: (a) low pass (b) high pass.

Chebyshev   Chebyshev filters are characterized by an equiripple function in the pass band and a monotonically decreasing function in the stop band. Also referred to as equiripple filters.

Elliptic   Elliptic filters are characterized by a magnitude function with equal ripples in both the pass and stop bands.

Active analogue filters are based around the use of operational amplifiers to enable 'ideal' components to be simulated and to allow for a proper impedance matching at both the input and output of the filter. This enables the performance of the active filters to more closely approximate to the ideal but at the expense of increased complexity and of problems, such as drift, associated with the use of operational amplifiers. Figure 8.27 shows realizations of simple active low pass and high pass filters.

Active analogue filters can be used to realize a range of characteristics including the Butterworth, Chebyshev and elliptic forms referred to above.

8.6.2   DIGITAL FILTERS: THE SAMPLING THEOREM

Digital filters take in information as a series of discrete sampled data, as illustrated by Fig. 8.28. These quantized levels are then processed to produce an output which is filtered with respect to the input. The design of digital filters is primarily based on the use of the z-transform, the detailed discussion of which is outside the scope of this text, to express the transfer function of the sampled data system. A fuller description of the z-transform and its use in the design of digital systems will be found in appropriate works in the Bibliography.

The principle of a digital filter can, however, be illustrated by reference to the schematic block diagram of Fig. 8.29 and to the simple, single pole recursive filter of Fig. 8.30a, corresponding to the low pass RC filter of Fig. 8.30b. The performance of this filter with constants $\alpha = 0.15$ and $\beta = 0.85$ is expressed by equation 8.34, and it produces the output given in Table 8.1 with the offset

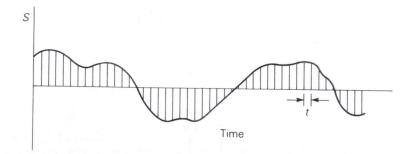

Figure 8.28   Sampled data.

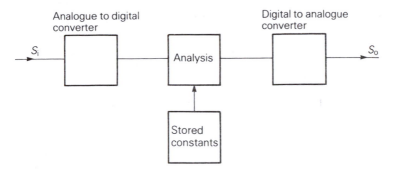

Figure 8.29   Block diagram of a digital filter.

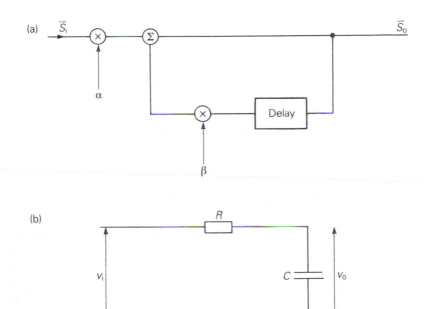

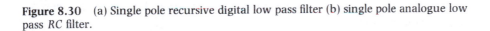

Figure 8.30   (a) Single pole recursive digital low pass filter (b) single pole analogue low pass RC filter.

triangular waveform of Fig. 8.31 as input:

$$S_{o(n)} = \alpha S_{i(n)} + \beta S_{o(n-1)} \tag{8.34}$$

Digital filters have a number of advantages over analogue filters:

1. Frequency response characteristics approximate closely to the ideal;
2. There is zero drift;
3. Low frequency signals can easily be accommodated;

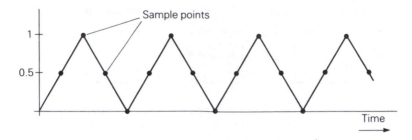

Figure 8.31   Input waveform and sample points for digital low pass filter of Fig. 8.30a.

Table 8.1   Output of single pole recursive digital filter with triangular wave input

Data input	0.5	1.0	0.5	0	0.5	1.0	0.5	0
Data output	0.075	0.214	0.257	0.218	0.260	0.371	0.391	0.332
(repeated)	0.357	0.454	0.461	0.391	0.408	0.497	0.497	0.423
	0.434	0.519	0.516	0.439	0.448	0.531	0.526	0.447
	...	...	...	...	...	...	...	...
	...	...	...	...	...	...	...	...
	0.463	0.544	0.537	0.456	0.463	0.544	0.537	0.456

4. There are no insertion losses;
5. Linear phase characteristics can be achieved;
6. Adaptive filtering is possible.

In any digital system there exists a finite number of quantized levels, in contrast to the infinite number of levels in an analogue system. The achievable accuracy of a digital filter is therefore a function of the number of bits used to represent the signal in digital form. However, as the number of bits increases so does the cost of the associated hardware, and there must be a trade-off in the filter design between accuracy and cost which will affect the overall performance characteristic.

The operation of a digital filter is also constrained by the sampling theorem, which requires that:

To be able to fully recover the continuous signal from the sampled data, the sampling frequency must be at least twice the frequency of the highest frequency component in the original signal.

If this criterion is not met then it is possible, as shown in Fig. 8.32, for high frequency signals to mimic low frequency signals, a condition known as aliasing. To prevent this an anti-aliasing filter is sometimes incorporated as a pre-processing stage before sampling takes place to suppress any high frequency components.

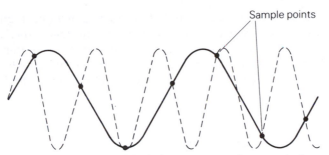

Figure 8.32   Aliasing: both low and high frequency signals generate the same sampled data.

The development of sophisticated, high performance digital filters is fundamental to the operation of complex systems such as compact disc players where the required performance is unachievable by analogue means. As microelectronic hardware continues to increase in power at reducing cost it is likely that digital filters will become increasingly more common, and will gain preference over and replace analogue filters for a wider range of applications in less complex environments.

8.6.3   PRE-PROCESSING AND POST-PROCESSING FILTERS

Pre-processing and post-processing filtering can be used to improve the noise performance of systems, particularly in relation to high frequency noise. Consider the arrangement of Fig. 8.33a. The input signal is first passed through the pre-processing filter, the characteristic of which is shown in Fig. 8.33b, which

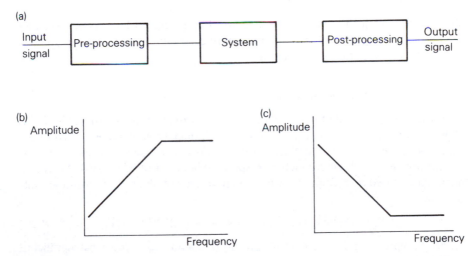

Figure 8.33   Pre- and post-processing of data: (a) system (b) pre-processing filter characteristic (c) post-processing filter characteristic.

has the effect of increasing the influence of the lower amplitude, high frequency components. The output signal from the pre-processing stage is then acted upon by the main system, during which stage noise is added. The output from this main stage is then taken to the post-processing filter, the characteristic of which is shown in Fig. 8.33c, and has the effect of de-emphasizing the high frequency components, including any added noise.

8.7   Digital signal processing

Digital signal processing makes use of the computational power of the microprocessor to analyse the incoming data. Digital techniques enable the introduction of compensation for temperature effects and signal linearization, as well as filtering and complex analytical operations such as the use of Last Fourier transforms. Digital systems also have a greater long term stability and are not prone to factors such as drift. However, the time required for many operations is longer than that required by the equivalent analogue system, and care must be taken to ensure that the inherent delays do not introduce problems such as instability into system operation.

Digital signal processing systems, like digital filters, begin by converting the continuous analogue signal into a series of discrete sampled data at regular intervals. The rate at which these samples must be taken to avoid aliasing is again defined by the sampling theorem (see section 8.6.2).

8.7.1   ANALOGUE TO DIGITAL AND DIGITAL TO ANALOGUE CONVERSION

The conversion of the analogue signal into binary digital form is achieved by means of an analogue to digital converter, while the digital to analogue converter is used to perform the reverse operation.

(a) DIGITAL TO ANALOGUE CONVERTERS

The simplest form of digital to analogue (D/A) converter is shown in Fig. 8.34. This consists of a series of graded resistors in binary sequence connected to the input of an operational amplifier. The reference voltage is connected to the resistors by means of switches which respond to binary 1 or a binary 0. Assuming a switch is closed on receipt of a 1 and open on receipt of a 0, the output voltage becomes

$$V_o = -(R_f/R)(s_0 + 2s_1 + 4s_2 + \cdots + 2^{n-1}s_{n-1})V_{ref} \qquad (8.35)$$

where s_0 represents the least significant bit (LSB) and s_{n-1} the most significant bit (MSB).

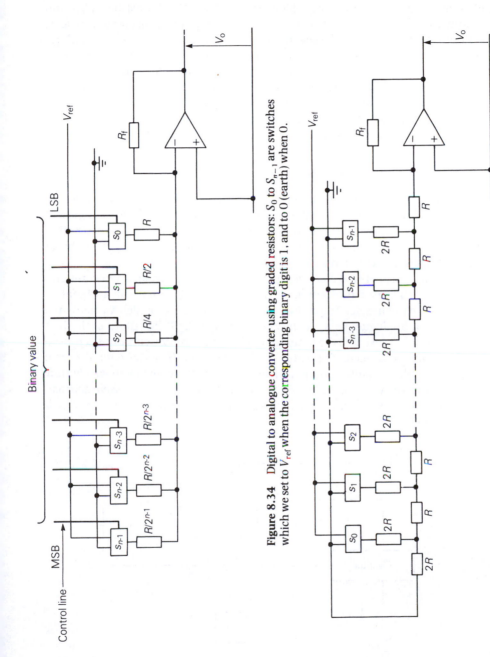

Figure 8.34 Digital to analogue converter using graded resistors: S_0 to S_{n-1} are switches which we set to V_{ref} when the corresponding binary digit is 1, and to 0 (earth) when 0.

Figure 8.35 Ladder-type digital to analogue converter.

The disadvantage of this type of D/A converter is in the resistors used to obtain the binary sequence. To obtain a high performance these must be stable and carefully matched within close tolerances; these requirements create problems in manufacture. The range of resistance values required is also large, particularly where high bit numbers are involved. Finally, the input impedance of the converter is low and varies with switching, introducing offset errors.

A better solution is the ladder network of Fig. 8.35, which uses only two values of resistance and presents the same value of impedance (R) at each node. By using laser trimming the individual resistors can be adjusted to within close tolerances, enabling high performance devices to be produced.

(b)   ANALOGUE TO DIGITAL CONVERTERS

The function of an analogue to digital (A/D) converter is the conversion of an analogue signal into a digital format. A number of different types of A/D converters are available offering a range of performance characteristics. Selection of an A/D converter for a particular application must then be in terms of these characteristics.

The simplest form of A/D converter is the successive approximation converter of Fig. 8.36. This uses the output of a counter as the input to a digital to analogue (D/A) converter to produce an analogue value, which is compared with the input analogue value by the comparator. When the output of the D/A converter equals or exceeds the input analogue value the conversion is complete and the count is stopped. The value of the count is then the desired digital value. If a simple sequential count from zero is used, the conversion rate is slow and varies with the value of the analogue input. The switched successive approximation converter uses a strategy which successively halves the range in which

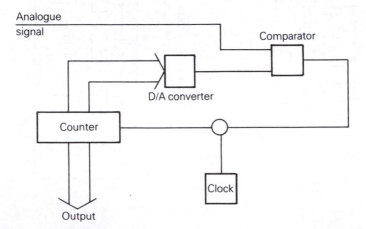

Figure 8.36 Successive approximation analogue to digital converter.

the solution must lie to obtain a result in n trials, where n is the number of bits in the solution.

The integrating A/D converter of Fig. 8.37 uses an integrator connected to a reference voltage to generate an analogue value which is compared with the input analogue value by a comparator. The time taken for the output ramp of the integrator to reach the input signal level then gives the binary solution.

In the dual slope converter of Fig. 8.38 the input analogue signal is first

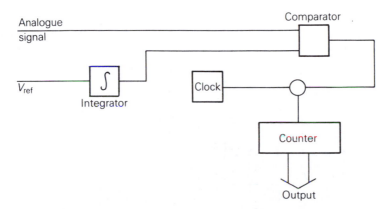

Figure 8.37   Integrating analogue to digital converter.

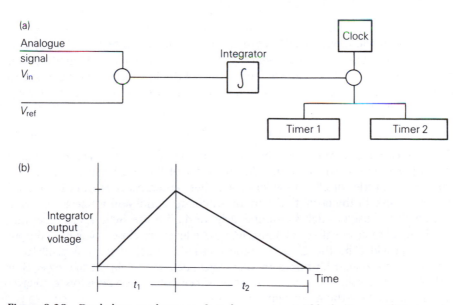

Figure 8.38   Dual slope analogue to digital converter: (a) block diagram (b) integrator output.

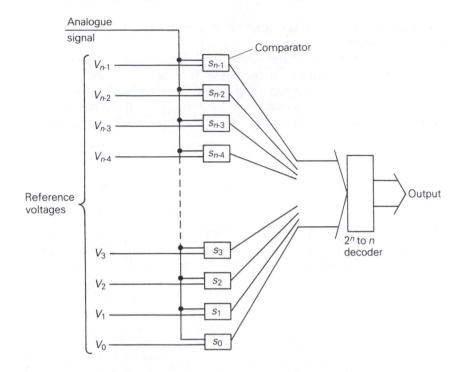

Figure 8.39   Flash analogue to digital converter.

connected as the input to an integrator for a fixed time interval. This input is then disconnected and a negative reference voltage is connected as the input to the integrator, and the time for the output of the integrator to return to zero is recorded. The ratio of the two times is then used to generate the binary solution, since

$$V_{in} = (t_2/t_1)V_{ref} \tag{8.36}$$

The A/D converters considered thus far are all relatively slow. Where increased speed is required then the flash A/D converter of Fig. 8.39 can be used. This consists of a series of 2^{n-1} individual switches representing each of the binary levels possible in the output. The input analogue signal and a reference signal are applied to each switch simultaneously, and all those switches representing levels equal to or less than the input analogue level are closed. The final output is then provided by the 2^n to n decoder circuit. Flash converters tend to be expensive; for higher bit numbers they would be used to obtain the lower 8 or 10 bits of the solution, and a switched successive approximation converter would be used for the more significant bits.

(c)   ERRORS IN ANALOGUE TO DIGITAL CONVERSION

In addition to the device errors such as linearity and monotonicity discussed in Appendix A (definitions and terminology), the operation of the A/D converter will itself introduce errors.

The process of converting the continuous analogue signal into a time sampled digital form results in the stepped output from the A/D converter of the form of Fig. 8.40. The resulting error is referred to as the quantization error and varies between plus and minus 0.5 LSB (least significant bit). The output of the A/D converter can therefore be considered as comprising the input analogue signal together with an added quantization noise component resulting from the transfer of the information from the analogue domain to the digital domain.

For a sinusoidal input signal the quantization noise can be shown to have an RMS value of 0.5 LSB/3, with a signal-to-noise ratio for an n bit converter of $6n + 1.8$ dB. Increasing the resolution of the A/D converter therefore reduces the level of quantization noise.

Performance can also be improved by the introduction into the system of a dither signal of the form

$$\text{dither} = (-1)^n d \tag{8.37}$$

where n is the sample number and d is approximately half the quantization level. The effect is to cause the output to oscillate about the quantization threshold, aiding the rounding of the signal to the appropriate level.

The conversion of the analogue signal into digital form requires a finite time. If the output of the A/D converter is not to be in error by more than ± 0.5 LSB then the conversion must be completed before the analogue signal can change by one LSB. The aperture time t_a for A/D converter is expressed as the period over which the analogue signal must be applied in order to

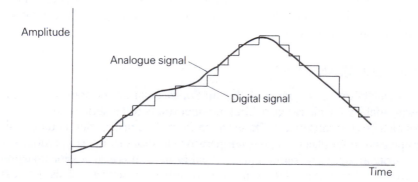

Figure 8.40   Output of an analogue to digital converter.

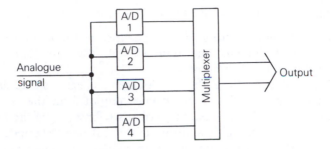

Figure 8.41 Analogue to digital converters with multiplexed output to increase the effective sampling rate.

enable the conversion to take place. The slew rate, defined by the highest frequency of a sine wave whose peak-to-peak amplitude corresponds to the full range of the A/D converter that can be converted correctly, may then be obtained in terms of the aperture time and the number of bits in the conversion.

For A/D converter of range S, the maximum rate of change of the corresponding full amplitude sine wave is then $\omega S/2$. The magnitude of one LSB is given by

$$\text{LSB} = S/(2^n - 1) \tag{8.38}$$

Hence the highest frequency that can be accommodated is

$$f_{\text{slew}} = 2/[\omega t_a(2^n - 1)] \tag{8.39}$$

If a higher slew rate than that which the A/D converter itself can provide is required, then a high speed sample and hold network can be used to sample the analogue signal over a shorter period. The output of the sample and hold circuit is then presented as the input to the A/D converter. By using a series of A/D converters taking input samples in sequence, as illustrated by Fig. 8.41, an effective sampling rate can be achieved which is much higher than that which could be achieved by a single converter, though at the expense of increased conversion delay.

8.7.2 SIGNAL ANALYSIS

The availability in the form of the microprocessor of substantial processing power enables a wide range of analytic operations to be performed on the data provided by the transducers. These range from the linearization of data and the compensation for effects such as temperature to more complex operations such as Fourier analysis and self-test and self-calibration. It is also possible to combine in the processor information from a number of sources. This process of information analysis and data fusion combines the measured data from all the available sources to build up a detailed picture of a complex environment.

(a) LINEARIZATION

It is possible that the output of a transducer or sensing system will have a complex relationship with the parameter being measured which could not readily be accommodated using analogue signal processing techniques. The ability of the microprocessor to obtain the required parameter value by either calculation or the use of lookup tables has enabled the increasing use of inherently non-linear transducers.

(b) COMPENSATION

Some transducers, such as those using silicon as the active material, are sensitive to a range of environmental conditions such as temperature variation which can distort their output. By separately sensing these external conditions, appropriate compensation can be provided by the microprocessor.

(c) SIGNAL AVERAGING

Where a repetitive signal is subject to random noise, signal averaging can be used to recover the signal from the noise. Consider such a signal to which has been added a Gaussian noise signal with a mean value of zero and a standard deviation of σ_n. If N successive sets of data each made up of n samples and containing the signal and noise are summed on a sample by sample basis, the average value of the rth sample is then

$$r_{av} = (1/N) \sum_{i=1}^{N} r_i \tag{8.40}$$

The noise signal still retains its Gaussian distribution but the standard deviation has become σ_{nN}, where

$$\sigma_{nN} = \sigma_n / \sqrt{N} \tag{8.41}$$

Therefore r_{ave} becomes more closely the actual value of the signal at that point.

(d) FOURIER ANALYSIS

Fourier analysis is a powerful analytical tool which is widely used to extract information about system behaviour. Based on the forward and reverse Fourier transform equations shown below, Fourier analysis translates a signal in the time domain into the frequency domain. Once in the frequency domain, information such as the power spectral density can be obtained and analytic tools such as correlation and cepstrum analysis can be applied.

$$X(f) = \int_{-\infty}^{\infty} x(t) \exp(-j2\pi ft)\, dt \qquad \text{forward} \tag{8.42}$$

$$x(t) = \int_{-\infty}^{\infty} X(f) \exp(j2\pi ft)\, df \qquad \text{reverse} \tag{8.43}$$

Depending on whether the signal in either the time or frequency domains is of periodic, continuous or discrete (sampled) form, the forward and reverse transforms are connected as shown by Fig. 8.42 and Table 8.2.

In digital analysis the form of the Fourier transform used is the Discrete Fourier Transform (DFT) in which both the time domain and frequency domain signals are of periodic and discrete form. The operation of the DFT defines a frequency, referred to as the Nyquist frequency, with a value of half the sampling frequency. Frequencies above the Nyquist frequency are capable of aliasing frequencies below the Nyquist frequency. This is in line with the conditions established by the sampling theorem referred to in section 8.6.2.

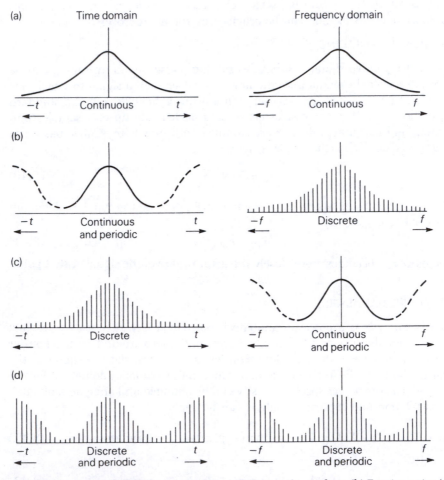

Figure 8.42   Forms of the Fourier transform: (a) integral transform (b) Fourier series (c) sampled (d) discrete Fourier transform.

Table 8.2 Forms of Fourier transform

Integral transform

$$X(f) = \int_{-\infty}^{\infty} x(t) \exp(-j2\pi ft)\, dt$$

$$x(t) = \int_{-\infty}^{\infty} X(f) \exp(j2\pi ft)\, df$$

Fourier series

$$X(f_k) = \frac{1}{T} \int_{-T/2}^{T/2} x(t) \exp(-j2\pi f_k t)\, dt$$

$$x(t) = \sum_{k=-\infty}^{\infty} X(f_k) \exp(j2\pi f_k t)$$

Sampled

$$X(f) = \sum_{n=-\infty}^{\infty} x(t_n) \exp(-j2\pi ft_n)$$

$$x(t_n) = \frac{1}{f_s} \int_{-f_s/2}^{f_s/2} X(f) \exp(j2\pi ft_n)\, df$$

Discrete Fourier transform

$$X(f_k) = \frac{1}{N} \sum_{n=0}^{N-1} x(t_n) \exp(-j2\pi nk/N)$$

$$x(t_n) = \sum_{k=0}^{N-1} X(f_k) \exp(j2\pi nk/N)$$

The DFT is usually implemented by means of Fast Fourier transform (FFT) algorithms which reduce the total calculation time required to obtain a solution.

The performance of the DFT/FFT is significantly influenced by the format in which information about the original signal is obtained. The original analogue signal may be considered as being viewed through a time window, the shape of which determines the weighting applied to the sampled data points within the window; this process is referred to as windowing. It has the effect of imposing, via the DFT/FFT, a series of filters centred at intervals of $1/T$ (where T is the duration of the window) on to the time domain signal. The characteristics of these filters are in turn determined by the shape of the window.

Figure 8.43 and Table 8.3 describe a number of common window functions. A true Gaussian window would extend from $-\infty$ to $+\infty$, and it is therefore truncated at three times the half amplitude width or 7.06 times the standard deviation.

Errors in the DFT are associated with the spread of energy between adjacent

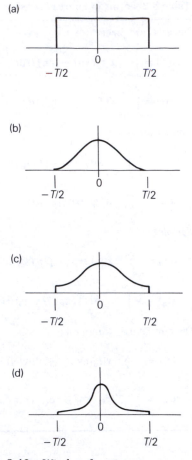

Figure 8.43   Window functions
(a) Rectangular window:

$$W(t) = \begin{cases} 1 & \text{for } -T/2 \text{ to } T/2 \\ 0 & \text{otherwise} \end{cases}$$

(b) Hanning window:

$$W(t) = \begin{cases} \frac{1}{2}[1 - \cos(2\pi t/T)] & \text{for } -T/2 \text{ to } T/2 \\ 0 & \text{otherwise} \end{cases}$$

(c) Hamming window:

$$W(t) = \begin{cases} 0.54 - 0.46\cos(2\pi t/T) & \text{for } -T/2 \text{ to } T/2 \\ 0 & \text{otherwise} \end{cases}$$

(d) Gaussion window:

$$W(t) = \begin{cases} \exp(-t^2/2\sigma^2) & \text{for } T = 7.06\sigma \\ 0 & \text{otherwise} \end{cases}$$

where σ is the standard deviation.

Table 8.3   Window function characteristics (T is window period)

Window function	Main lobe bandwidth	First sidelobe
Rectangular	$1/T$	$-13\,dB$
Hanning	$1.5/T$	$-32\,dB$
Hamming	$1.4/T$	$-42\,dB$
Gaussian	$1.9/T$	$-44\,dB$

frequencies (spectral leakage) and the influence of any frequency components of the original signal which lie between the filters.

8.8   Smart sensors

The integration of the sensor and transducer stages and their associated electronics in a single package has led to the development of the concept of the smart sensor. Such devices would typically provide some or all of the following features:

1. Error compensation covering external environmental factors such as temperature variation and internal effects such as offset and drift;
2. Signal processing using both analogue and digital techniques;
3. Self-test and self-calibration including diagnostics;
4. Decision making;
5. Bidirectional communications with other devices and with a central controller.

The first smart sensors to be introduced were silicon pressure transducers which featured a 4–20 mA interface that enabled the measurement range and scale to be remotely reset. These are now being used in environments where normal maintenance and checking procedures would be difficult to apply.

As yet, smart sensors are only finding limited application in support of conventional sensing systems. The emphasis in their development has also been one of improving existing techniques rather than introducing new forms of sensor and transducer. However, as the technology develops, including the increased integration of signal processing elements on to the same piece of silicon as the sensing element and the use of thick film hybrid circuits, the availability of smart sensors at reducing cost levels will undoubtedly lead to their wider application.

8.9   Expert systems, artificial intelligence and measurement

Expert systems are currently available for the analysis and interpretation of data in environments as diverse as medical diagnosis (MYCIN) and the design of computer systems (R1). It is likely that they will increasingly perform a similar role within mechatronics in areas such as condition monitoring, fault analysis, vision systems, autonomous robots and the presentation of information. Movements in this direction, though not at the expert system level, can already be seen in a simple form in advanced automatic cameras, where the optimum settings are determined by the camera with minimum operator intervention; in vehicle engine management systems, where the engine is continuously retuned to improve efficiency and economy; and in aircraft cockpit displays, where the nature and format of the information displayed is controlled to maximize its value to the pilot.

In complex and unstructured environments in which large amounts of data are available from a number of sources, the concept of data fusion is currently under development, particularly in areas such as vision systems for mobile robots and military threat analysis. In this process, artificial intelligence and expert systems techniques are used to analyse and interpret the available data. Although such techniques currently require large amounts of computing power, the probability is that they will become viable for a range of systems in the foreseeable future.

Part Two
Embedded Microprocessor Systems

Chapter 9

Microprocessors in mechatronic systems

Mechatronic systems contain elements of control and information processing which are typically performed by microprocessor systems. The prevalence of the microprocessor as opposed to any other form of control or processing is due to a range of factors, the most important of which are as follows:

Stored programmed control A microprocessor is essentially a minimally configured computer system and is capable of executing instructions from a program stored in memory at very high speed. This program is a logical sequence of operations which is both deterministic and repeatable, and is ideal for the implementation of control and signal processing functions based on mathematical or logical algorithms.

Digital processing Information in the microprocessor system is represented by binary numbers. These do not suffer from analogue noise and can have a variable resolution chosen according to the requirements of the applications.

Speed of operation Although a typical microprocessor is constrained to execute instructions in a sequential order, the speed at which it does so often means that it is capable of performing many sequential tasks in a very short time; thus within the response time of the observer the tasks appear to be executed concurrently. This observer may be a machine, device or system which the microprocessor is controlling.

Design flexibility A microprocessor system offers a designer a very flexible system solution as the same hardware can be made to perform a large number of different functions. This is possible within the physical bounds provided by the hardware, such as the range of input and output facilities of size of memory, by changing the program it executes. This amounts to simply changing the contents of a memory device. More importantly, the ability of the microprocessor to conditionally execute programs allows operational flexibility to be designed into the system. Together with its ability to communicate with other microprocessor systems, and a human operator, these features have been compared to a basic form of intelligence and have resulted in words such as 'intelligent' and 'smart' being used for devices containing microprocessors.

Integration   Integration is a major benefit of any microelectronic system in a mechatronic application, where physical space and the supply of power may be restricted. With the improved VLSI techniques that have enabled more hardware to be incorporated on the same area of silicon, a complete microprocessor system can be placed on a single silicon chip, allowing a high level of integration to be achieved.

Cost   An extremely important commercial consideration is the cost of the microprocessor, which with improved VLSI techniques has continued to decrease to a level such that it has now become viable to apply microprocessor technology to a wide range of low cost applications. As the cost continues to decrease, the number of viable applications grows exponentially.

9.1   Embedded real-time microprocessor systems

The requirements of a typical mechatronic application can be met by a class of microprocessor system referred to as embedded real-time microprocessor systems. The term 'embedded' refers to the fact that the microprocessor system is itself a subsystem of a much larger system which, in the case of mechatronic applications, is based on a mechanical product or process. For example, the microprocessor system which implements the control in a washing machine or in an engine management unit would be referred to as an embedded microprocessor system.

Embedded real-time microprocessor systems form one of the two major categories of microprocessor applications. The second category comprises what can be broadly referred to as data processing applications. Here, the microprocessor is built into an environment designed for typical computing applications, such as a personal computer or a workstation.

It is useful to compare the characteristics of microprocessor systems within these categories in order to clarify what constitutes an embedded microprocessor application. The basic characteristics of each category of application are summarized in Table 9.1, which compares them in terms of memory requirement, central processing unit (CPU) performance, and input/output structures.

That there is considerable difference in memory requirements between embedded and computing applications is due to the fact that an embedded system typically operates with a fixed program while a computer system must be capable of being reloaded with a wide range of different programs. In each case, the term 'system memory' is used to refer to that part of the memory which is directly accessible to the microprocessor, as opposed to storage media such as a magnetic disc or tape drive.

In an embedded system, the memory consists of varying amounts of non-volatile memory, the contents of which will not be lost when its power supply is removed, and volatile memory, which loses its contents on the removal of

Table 9.1 Comparison of microprocessor system requirements for embedded real-time applications and data processing applications

	Real-time embedded application	Data processing application
Memory	Small amounts of both non-volatile read only memory for program storage, and read/write memory for data storage	Large amounts of read/write memory which can be loaded from a data storage device such as a floppy disc, and only a small amount of non-volatile program memory used during initialization of the system
Input/output	Both digital and analogue interfaces to sensors and actuators	Mainly digital I/O to keyboard, visual display unit, disc drives, printers, and communications with other computers
Processor	Wide range of performance to suit cost and application; fast response required to external events	High performance required to improve the productivity of the user; response time is generally not critical

its power supply. A characteristic of non-volatile memory is that it can only be written to using a procedure which is incompatible with normal microprocessor write operations. Once information has been programmed into non-volatile memory it can be considered as permanent, and so would be used for holding fixed programs and constant data (firmware); it is referred to as read only memory (ROM). Conversely, volatile memory can be both read from and written to by the microprocessor and is normally used for data storage; it is referred to as random access memory (RAM) or read/write memory.

The programs in embedded systems are typically small; for example, a washing machine control program may require only 2 kilobytes of memory. For more demanding embedded applications, for example in communication controllers, several hundreds of kilobytes of non-volatile program memory may be required together with a corresponding amount of read/write memory to store the communications data.

For a general purpose computer, the system memory consists mainly of volatile read/write memory which can be reloaded with different programs from a non-volatile storage memory device such as a floppy disc. However, there will always be a small amount of non-volatile system memory to contain a simple boot program which allows access to the operating system when the system is powered on. Large amounts of read/write system memory are required in these systems to accommodate the large and complex programs required for computer aided design and simulations. Consequently, the microprocessor must be capable of addressing large amounts of system memory, of the order of tens of kilobytes to several megabytes, and ideally provide some means of memory management support.

The primary input and output devices for a computer system are the keyboard and visual display unit (VDU), and obviously the main controlling influence over the operation of the system is the human operator. The computer system therefore requires a man–machine interface which has to be matched to the operator in order to achieve a high level of operational efficiency. As it is impractical to change the physical attributes of a human being, the keyboard and VDU must be designed to transfer information efficiently between the computer and its operator. The standard keyboard is generally regarded as an inefficient man–machine interface, and efforts to improve this have resulted in the use of a mouse together with pull-down menus which to a large extent bypass the keyboard.

The major sources of data input to an embedded microprocessor system are the sensors connected to the machine or device it is controlling. Similarly, the outputs from the microprocessor system will operate actuators, which will in turn control the machine or device. The primary interface for an embedded microprocessor system is therefore a machine–machine interface and the characteristics of the signals which flow across this interface are quite different from those for a man–machine interface, the latter being based on a conversational dialogue which is interpreted by the operating system of the computer.

Embedded real-time systems may also make use of an operating system called a real-time multitasking executive. Instead of being driven by commands from a keyboard, the executive will be driven by events external to the system called interrupts, and will invoke the execution of special programs to service the requests that generated the interrupts. The executive may also have an internal scheduling mechanism to enable it to control the execution of several individual programs, thereby creating a multitasking system. A real-time multitasking executive is therefore seen to provide a set of services on to which systems can be built from tasks which may have complex interactions with one another and with external devices. In an embedded system, the executive would reside together with the application programs in program memory, whereas in a computer system it would normally be loaded from a disc.

Microprocessor performance is an important issue in all applications, embedded or otherwise. As in any design procedure, the performance of the microprocessor must be properly matched to the application as there normally exist significant cost penalties in providing too much performance for a given application. Conversely, too low a performance specification may render the system inoperable or unable to match its specifications.

There are many measures of microprocessor performance; that most usually quoted is a measure of the rate at which the processor can execute instructions.

For an embedded system an important performance measure is its response time. For a typical computer system, its response time to commands entered by the operator is not critical to the interaction between the operator and the computer. An operator may complain when the computer takes several seconds to draw a picture on a graphics screen, but this does not cause a catastrophic failure of the system. On the other hand, an embedded system, such as that responsible for a flight control system, may be required to output control signals to actuators within a certain time deadline after receiving event data from its sensors. The consequence of this deadline not being achieved could be the loss of control of the aircraft. Such a control system is said to be a hard real-time system because the time deadline is a strict qualifier to the correct operation of the system. Systems that are similar, but where the failure to meet time deadlines causes a degradation in performance rather than a catastrophic failure, are referred to as soft real-time systems. The term 'real-time' is used to describe any system whose response to external events is required to meet some function of time.

For this reason, in environments where the ability of a system to function is of prime importance, a distributed configuration is adopted which incorporates some form of polling or voting procedure between a number of equivalent systems. For further security, and to prevent both hardware and software faults, such safety critical systems are often built using different processors with software from different sources for each voting element.

9.2   The mechatronic system

Figure 9.1 shows a block diagram of a typical microprocessor system used to control a machine. In order to add some reality to the general picture it can be assumed that the machine is a robot arm and that the microprocessor system is its controller.

The microprocessor system consists of its central processing unit (CPU), program and data memory and a variety of input and output (I/O) functions, all of which are interconnected by a microprocessor system bus. The amount of memory required for a robot control application would perhaps be of the order of 0.5 megabytes of program memory and 128 kilobytes of read/write memory. In addition there may also be a small amount, in the region of several kilobytes, of non-volatile read/write semiconductor memory to store configuration parameters.

The primary function of the microprocessor based robot controller is the control of the axes of the robot arm. Each of these axes forms a separate control loop, and hence is considered as an individual task or program module. In addition, there will be the coordinating task of calculating the speeds and directions of individual axes in order to move the robot arm along any desired trajectory. It is common in such applications to control the execution of each

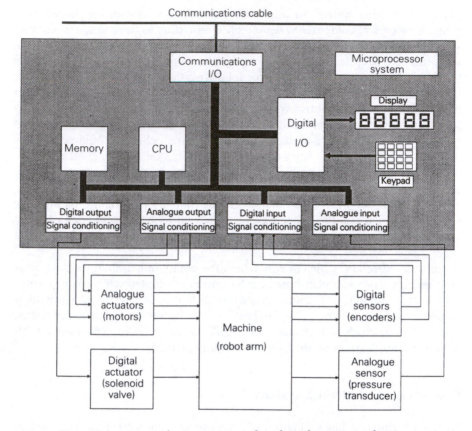

Figure 9.1   A typical microprocessor based mechatronic application.

of the tasks using the real-time multitasking executive referred to above.

The calculation of the robot axis speeds, directions and control algorithms will involve the CPU in a large amount of arithmetic computation. This requires the CPU to have an adequate performance as defined by the processing requirements of the application and its real-time requirements. In robot control applications, 16 or 32 bit CPUs are normally required as these are higher performance devices. In many applications, the computational ability of the CPU may be supplemented by the addition of a coprocessor dedicated to the fast execution of arithmetic calculations.

The robot arm itself is interfaced to the microprocessor system by means of both analogue and digital sensors and a range of actuators. In the system under consideration, each x, y and z Cartesian axis is driven by an electric motor, the speed of which is controlled by an analogue output signal from the microprocessor system. The speed of the motor is sensed by a digital output encoder. Secondary functions, such as a pneumatic gripper, require control of the operation of a digital solenoid valve, while the resulting pressure on the gripped object is measured by pressure pads in the gripper producing an analogue signal which is used as an input to the microprocessor system. All these different types of signals require signal processing in order to match the interfacing requirements at the microprocessor system and the robot arm.

A system such as a robot will seldom operate as an isolated machine, but is more likely to be part of a larger controlled system consisting of other robots and controllers with which it has to cooperate. The robot controller is therefore equipped with both a human operator interface in the form of a keypad and display, and a communications interface to a communications network which will link together a number of machines and systems. Although these machines and systems may be physically located in different areas, the communications network provides a means by which they can communicate with one another and with computers which provide global supervision of the control system. This type of system is called a distributed control system, the effectiveness of which lies in matching the design of the communications system to the application.

Chapter 10

The microprocessor system

A microprocessor system can be described at a number of different levels of complexity. The least complex form is that of a simple block diagram describing the interconnection and flow of information between functional blocks and will be used in this chapter to examine the operation of a microprocessor system.

10.1 The system components

All microprocessor systems contain a central processing unit (CPU), program and data memory and input and output (I/O) devices. Figure 10.1 shows a block diagram of a typical embedded microprocessor system in which each block corresponds roughly to the individual integrated circuit (chip) used in the system.

The memory section contains both non-volatile read only memory (ROM) as program memory and volatile random access memory (RAM) as read/write data memory. For each type of memory there are a number of different types of devices, such as erasable ROMs and static or dynamic RAMs, each of which is chosen for an application based on its cost and function. These different types of semiconductor memory are reviewed in section 12.1.

Four different I/O functions are shown in Fig. 10.1. An analogue input channel to the microprocessor system is provided by the analogue to digital (A/D) converter and may be used to connect a device such as an analogue sensor. An analogue output channel is provided by the digital to analogue (D/A) converter and could be used to control an output transducer such as an electric motor. The parallel I/O device provides a number of individual lines. In output mode these can be programmed to provide logic levels 1 or 0 to activate binary (on/off) devices such as lamps. In input mode it allows the microprocessor to read the state of switches and other binary devices. The serial I/O device is used to provide communications with other microprocessor systems or with an operator console used to configure the system for various operational modes.

The level and power specifications of the interfacing signals of the micro-processor system are frequently incompatible with the signal specifications of

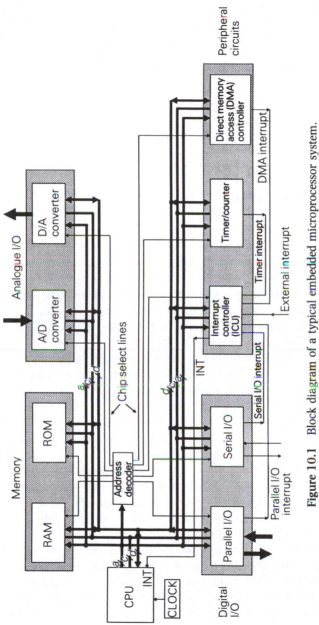

Figure 10.1 Block diagram of a typical embedded microprocessor system.

the devices which are to be interfaced to it. For example, the output voltage of a D/A converter may typically be in the range 0–5 volts and be capable of supplying only a few milliamperes of current, while the electric motor may require a control voltage range of plus and minus 12 volts at a maximum current of 1 ampere. Consequently, additional analogue interface circuitry is often necessary to perform functions such as signal level shifting, amplification and filtering.

Figure 10.1 also shows three peripheral circuits: an interrupt control unit (ICU), a programmable counter/timer, and a direct memory access unit (DMA). These devices provide additional interfacing functions to the system and are described in more detail in section 12.3.

10.2   The system bus

All of the above devices are interfaced to the CPU by means of a system bus which is itself made up from an address bus, a data bus and a control bus. Physically, a bus is simply a collection of parallel interconnections between two or more devices. The number of lines contained in each bus is dependent on the type of microprocessor used in the system and the function of the bus. In Fig. 10.1 we assume that the address bus has sixteen lines, the data bus has eight lines, and the control bus contains an arbitrary number of lines depending on the control functions provided by the CPU.

The concepts of address and data are fundamental to the operation of a stored program computer and form a feature of all microprocessors and computers. The memory will consist of a number of memory locations capable of storing data written to them by the CPU over the data bus. Each memory location is uniquely identified to the CPU by a number called its address. The CPU controls the address and control bus lines in order to write or read information to or from the memory or I/O devices. For example, if the CPU wished to write the binary number 01010101 into a memory location which had the address 0000000000001111, the CPU would first place the address on to the address bus, then place the number 01010101 as data onto the data bus. Control lines in the control bus would then be activated to cause the data to be loaded into the appropriate memory location.

A similar procedure would be used if the CPU was then to read a memory address, except this time the flow of data would be from the memory to the CPU. After the CPU had placed the address of the required memory location on to the address bus, it would indicate to the memory that it wished to read the value by activating the relevant line in the control bus. The memory would respond by placing the contents of the memory location as data on to the data bus, and this would then be read by the CPU.

Within the system bus, the address bus is an output bus from the CPU and an input bus to the other devices. The control bus consists of a number of lines, each of which may be either a control output from the CPU or a control input to the CPU. The data bus however acts as both an input bus and an output bus depending on whether the CPU is reading or writing data. Figure 10.1 shows that all devices in the system are connected together by the data bus and this means that, potentially at least, the outputs of all the memory and I/O devices are connected. If this were in fact to happen it would cause the destruction of several or all of the connected devices, because some devices would be trying to drive the bus to a logic 1 state while others were trying to drive it to a logic 0 state. To avoid this problem, the data bus connections of each device are capable of being placed into a third, high impedance state where the device no longer has any loading effect on the bus. This allows other devices connected to the data bus to output their data on to this bus when they are correctly enabled, which in turn means that only one device should be enabled at any one time to the data bus. The ability of a device to be either at a logic 1 or at a logic 0 or in a high impedance condition in relation to the data bus is called a tristate condition, and is an essential feature of devices which share a common data bus.

10.3   The memory map

In the example of Fig. 10.1 the data bus has eight lines, and hence the range of values which a single item of data can take is restricted to that which can be represented by 8 binary digits or bits. Eight bits are referred to as a byte, and can represent a decimal number from 0 to 255 ($2^8 - 1$). Likewise the address bus, consisting of sixteen lines, can represent an address number in the range 0 to $2^{16} - 1$ or 65 535. This number is usually abbreviated to the binary equivalent of the decimal number and expressed as 64K, where K is equal to 1024 in the binary number system. To the CPU, the system appears as a series of 64K consecutive memory locations, each capable of storing an 8 bit binary value.

The CPU will contain a number of registers which are used to manipulate the data and its addresses. In the example chosen, these data registers will be 8 bit registers and all data manipulations will be performed on 8 bit quantities. The CPU is therefore referred to as an 8 bit CPU. However, registers which support address manipulations need to be 16 bit registers because of the 16 bit address bus. The size of the address bus is independent of the size of the data bus, so that 16 bit or 32 bit CPUs may typically have a 16 bit, 24 bit or 32 bit address bus.

It is normal when working with microprocessors to represent binary numbers as hexadecimal (base 16) values, because a single hexadecimal (hex) digit corresponds to a group of four consecutive binary digits. The hexadecimal number is easier to read and write than its binary equivalent, and it requires only simple mental calculation in order to translate from hexadecimal to binary and back again. Hexadecimal numbers and digits are identified in the text by prefixing them with 0x, which is the convention adopted in the C programming language. For example, the 16 bit binary number 1111110100111001 would be written in hexadecimal form as 0xFD39.

It is not necessary for the hardware designer to make use of the entire address range of the CPU, and the physical memory required can be implemented anywhere within the address space of the CPU. In addition, the I/O devices will contain registers through which the I/O devices may be configured, and which provide addresses through which the CPU can read and write data into and out of the system. These registers appear to the CPU as normal memory locations, and so occupy parts of the address space. The arrangement of the memory and I/O addresses within the address space is described by the memory map of the system, an example of which is given in Fig. 10.2a.

The memory map will be designed to meet the requirements of the application, and will be used by the hardware designer to partition the address space so that the address range of the memory devices in the system corresponds to the address range specified by the memory map. This is achieved by means of the address decoder. Shown in Fig. 10.1, this uses lines from the address bus as its input and produces individual chip select signals for each chip which contributes to the memory map of the system. Figure 10.2a shows that the particular system

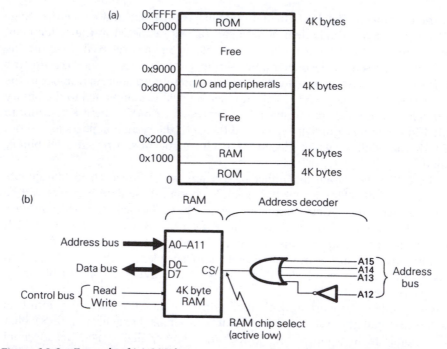

Figure 10.2   Example of (a) 64 K byte memory map (b) RAM and address decoder for the address range 0x1000 to 0x1FFF.

under consideration should have RAM from memory location 0x1000 to memory location 0x1FFF. This address range represents 4K bytes (4096) of memory, requiring the lower twelve address lines of the address bus, A11–A0, to be connected to a 4K RAM device. The remaining four upper address lines, A15–A12, are taken to the address decoder which outputs a chip select signal to the RAM whenever the lines A15–A12 have the value 0001. The address decoder therefore implements the Boolean equation

$$\text{RAM enable} = \overline{\text{A15.A14.A13.A12}} = \text{A15} + \text{A14} + \text{A13} + \overline{\text{A12}} \qquad (10.1)$$

as in Fig. 10.2b.

In practice, the design of the address decoder can be simplified by partitioning the memory map into equal sections which are themselves convenient binary sizes, such as 4K, 8K or 16K. This will normally result in an address decoder which may be implemented with simple logic gates. It can easily be seen that similar decoding logic would be required for each chip in the system that occupied the same size of memory, and so this circuitry can be found in logic decoder integrated circuits such as the 74138 and 74139. Where more complicated memory mapping schemes are required, the decoder equations can be implemented with a programmable logic array (PLA). PLAs are described in more detail in section 13.2.2.

The CPU will normally place some constraint on the position of the RAM and ROM within the memory map. For example, when the system is powered on, the CPU will start to fetch its first instruction from a predetermined memory location. It is therefore necessary for the memory map to have some non-volatile memory at this location containing the start-up program to be executed.

10.4   The microprocessor bus operation

The CPU is responsible for the control of the address, data and control buses and as such is referred to as the bus master, while the other components, such as the memory and I/O, are the bus slaves. The CPU executes a program which is stored in memory by performing sequences of fetch and execute operations. These operations consist of a number of clock cycles which make use of the system bus in order to read and write information to and from the memory or I/O. For instance, to fetch an instruction may cause several memory read cycles to occur as the instruction may be several bytes long, while an execute operation may consist of a read or write cycle or a combination of both. The execution may also result in an operation which occurs entirely within the CPU such as execution of a calculation, in which case the system bus is not used.

It can be seen in Fig. 10.1 that the CPU receives an input from a clock generator. The clock signal is simply a logic square wave which is used to drive the internal circuitry of the CPU. All actions within the CPU, and all actions which take place on the system bus, occur synchronously with respect to the

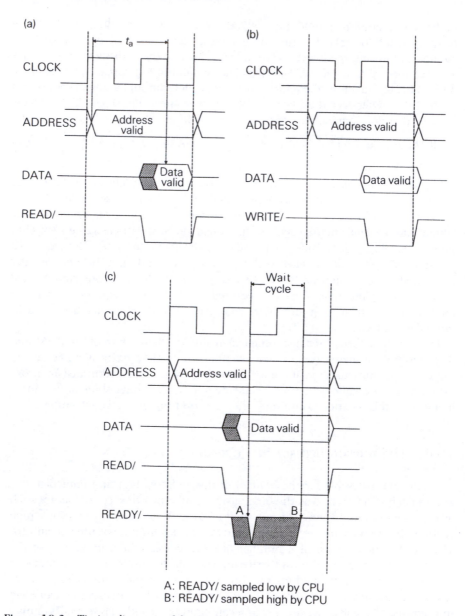

A: READY/ sampled low by CPU
B: READY/ sampled high by CPU

Figure 10.3 Timing diagrams of the CPU bus cycles: (a) read cycle (b) write cycle (c) read cycle with one wait state.

CPU clock signal. Changes in the state of the signals of the system bus are controlled by the CPU and take place at the edges of the clock pulses. Figure 10.3a shows the relationship between the clock signal and the system bus signals during the read cycle of a typical microprocessor. The exact details of such operations will differ between different microprocessors, but the principle remains the same.

The read cycle shown in Fig. 10.3a lasts for two cycles of the clock signal. At the rising edge of the first clock cycle, the address of the memory location to be read is placed on the address bus by the CPU. As soon as the address appears on the address bus, the address decoding logic will select the device which contains the memory or register corresponding to that address. During this first clock cycle, the status of the data bus is undefined because all of the outputs on to the data bus from the memory and I/O devices in the system will be in a high impedance state (or tristated), even though a device will have been selected by the address decoding logic. In order for the selected device to output its data on to the data bus, it must receive an active read signal which will cause the data bus buffers within the device to come out of their high impedance state and to drive the data bus. This occurs at the beginning of the second clock cycle, when the read control line goes from a high (inactive state) to a low (active state) level. Then, after a time delay determined by the internal delays of the selected device, the valid data will be placed on to the data bus. The diagram then shows that the levels on the data bus are sampled by the CPU at the falling edge of the second clock cycle, at which point the data is read into the CPU.

Figure 10.3b shows the relative timing of the system bus signals for a CPU write cycle. It can readily be seen that the read and write cycles are very similar, although there are some subtle differences which are important to the write operation. As in the case of the read cycle, the CPU places the address of the memory location to be written to on to the address bus at the beginning of the cycle, and the address decoding logic then selects the relevant device. For the write operation, the CPU is responsible for placing the data on to the data bus, and this occurs at the same time as the write signal becomes valid at the beginning of the second clock cycle. The active write signal enables the selected device to read in the value on the data bus by enabling a data buffer between the data bus and its internal registers or memory. However, memory devices and many I/O components do not rely on a clock for their internal operation and are instead based on a completely static logic design. A consequence of this is that the value on the data bus must be valid at the time the write signal is made inactive by the CPU, as this has the effect of disabling the data bus from the internal workings of the device and means that the value of the data bus present during the rising edge of the write signal will be the value retained by the memory or register. The active data bus condition therefore extends slightly into the beginning of the next cycle.

The speed of many of today's microprocessors is often too fast for them to be able to read and write data to some of the slower devices. This is especially true in the case of I/O devices, and may also be true for some memory devices as the cost of fast memory devices is significantly more than that of normal speed components. The speed of the memory or I/O device refers to the read or write access time which, in the case of the read access time, is the time between a valid address being applied to its address input and the data at that address appearing at the data bus output, given that the device has been selected and that the read control line is active. This corresponds to the time t_a shown in Fig. 10.3a, assuming that the time taken to generate a chip select signal for the memory device is negligible. The write access time is a little more difficult to define, but could be stated as the minimum time required from a valid address receiving an active write signal to the time that the write signal becomes inactive, in order to store the contents of the data bus in the address location. Normally the worse case is the read access delay, and so the term 'access delay' is taken as a value which will meet both read and write timings.

It is possible to accommodate slower devices into a microprocessor system by inserting wait states into the read and write cycles. This may be achieved by using an external control signal which indicates to the CPU when the slow device has had sufficient time to read or write the data. With the aid of extra logic circuitry, the wait control signal will be activated whenever a slow device is accessed by the CPU. This will indicate to the CPU that a wait state is requested which will extend the read or write cycle by inserting extra clock cycles after the read or write line has become active. This is shown in Fig. 10.3c for the read cycle with an added wait state. The wait control line is sampled by the CPU in the middle of the wait cycle and its state will determine if another wait cycle is to be added or if the cycle is to finish. In the case shown, the wait control line has returned to an inactive state and so the cycle terminates.

A disadvantage with the scheme outlined above is that extra logic is required to service the wait control line. Some microprocessors avoid this by having a wait control line which is permanently active to always insert a fixed number of wait states on each external access. This is less flexible than the previous scheme as it inserts wait states regardless of whether an access is being made to a fast or slow device.

It may seem that the use of wait states is an overcomplicated scheme when the bus timings can be more easily slowed by choosing a slower clock frequency. However, the slower clock frequency would mean that the internal operation of the CPU would be slowed down, with a similar effect on the internal execution of instructions. By using wait states, only external system bus accesses cause the system to slow down, while allowing less expensive memory devices to be used.

So far, only the system bus and the way in which it provides a flow of information between the CPU and its memory and I/O have been examined. The operation of the CPU must be considered in order to gain an understanding of how it performs the operations defined by the instructions that it reads from memory. Examples in the following chapters assume that the CPU is a simple 8 bit processor with a 16 bit address bus, as used to describe the system bus operations above. The principles remain the same for both 16 bit and 32 bit processors.

Chapter 11

The central processing unit

Although many different types of microprocessor exist, they all have a common set of functions which are fundamental to their operation. In Chapter 10 the operation of the microprocessor system was considered at the block level in terms of the interconnections and the operation of the interfaces within the system. The next level of detail is obtained by expanding the blocks at a functional level, and in this chapter the operation of the central processing unit (CPU) is described.

A basic element within the CPU and other components within the system is the register. A register is similar to a memory location in that data can be written to and read from the register by the microprocessor. In addition, the contents of a register are responsible for the control of the hardware with which it is associated.

The microprocessor is a CPU on a single chip, and contains all the necessary features which allow it to fetch instructions from memory, decode the instructions, and execute them in a sequential fashion. In order to better understand the functions provided by a typical CPU, it is instructive to look at the features that must be supported by the CPU in order for it to be able to execute a program. To achieve this we can list some of the characteristics of a typical program as follows:

1. Instructions are executed sequentially one after the other, unless a specific instruction, such as a jump, transfers the operation to another point in the program.
2. Sections of a program which are frequently repeated are separated out from the main program to form a subroutine which can be called from the main program. Calling the subroutine has the same effect as jumping to the beginning of the subroutine. However, when the execution of the subroutine is completed, control automatically reverts back to the main program at the instruction immediately after the point at which the subroutine was called. Subroutines can normally be nested, enabling a subroutine to call other subroutines.
3. A program may contain groups of instructions which are to be conditionally executed depending on the result of previous instructions.

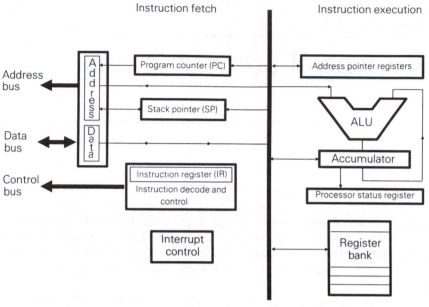

Figure 11.1   The simplified internal organization of a CPU.

4. The program will contain assignment and basic arithmetic functions for the manipulation of data.

In operation, the program instruction and any associated data are recovered from memory by the CPU during the fetch phase by means of a series of read cycles. This is followed by the execution phase in which the instruction is decoded and the appropriate actions initiated. The functions within the microprocessor can be conveniently partitioned into those that support the instruction fetch phase and those that support the instruction execution phase. Figure 11.1 illustrates this by showing the functions of a typical CPU as partitioned either side of the internal bus of the CPU. Those functions to the left in the diagram support the fetch phase, and those to the right the execution phase. Each function is described in the following sections.

11.1   CPU operation: the fetch phase

11.1.1   THE PROGRAM COUNTER

The sequential fetching of instructions from memory is achieved using a special register called the program counter (PC) which contains the address of the next instruction to be fetched from memory. At the beginning of the read cycle the contents of the PC are latched into the the address buffers and placed on to the address bus. The contents of the PC are then incremented in preparation for fetching the next item from memory. If the maximum addressing capability of the CPU is 64K bytes, requiring 16 lines on the address bus, then the program counter register must be 16 bits long.

As instructions are read into the CPU from memory, they are placed into the instruction register which is part of the instruction decode and control section of the CPU.

Assuming the data bus to be 8 bits wide, each data transfer operation between the CPU and memory transfers a single byte. However, instructions can be more than one byte long as they may refer to an operation to be performed on data which is in memory, in which case more than one read operation is necessary to read both the instruction and the data. The length of an instruction is typically encoded as part of its first byte, so that the CPU then knows how many more read operations are required to complete the instruction fetch phase.

As an example, consider the PC to be currently pointing to address location 0x5000 where it finds the first byte of the jump instruction:

JMP 0x2000

This indicates that the next instruction to be executed is located at address 0x2000. The sequence of execution of this instruction is then as follows:

1. The byte stored at address 0x5000 corresponding to the command JMP is read into the CPU and stored in the instruction register.
2. The PC is then incremented by 1 to 0x5001.
3. The value in the instruction register is then decoded by the instruction decode and control section of the CPU to determine that it is a jump instruction and as such requires the destination address to be read before it can be executed.
4. As the destination address is a 16 bit value, the CPU will know that it has to make two more read operations, each time incrementing the value of the PC by 1. By the time it has read the whole instruction the value of the PC has reached 0x5003.
5. The instruction decode and control section then executes the jump instruction by transferring the value 0x2000 to the PC. The program will continue sequentially from address 0x2000, with the next fetch cycle transferring the byte at this location into the instruction register.

11.1.2   THE STACK POINTER

Jumping within a program causes program control to be transferred from one part of the program to another. However, it is frequently found that a section of a program may be used many times, such as a routine used to write or read data to or from an I/O port, or a routine to multiply two numbers together. Instead of duplicating the code each time the function is required, it may be

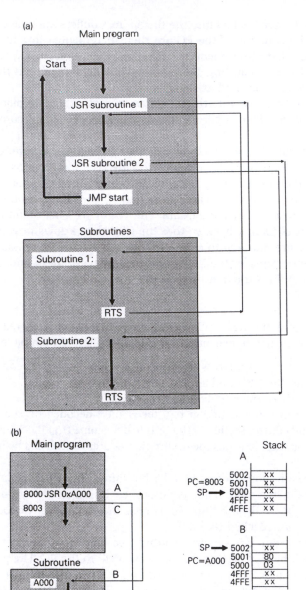

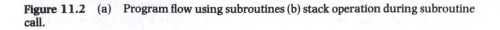

Figure 11.2   (a)   **Program flow using subroutines (b) stack operation during subroutine call.**

implemented only once as a subroutine with special instructions which allow it to be called from the main program whenever necessary and, when it is finished, to return control to the main program, as illustrated in Fig. 11.2a. The instruction JSR or 'jump to subroutine' transfers program control to the subroutine, and the instruction RTS or 'return from subroutine' automatically returns control to the main program.

It can be seen from the above that once the subroutine has been called, the contents of the program counter will have been changed so that the CPU fetches instructions from the subroutine. As control is automatically to be returned to the main program on completion of the subroutine, the CPU must keep a copy of the return address. When the 'return from subroutine' (RTS) instruction is executed, the return address is reloaded into the PC and execution of the program continues from the instruction following the 'jump to subroutine' (JSR) instruction in the main program.

One way of storing the return address is to provide a dedicated register within the CPU. However, if the CPU is to support nested subroutines, where a subroutine may itself call another subroutine, then a number of registers must be provided depending on the depth of the nesting.

The most common method of retaining the return address is by means of a stack. This is an area of read/write memory which is treated like a first in, last out buffer. The address of the top of the stack is held in a register in the CPU called the stack pointer (SP), so named because its contents point to the location of the top of the stack in memory. Each time an instruction writes a data byte to the stack the SP is either incremented or decremented according to whether the stack grows upwards or downwards in memory. Similarly, when the top of the stack is read by the CPU, the SP is changed in the opposite direction.

As an example, consider the operation of the stack during execution of the JSR instruction. For an 8 bit CPU with a 16 bit address bus, the return address must be written to and read from the stack as two bytes. Assuming that the stack grows upwards in memory, when the JSR instruction is executed, the contents of the PC, which is the address of the following instruction, are written to the top of the stack using two write cycles. After each write cycle the stack pointer is incremented to point to the top of the stack. When the RTS instruction is executed at the end of the subroutine, the exact reverse procedure is followed in order to restore to the PC the address of the instruction following the JSR instruction. This time, two read cycles are required and the contents of the SP are decremented before each read cycle.

This is illustrated in Fig. 11.2b for the execution of the instruction JSR 0xA000. When this is executed, the PC will contain the address of the following instruction, 0x8003, which is stored on the stack in memory at addresses 0x5000 and 0x5001 pointed to by the SP. The SP is incremented accordingly to point to the top of the stack at address 0x5002. The PC is loaded with the subroutine address, 0xA000, and is executed. At the end of the subroutine, the RTS instruction causes a reverse procedure to be followed which results in the

PC being loaded with the last two bytes written to the stack, and causing the stack pointer to be decremented by two. In this example the stack has been chosen to grow upwards in memory, such that as data is written to the stack it occupies higher addresses. Some CPUs implement stacks in memory which grow downwards in memory, but have the same function as described above.

The stack can also be used as a storage buffer for CPU registers. The operation which causes a value to be put on to the top of the stack is called a push, and the operation which reads the value from the top of the stack into a register is called a pop. These operations are invoked by special instructions which are part of the instruction set of the CPU (see section 11.5.7).

11.1.3 INSTRUCTION DECODE AND CONTROL

Within the microprocessor system both instructions and data exist simply as bytes of information, and the CPU has no way of knowing by reading the byte if it is part of an instruction or a piece of data. The CPU will always interpret the first block of information read during the fetch phase as an instruction. A consequence of this is that the microprocessor system can 'crash' when, for example, noise causes the address lines to be corrupted such that the CPU reads data instead of an instruction and thus operates unpredictably. In an embedded real-time system a microprocessor crash is highly undesirable. It cannot be guaranteed not to happen, especially in electrically noisy environments. Such applications normally employ recovery mechanisms which reset the CPU to a known state; these are described in section 11.3.6.

As instructions are read from the program memory, the CPU has first to decode those instructions in order to generate the correct control signals. The instruction decode and control structure is the heart of the CPU, as it defines exactly how the instructions are to be executed. To do this, it controls the flow of data between registers and the bus interface, generates external control signals for the control bus, and controls the operation of the arithmetic and logic unit.

Encoding of the instructions, such that a particular value of byte will be interpreted as a specific instruction, is designed to make the decoding of the instructions as efficient as possible so that the instruction decode and control unit is fast, and as small as possible to reduce the cost of the CPU. There are two ways of implementing the instruction decode and control unit. One is to use microcode to define the function of an instruction, and the second is to use fixed logic.

11.1.4 MICROCODED INSTRUCTION DECODE AND CONTROL

In a microcoded instruction decode and control unit, a block of ROM within the CPU contains the state of the signals used to control the internal and external functions of the microprocessor system. The contents of the ROM are referred

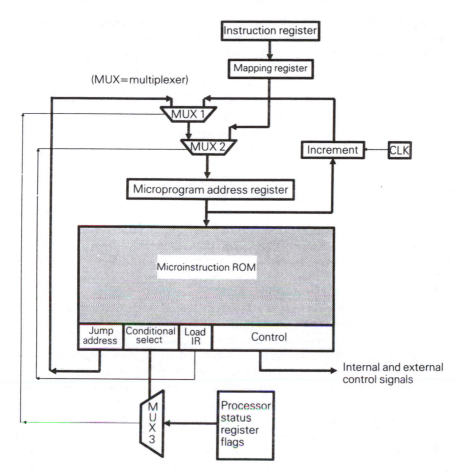

Figure 11.3   The microprogrammed control unit.

to as microinstructions and form a microprogram that defines the operation of the instructions read by the CPU from program memory. An example of the arrangement of a microcoded control unit is shown in Fig. 11.3. This diagram has been simplified in order to convey the principle of its operation. To avoid possible confusion, the instructions of the microinstruction will be referred to as microinstructions and the instructions read by the microprocessor from its external program memory will be referred to as machine instructions.

The principle of its operation can be understood by considering the ROM, referred to as the microprogram ROM, and the microprogram address register and counter in Fig. 11.3. Assume that this address register contains a binary number X and that multiplexers 1 and 2 are selected so that the microprogram address comes from the increment register and back into the microprogram address register. (The multiplexer is a simple logic function which, depending on its control inputs, will select one of its inputs to become connected to its output.) The output of the microinstruction ROM is the microinstruction held in location X; it will determine the state of control lines, such as the external read and write lines and the internal ALU control lines. Normally, the content of the address register is incremented from the increment register by the clock signal so that the microinstructions are executed in a sequential manner.

Each time a new machine instruction is loaded into the instruction register, it defines the start address in the microinstruction ROM of the microinstructions which implement the machine instruction. This is achieved by first mapping the relevant part of the machine instruction into an address which is compatible with the address bus of the microinstruction ROM. The address is then selected in multiplexer 2 by the 'load IR' bit in the microinstruction and is loaded into the microprogram address register.

Multiplexer 1 allows either the contents of the incremented microprogram address register or the value of the 'jump address' field in the microinstruction to be passed to the microprogram address register via multiplexer 2. The microprogram may therefore jump to another microinstruction when the 'jump address' field is enabled through the multiplexers to the address register.

Conditional jumping to other microinstructions is also possible using the status signals generated by the arithmetic and logic unit (ALU). For example, if the result of an ALU operation was zero, a signal line will be set at the input of multiplexer 3. If the current microinstruction is then required to jump conditionally on the status of this signal line, the 'conditional select' bit of the microinstruction is selected. Then either the jump address or the incremented address register is selected according to the state of the relevant status signal.

The size of the instruction decode and control unit, and in particular the microinstruction ROM, is dependent on the number of different machine instructions defined in the instruction set of the microprocessor. In addition, the width of the microinstruction ROM, or the number of bits in its data output, will depend on the number of control lines that have to be provided for internal and external functions. This can typically be in the range of 50 to 200 control lines. Because the control unit is based on a ROM, it provides a very flexible way of implementing the microprocessor instruction set as only the contents of the ROM need to be defined. Hence, microcoded instruction decode and control units are also an effective way of implementing large and complex instruction sets. In some microprocessors the microinstruction memory is accessible by the programmer, and this allows him to implement his own instruction set. However, the microcode in most CPUs is made permanent at the time of its manufacture, and therefore may only be changed by the designers of the CPU.

11.1.5 HARD WIRED CONTROL UNITS

The alternative to a microcoded control unit is to implement the instruction decoding and control by means of hard wired logic functions. In this case, the internal and external control signals are generated by combinational logic directly from the instruction register. The major benefit of this type of control unit is its higher speed due to the low propagation delay of signals through logic gates compared with the access time of a microinstruction ROM. Its disadvantage is that once an instruction decode and control scheme has been implemented, it is comparatively more difficult for its designers to make modifications or changes. Also, the size and complexity of the circuitry tends to grow exponentially with the size and complexity of the instruction set. Consequently, hard wired control units are mainly used to implement small instruction sets for fast CPUs.

11.2   CPU operation: the execution phase

11.2.1   THE ARITHMETIC AND LOGIC UNIT AND THE ACCUMULATOR

The main function of the instruction decode and control unit is to decode machine instructions and issue control signals to move information between internal registers and external memory, and to control the functions in the CPU which process the information. All CPUs contain a functional block which is capable of receiving data from registers and performing arithmetic and logic operations on them. This functional block is called the arithmetic and logic unit (ALU).

The operation of the ALU is associated with a special register called the accumulator, which may be regarded as the key register within the CPU. The close relationship between the ALU and the accumulator can be explained by considering the way in which the microprocessor would execute an arithmetic or logic instruction.

Consider the arithmetic expression

$$a = b + c$$

where b and c are single byte numbers stored in memory. In order to perform the addition, the CPU must present both b and c to the ALU together with an instruction to add the numbers together. It must then be able to access the result, either to store it in memory or to use it to perform further calculations.

Because the CPU can only access one byte at a time, it must load the numbers sequentially from memory. This requires temporary storage for one of the numbers inside the CPU while the other number is being fetched. The problem is solved by the arrangement of the ALU and accumulator shown in Fig. 11.1, in which the accumulator acts both as the source of one of the operands for the ALU and as the destination for the result. The other ALU operand can normally be sourced from other CPU registers or external memory. Hence, for the example given, the microprocessor could implement the calculation with the two instructions:

1. Load the accumulator from the memory location that contains b;
2. Add to the accumulator the memory location that contains c.

The second instruction assumes that one of the operands has already been loaded into the accumulator and will store the result back into the accumulator, overwriting the value b stored there as a result of the first instruction.

Most ALUs offer a variety of logical and arithmetic operations on one or two operands. Some of the typical operations are as follows.

(a) LOGICAL OPERATIONS

logical AND	b AND c
logical OR	b OR c
exclusive OR	b XOR c
complement	$b = $ NOT (b)
logical shift left or right	bit pattern shifted by one bit position, normally with zero infill
logical rotate left or right	bit pattern shifted and wrapped around on itself

Logical operations normally operate on a bit-by-bit basis.

(b) ARITHMETIC OPERATIONS

arithmetic addition	$b + c$
arithmetic subtraction	$b - c$
arithmetic multiply	$b \times c$
arithmetic division	b/c

With the exception of multiply and divide, these operations can be implemented using simple logic gates. To add hardware multiply and divide functions to the ALU significantly increases its size and complexity and hence the cost of the CPU. An alternative approach is for these operations to be implemented by means of microcode additions and shift operations in a similar way to a long multiplication calculation. The penalty of microcoded multiply and divide operations is the time required for their execution because of the larger number of instructions that have to be performed. Hard wired multiply and divide is normally implemented in an ALU where these instructions are used frequently and speed is critical, for example as in digital signal processing and high speed control applications.

11.2.2   THE PROCESSOR STATUS REGISTER

In the arithmetic operation

$$a = b + c$$

each of the *a*, *b* and *c* values is stored in byte registers or memory. However, it is apparent that the result of the calculation can exceed the maximum value that can be represented by a single byte. In fact, for addition, one more bit is required. This bit can be considered as the carry flag when additions are performed on individual bytes. So, to add two 16 bit numbers together using byte operations, the program must first add the two least significant bytes together, and then the two most significant bytes together with the value of the carry flag generated by the previous addition.

Similarly, a borrow flag can be generated by subtractions where the result may be negative. Negative numbers are normally represented in a binary system in two's complement form. This has the advantage that the sign and value of the result of arithmetic operations involving mixtures of positive and negative numbers is inherently correct.

The carry and borrow flags are generated by the ALU to provide status information about the last addition or subtraction performed. Most CPUs have a range of flags which reflect its general status and the status of the last ALU operation. These flags are contained in a register known as the processor status register. Other popular names for this register are the condition code register or flag register.

In the microcoded control unit, the status of the flags is used for the conditional execution of microinstructions, which in turn are used to implement the conditional execution of machine instructions. The following is an example of how the microprocessor would execute a simple delay loop which uses the accumulator as a loop counter initialized with the value 10:

	load accumulator with 10	; set initial contents of accumulator to 10
label:	decrement accumulator	; accumulator = accumulator − 1
	if the contents of the accumulator have not reached zero then jump to 'label'	; conditional jump to 'label'

The last instruction is a conditional instruction which will execute a jump if the value of the accumulator has not reached zero. The microinstruction which implements the machine level conditional jump instruction must first select the zero condition flag from the processor status register in Fig. 11.3. This flag has a value of 0 if the content of the accumulator is not zero, and a value of 1 when the accumulator is zero. If the flag is not set (has a value of 0) the control unit executes a microinstruction to reload the program counter with the address of 'label'. If the zero flag is set (has a value of 1) no jump will take place and the control unit will fetch the next machine instruction after the loop.

11.2.3 THE REGISTER BANK

Many microprocessors contain a set of general purpose registers which can be used, for example, to store the intermediate values generated in long calculations. The reason for having available such a register bank is the improvement in speed that can be achieved by having data already available inside the microprocessor instead of having to fetch it from memory. In general, a microprocessor can always access an on-chip register faster than an external memory location because an external access must operate through the bus interface of the CPU and external logic, as well as being dependent on the speed of the external memory. Also, the machine instructions which access on-chip registers are typically single byte instructions, whereas an instruction which accesses external memory must be two or three bytes long in order to specify the data itself or the address of the data. Hence the fetch phase of an instruction which access an on-chip register is also faster.

In addition to general purpose registers, the CPU may also contain special registers which are intended to be used as address pointers. Instead of referring to a memory address explicitly, which means that the instruction would contain the address and therefore consist of two or three bytes, the address of the operand is specified as the contents of the appropriate pointer register. As in the case of general purpose registers, this means that the CPU has rapid access to the pointer register because it is on-chip and the instruction can be encoded as a single byte. This is explained in more detail in section 11.6.

11.3 Interrupt processing

The normal execution of a program in a microprocessor system follows a sequential and deterministic process. In a real-time environment, the microprocessor system will interact with a process or machine which will generate events, such as a limit switch being activated, that occur asynchronously to the operation of the microprocessor. During normal operation, the microprocessor system must be capable of being interrupted so that it can execute a program routine dedicated to servicing the event.

Physically, an interrupt takes the form of a single input signal to the CPU which, when active, causes the CPU to start executing a program from a predefined address. This program will normally take any necessary actions to service the cause of the interrupt and is referred to as an interrupt service routine (ISR). The ISR can be thought of as a subroutine which is invoked by a hardware signal to the CPU instead of a software instruction.

11.3.1 REGISTER STACKING AND CONTEXT SWITCHING

The interrupt will cause the CPU to transfer program control to its interrupt service routine which, like the subroutine, must have a return address to the main program stored on the stack. In addition, because the occurrence of the interrupt is asynchronous to the execution of the CPU, it is not possible for the CPU to determine at what point the main program will be interrupted. It is therefore also necessary to save the status of certain registers in the CPU which may be changed during the execution of the ISR. The most important registers that are typically stacked automatically are the accumulator and the processor status register.

There are three common approaches to saving the remaining CPU registers. The first of these involves the CPU automatically saving the contents of its programmable registers at the beginning of the ISR, and automatically restoring them on execution of the 'return from interrupt' instruction. This method may be convenient for the software programmer in that saving and restoring the CPU registers is done automatically, but it also has the disadvantage that it takes longer for the CPU to start executing instructions in the ISR immediately after the interrupt. This interrupt latency occurs because, for each register stored on the stack, the CPU must perform a write cycle to the stack in memory. The more registers the CPU has, the longer it takes to stack the registers and start executing the instructions in the ISR. In real-time systems, a short interrupt latency is required to allow a fast response to interrupts.

An alternative method is to provide no automatic register stacking, but instead to place the responsibility for stacking the registers in the hands of the programmer, who can then use special stack instructions to save the registers on the stack. With this method, only those registers which are modified by the ISR need to be stacked, and so the minimum time is spent performing stack operations. However, it also means that the programmer must explicitly restore the registers before returning from the ISR. Because the stack operates on a first in, last out principle, the order in which registers are unstacked should be exactly the reverse of the order in which they were stacked.

The third method avoids the use of stack operations. Instead, the CPU contains duplicate sets of registers to which it can switch whenever an interrupt occurs. For example, the CPU may have four sets, numbered 0 to 3, of eight 8 bit registers, where the active register bank can be selected by two bits in the processor status register. When there is no interrupt present and the main program is executing, register bank 0 is in use. When an interrupt occurs, the PC and processor status register are stacked in the normal way, and the first instruction of the ISR selects register bank 1 by writing to the processor status register. This has the effect of freezing the contents of register bank 0. It also provides a means of automatically selecting the previously enabled bank when the 'return from interrupt' instruction is executed, as the previous contents of the processor status register will be restored from the stack. If during the ISR another interrupt were to occur, then a similar procedure would be used to select the next register bank, bank 2, and so on until three interrupts had been nested.

The values of the registers in the CPU at any instant are sometimes referred to as its context, as they reflect the status of the CPU at that time in the context of the program that it is executing. The consequence of an interrupt is that the current program context is saved such that it can be restored after the execution of the ISR. The freezing of one program and the continuation of another program therefore involve a context switch.

11.3.2   SYSTEMS WITH MULTIPLE INTERRUPT SOURCES

Within an embedded microprocessor system, interrupt requests may be generated by its input and output devices. A simple device such as a switch can generate an interrupt by direct connection to the CPU such that when the switch is closed it activates the interrupt. It is normal for there to be a number of different sources of interrupts within a system, and unless the CPU has many separate interrupt inputs, some form of interrupt controller is required to arbitrate the access of several interrupt requests to the CPU, and also to identify their source to the CPU. A typical arrangement was shown in Fig. 10.1, where the CPU has only a single interrupt input connected to the output of an interrupt controller. The interrupt controller interfaces the interrupt lines from the serial port, the parallel I/O port and the A/D converter to the CPU.

When an interrupt is recognized by the CPU, it starts to execute an ISR at a predefined address in program memory. The CPU makes use of either a vectored address or a non-vectored address method to determine this start address.

11.3.3   NON-VECTORED INTERRUPTS

A non-vectored interrupt uses a fixed address within the memory at which the start of the ISR is located. When the interrupt occurs the program counter will be loaded with this fixed address. The return address to the main program will be stored on the stack in the same way as a JSR (jump to subroutine) instruction stores the return address. The end of the interrupt service routine is defined by a 'return from interrupt' instruction which returns control to the program at the point at which it was interrupted. If an interrupt occurs halfway through the execution of an instruction, the CPU will complete the execution of the instruction before servicing the interrupt. The operation of non-vectored interrupts is shown in Fig. 11.4a. At the point where the interrupt occurs in the execution of the main program, the PC is loaded with a fixed address, 0xF000, which is the start address of the ISR.

11.3.4   VECTORED INTERRUPTS

An interrupt vector is a pointer register or memory location whose content is the start address for the ISR. Although the address of the vector itself is fixed, its contents may be any address value; therefore a programmer may place the start address of an ISR anywhere in memory. When an interrupt occurs, the

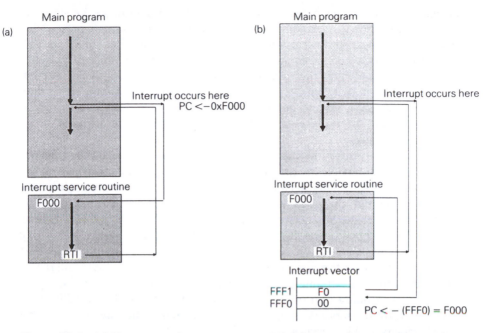

Figure 11.4   (a) Non-vectored interrupt processing (b) vectored interrupt processing.

current instruction execution is completed, and the interrupt vector is read from a fixed memory location, or register, which has been loaded with the start address of the ISR. This value is loaded into the program counter and the remaining operation then follows the sequence described in section 11.3.1. The operation of vectored interrupts is shown in Fig. 11.4b. When the interrupt occurs, the PC of the CPU is loaded with the contents of the two consecutive bytes starting at 0xFFF0, which in this case is the address 0xF000.

11.3.5   MULTIPLE INTERRUPT PROCESSING

An embedded real-time microprocessor system may have many sources of interrupts. Many of these interrupts will be generated by the I/O devices within the system, and the CPU then faces the problem of having to identify the source of the interrupt in order to be able to execute the correct ISR for the interrupting device. If the CPU uses non-vectored interrupts, only a single ISR address is available, and so the ISR must identify the device which requested the interrupt. To achieve this, either the device that requested the interrupt, or the interrupt controller, must have an interrupt status register which will indicate to the CPU the source of the interrupt. The ISR must read this interrupt status register and, having identified the interrupt source, execute the correct interrupt service routine. In cases where each device contains an interrupt status register, the

ISR will have to read each device in turn in order to determine the status of each interrupt. This is referred to as polling the interrupt status registers.

For a microprocessor system which depends on several interrupts and ISRs to transfer data between itself and the real world, it is possible that an interrupt will be generated during the execution of an ISR from an earlier interrupt. Should the interrupts be enabled at that time, then the executions of the ISRs become nested. Each time an interrupt occurs, the return address is stored on the stack, so that as the level of nesting becomes deeper, the contents of the stack become larger.

In many cases it is not necessary to allow nested interrupts, and is sufficient to process each interrupt in turn. In this case the interrupt controller or the CPU should have the ability to enable and disable its interrupt inputs. However, during the time that interrupts are disabled, it is still necessary to recognize the fact that an interrupt has occurred, such that the CPU can respond to it as soon as it has finished processing the current ISR. An interrupt controller may therefore provide an interrupt pending register to indicate the interrupts that have occurred during the time that interrupts were disabled. When interrupts to the CPU are again enabled, the interrupt controller will determine if any interrupts are pending and generate an interrupt to the CPU according to a priority such that, if more than one interrupt is pending, the interrupt with the highest priority will be serviced first by the CPU.

A method used by some CPUs and their associated interrupt controllers to avoid the problem of identifying the interrupting device by reading a status register, or polling status registers of individual devices in software, is to read the interrupt vector from the interrupt controller. The interrupt vector is read automatically by the CPU from a register within the interrupt controller as the CPU responds to the interrupt. This value is then used by the CPU to generate the start address of the ISR which is loaded into the PC. When the interrupt occurs, the interrupt controller must generate the correct interrupt vector for that interrupt and place it in the register ready for the CPU to read it.

11.3.6   NON-MASKABLE INTERRUPTS AND CPU RESET

The interrupts considered so far are the general purpose interrupts to the CPU which can normally be enabled and disabled by the CPU. This process of enabling and disabling is sometimes referred to as masking, because when the register disables the interrupts it is masking off the interrupt function from the CPU.

Most microprocessors contain a non-maskable interrupt (NMI) which, as its name implies, cannot be disabled. Its function is identical to the general purpose interrupts, but it is normally reserved for emergency situations.

A typical emergency situation might be a power failure in a system. This would be indicated to the CPU through the NMI input by circuitry monitoring the voltage levels on the power supply to the microprocessor system. When the voltage drops to a level which is considered close to that below which the microprocessor system can no longer operate, an interrupt signal is sent to the NMI input. If the power supply had been suddenly cut, it would take a finite time for the voltage levels to decay to zero. Although this time may only be a matter of milliseconds, it may still be sufficient time for the CPU to execute important functions such as failsafe precautions which, once the power had been restored, would allow the system to recover gracefully from the power-down condition.

Because the NMI is intended to allow the CPU to respond to emergency situations, its interrupt latency should be as small as possible. Hence, for CPUs which employ an automatic stacking of a large number of CPU registers, the NMI may only stack the program counter and the processor code register.

All microprocessors have a reset input signal which, when active, resets the internal state of the CPU and initializes its registers. Its normal usage is during power-on of the microprocessor system in order to reset the CPU, and other circuits, and to begin execution of its program from the beginning. The operation of a reset is in many ways similar to that of an interrupt in that the start address of the program to be executed is either a fixed value or a vector obtained from a fixed address. However, instead of executing an ISR, the program executed on reset will be the main program, which will typically initialize functions such as the stack for subsequent subroutine and ISR execution.

Many systems also provide a manual reset facility to reset and restart the system should the execution of the program become lost or corrupted, which may occur due to electrical noise on the microprocessor's buses and control lines. For an embedded system, access to the reset button may not always be possible, and so a watchdog system may be implemented to automatically activate the reset input should the normal execution of the program be stopped. Most watchdog methods are based on a timer which is prevented from timing-out by instructions in the program which reset the watchdog timer. If normal program execution does not take place, the timer will not be reset, causing it to time-out and so reset the CPU.

11.4   The central processor unit instruction set

The microprocessor executes a program by continually fetching and executing machine instructions from memory. An instruction is normally composed of two parts: an operation code, called the opcode, which defines the operation to be performed; and the operand, which defines the data to be operated on. These are shown in Fig. 11.5.

The way in which the operand is accessed by the instruction, for instance by value or by address, is called the addressing mode of the instruction. If the instruction set of a microprocessor is defined such that all addressing modes are possible with all instructions, then the instruction set is said to be orthogonal or regular. Such an instruction set is considered to be advantageous as it

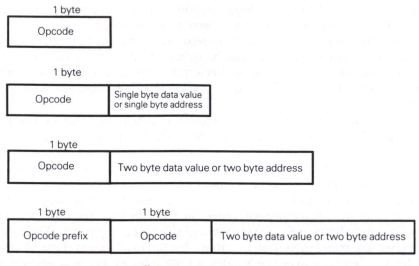

Figure 11.5   Different microprocessor instruction formats.

improves the flexibility of the instructions and consequently the ease of programming, since the programmer does not have to remember which addressing modes apply to which instructions. Not all microprocessors possess an orthogonal instruction set, but instead may optimize, in terms of size and hence speed, those instructions and addressing modes used most.

In an 8 bit microprocessor, it is convenient to be able to encode all possible opcodes with their addressing modes as a single byte. The use of a single byte opcode allows a possible 256 instructions, with their associated addressing modes, to be defined. Although this may sound a large number, it is only necessary to look at a typical microprocessor instruction set to see that a single byte opcode is often not enough to avoid either making some sacrifice to the orthogonality of the instruction set, or having some two byte opcode instructions. A common way of implementing two byte opcodes is to use an opcode prefix for less commonly used instructions and addressing modes, which informs the instruction decode and control section of the CPU that the next byte is also an opcode instead of an operand or an operand address.

In general, microprocessors make use of variable length instructions. Figure 11.5 shows instructions of one, two, three or four bytes in length if an opcode prefix is used. Although these require a more complex instruction decode and control unit, it increases the speed of their execution as only the minimal number of bytes have to be fetched from the memory. For fixed length instructions, its size must be equal to the largest variable length instruction. Hence only the minimum amount of memory is used with variable length instructions when compared with programs with fixed length instructions.

11.5   Addressing modes

The addressing modes of the instruction set refer to the different ways that the microprocessor accesses the data on which the opcode is operating. In this section the most commonly used addressing modes are described. The names given to these addressing modes and their descriptions are those common to a wide range of microprocessors, although differences can be found between specific devices. In the examples used to illustrate the addressing modes, mnemonics are used to represent the opcode of the instruction.

11.5.1   IMMEDIATE ADDRESSING

The immediate addressing mode is so called because the operand follows immediately after the opcode. The complete instruction therefore consists of the opcode followed by either an 8 bit or a 16 bit data value. An example of an instruction that uses immediate addressing is

LDA #0x12   ; load the accumulator with the value 0x12

This instruction would appear in program memory as two consecutive bytes. The first byte of the instruction read by the CPU would be the opcode for the 'load accumulator with 8 bit immediate value' instruction, and the second byte would be the value 0x12. The # symbol is used to indicate that the value 0x12 is an immediate value and not an address (see the next section, on direct addressing).

In the context of computer programming, the immediate addressing mode is used to load constant values into registers, and the constant is defined as part of the instruction.

11.5.2   DIRECT ADDRESSING

This addressing mode takes its name from the fact that the operand is specified in the instruction directly by its address. So, if the data value 0x12 was stored in read/write memory at address location 0x4000, we could achieve the same result as the instruction above with

LDA 0x4000   ; load the accumulator with the contents of
                ; address 0x4000

Notice that the instruction does not contain a # sign as in the immediate addressing mode because the value 0x4000 refers to an address and not an immediate data value. Although the address is defined as part of the instruction, its contents are not defined, and so this type of addressing mode is used to access variable values.

This instruction will consist of three bytes in program memory as shown. The opcode in this case will be different from that used in the instruction LDA

#0x12 because of the different addressing mode, and will indicate to the CPU that it has to read two more bytes which it is to interpret as a direct address.

A special form of direct addressing is the register direct addressing mode. Instead of occupying memory locations within the address space of the CPU, the registers of the CPU are normally addressed by specific bits within the opcode itself. This has the advantage that the instruction need only be a single byte in length. So, for example, the first five bits may define the type of instruction, such as a load or add operation that uses register direct addressing, and the next three bits define which CPU register is to be operated on. These three bits can be thought of as register address lines internal to the CPU which address its registers. The instruction will consist only of a single byte opcode and will have a short execution time because no external memory accesses are needed by the CPU.

11.5.3   PAGED ADDRESSING

Although direct addressing may provide the programmer with access to any memory location within the address space of the CPU, an instruction which uses memory direct addressing is three bytes long, and so requires the CPU to execute three read cycles before the instruction can be executed. One way of reducing the size of the instruction is to assume that the direct address lies within some fixed range of the address space such that the upper address byte can be assumed by the CPU to be a known constant value. Then only the lower address byte needs to be specified as part of the instruction.

This method is called paged addressing because the address space is split up into a number of pages, where the upper address byte is the page number and the lower address byte is the address within the page, as shown in Fig. 11.6. Some CPUs contain a special page number register which can be previously loaded with the page number. Then all subsequent instructions using the paged addressing mode would access memory locations within the same page, the advantage being that the instructions are only two bytes long and so do not take as long as the direct addressing instructions to fetch and execute, or take up as much memory.

Other CPUs may not have a fully paged addressing mode with a page number register, but instead assume that the upper address byte is zero, so that the effective address is in page number zero. This addressing mode is called zero page addressing.

11.5.4   INDIRECT ADDRESSING

Indirect memory addressing means that the operand is addressed indirectly through another memory location. Hence the instruction specifies the address of the address of the operand. This is illustrated in Fig. 11.7 for the instruction

LDA @0x1000   ; load the accumulator with the contents of
                  ; the address location stored at address 0x1000

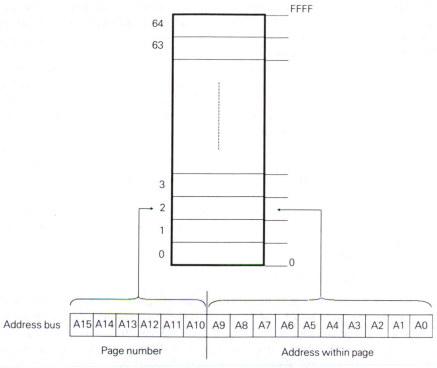

| Address bus | A15 | A14 | A13 | A12 | A11 | A10 | A9 | A8 | A7 | A6 | A5 | A4 | A3 | A2 | A1 | A0 |

Page number | Address within page

Figure 11.6 Paged addressing.

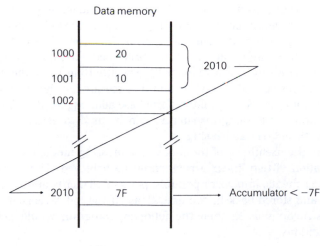

LDA @ 0x1000

Figure 11.7 An example of indirect addressing.

The two bytes stored at addresses 0x1000 and 0x1001 are taken to be the address of the operand, and the address locations 0x1000 and 0x1001 act as a pointer to the operand. The @ symbol is used in the description of the instruction to indicate that indirect addressing being used.

A benefit of this addressing mode is that it allows the same instruction to access any memory address indirectly through a fixed memory location. In effect, the instruction uses two memory locations as a pointer register to the operand. A disadvantage is that the indirect addressing mode takes longer to execute than the direct addressing mode because the CPU has to read the two consecutive byte values at the address in external memory specified in the instruction, and then use this as the address of the operand in the execution of the instruction. A solution to this problem is to implement special address pointer registers in the CPU to avoid the additional memory accesses and so speed up execution. When a CPU pointer register is used, this mode is referred to as register indirect addressing, and can be implemented in a single byte instruction.

11.5.5   INDEXED ADDRESSING

Information is often stored and processed as tables or records of data. For example, a two-dimensional matrix would have to be stored as a linear array of numbers within the memory of a microprocessor system, with each row or column stored consecutively. Any processing performed on the matrix would then need to take account of this linearization. A convenient way of addressing individual elements within the matrix is to have a pointer to the start of the row or column together with an offset to address the element within the row or column. The required memory address is the sum of the start or base address and the offset. This technique of addressing a memory location according to a start or base address plus an offset is called indexed addressing.

The CPU may provide a special pointer register called the index register, which can be used to hold either the base address or the offset address according to the way in which the information is to be processed. Although an instruction using indexed addressing will take longer to execute than the same instruction using direct addressing, because of the addition required in the addresses, it has the advantage that the effective address, the base address plus the offset address, is a variable and so the same instruction can be used to access any memory location, as with indirect addressing.

To illustrate the usefulness of the indexed addressing mode, Fig. 11.8 shows how the addition of two linear arrays could be achieved. These arrays each consist of 16 elements, and corresponding pairs of elements are to be added to one another and stored back in the first array. If the index register is referred to in the instructions as X, then the following program would perform the necessary additions:

; define the labels and addresses to be used in the program

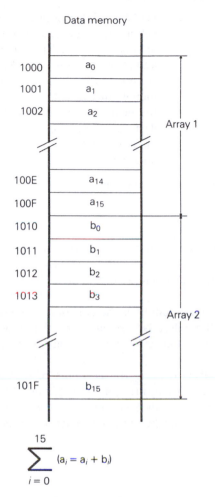

Figure 11.8   Two data arrays to be added with instructions that use indexed addressing.

COUNT = 0x0FFF	; define a variable to act as the loop counter
	; the address of the variable is defined as 0x0FFF
ARRAY1 = 0x1000	; define start address of ARRAY1
ARRAY2 = 0x1000	; define start address of ARRAY2
; the program starts here	
LDX #0x00	; initialize the index register
LDA #0x10	; load the accumulator with the
	; value 0x10 (16 decimal)
STA COUNT	; store value 0x10 in the COUNT variable
	; this initializes the loop counter with the
	; number of elements in the array

```
LOOP:                        ; LOOP is a label used for convenience to
                             ; define the beginning of the program loop
LDA #ARRAY1, [X]             ; load element of ARRAY1 into accumulator
ADD #ARRAY2, [X]             ; add element from ARRAY2 to accumulator
STA #ARRAY1, [X]             ; store result back to ARRAY1
INC X                        ; increment the index register to point to
                             ; the next element in each array
DEC COUNT                    ; decrement the loop counter value
JNZ LOOP                     ; jump if COUNT not zero to the label LOOP
                             ; and execute the program loop another time
```

The instructions using indexed addressing are the first three within the loop. For example, the instruction

LDA #ARRAY1, [X]

can be read as 'load the accumulator with the contents of the memory location pointed to by the effective address created by adding the contents of the index register to the start address of the array.'

In this example, the base address is the constant value defined as #ARRAY1 or #ARRAY2, and the offset is the value of the index register which is incremented each time the loop is executed to point to the next element in each array.

Incrementing and decrementing the contents of the index register are so frequently used in program loops that some CPUs provide indexed addressing instructions which automatically increment or decrement its contents. In the above example, the store and increment instructions would be replaced with the signal instruction

STA #ARRAY2, [X +]

11.5.6   RELATIVE ADDRESSING

In relative addressing, the effective address is calculated relative to the current contents of the program counter. The most common use of relative addressing is with the jump instruction. Taking the example used in section 11.5.5 for indexed addressing, the JNZ instruction (jumps if the contents of the accumulator are not zero) causes the program counter to be loaded with the address of the label LOOP when the contents of COUNT are not zero. If the program between LOOP and the instruction after the JNZ instruction was 15 bytes long, then the jump instruction using relative addressing would take the form

JNZ − 0x0F

This would have the effect of taking the current contents of the program counter when executing the JNZ instruction, subtracting 0x0F from it, and loading the result back into the program counter.

However, if the address of LOOP was 0x2000, and the JNZ instruction used absolute addressing, then the address 0x2000 would be specified as part of the instruction as

JNZ 0x2000

There are two advantages to be gained from relative addressing. Firstly, the relative offsets are typically small values and can be specified in a single byte, so that the instruction is reduced to two bytes instead of the three bytes required for an absolute jump. Secondly, the machine code is no longer constrained to a fixed location in memory as a result of specifying absolute addresses in instructions.

This can be better understood by again considering the example above where the label LOOP is the address 0x2000, and the jump instruction is implemented using absolute addressing and becomes JNZ 0x2000. If the machine code program is then moved such that LOOP corresponds to address 0x3000, the JNZ instruction would have to be modified to JNZ 0x3000 in order for the program to run correctly. However, no modification would be required if relative addressing were used because the jump would be specified relative to the contents of the program counter such that the offset remains the same, and so the program could be moved anywhere in memory, thereby creating relocatable machine code.

11.5.7 STACK ADDRESSING

Stack addressing is a special form of register indirect addressing. In addition to the normal attributes of register addressing, such as single byte instructions, the instructions which use the stack pointer to address data automatically manipulate the stack pointer register to maintain the first in, last out operation of the stack.

Two common stack instructions are PUSH and POP. The PUSH instruction puts the contents of a specified register on to the stack and then modifies the stack pointer to again point to the first free location at the top of the stack. The POP instruction does the opposite; it modifies the stack pointer to point to the first value on the stack and then loads this value into the specified register. For example, the instruction PUSH A would stack the contents of the accumulator, and POP A would load the accumulator with the last value that was stacked.

These instructions are most commonly used to maintain the status of registers and program variables during subroutines and interrupts.

11.6   CISC and RISC instruction sets

In the early days of the microprocessor, VLSI technology was less advanced and consequently the costs of the microprocessor and other components were much higher. This was particularly critical in the case of memory devices, as not only were they more expensive but more of them were needed because the technology had not yet been developed to produce the high density semiconductor memories in use today. As a direct consequence of this, the microprocessor instructions were designed to perform as many functions as possible in order to make their programs small and so require less memory. This trend resulted in the so-called complex instruction set computer (CISC), which was typical of most microprocessor instruction sets until quite recently.

As the instruction set of a microprocessor becomes more complex, it is only practical to implement it with a microprogrammed instruction decode and control unit, which because of its complexity can be inefficient at executing simple instructions, and its size makes it costly to implement.

If the usage of instructions in a large program is analysed, it is found that the majority of operations are simple instructions such as 'load register' and 'store register', while the more complex instructions are used less frequently. It has been argued that the instruction set of a microprocessor should therefore consist only of a small number of simple instructions, as this would significantly reduce the complexity of the instruction decode and control unit which could then be implemented as a hard wired logic design. This would have the benefit of decoding and executing instructions much faster than in a CISC, and the size of the CPU would be much smaller and hence cheaper to produce. These reduced instruction set computers (RISCs) are characterized by a small and compact instruction set where instructions are executed in a single cycle. These devices also tend to have a large number of general purpose on-chip registers so that the need to access external memory is not so frequent, which helps to maintain the speed benefits of a RISC CPU in a complete system by reducing the number of memory accesses.

While a RISC CPU may execute instructions faster and cost less than a CISC, its individual instructions are much simpler, so that it would have to execute several instructions in order to perform the equivalent function of a single CISC instruction. In addition, the RISC will require more memory to store the extra instructions, so that at the end of the day there may be little to choose between a RISC microprocessor and a CISC microprocessor. The trend today is to produce microprocessors that implement the benefits of both philosophies, so that an instruction set will consist of a core of frequently used instructions capable of executing in a single cycle, with additional more complex instructions to improve the functionality of the instruction set.

Chapter 12

Semiconductor memory, input and output, and peripheral circuits

12.1 Semiconductor memory devices

A single digital binary memory cell is a device or circuit which is capable of maintaining the state (high or low) of its last enabled input value. This state is maintained until the input to the memory is again enabled, and the state of the memory changes to match that of the input signal. This action is equivalent to a memory write operation. The memory cell will also allow access to its current state by circuits which, when enabled, will transmit the contents of the memory cell to other devices outside the memory. This is the memory read operation and does not alter the contents of the memory cell.

The function of a semiconductor memory device in a system is to provide an addressable array of contiguous memory elements, with each element containing a number of basic memory cells, which can be interfaced to the microprocessor system bus. Figure 12.1 shows a block diagram of a typical read/write memory device.

The array itself can be arranged in a number of different ways. For an 8 bit microprocessor with an 8 bit external data bus, the most obvious arrangement is to have a memory element consisting of 8 memory cells ($d = 8$ in Fig. 12.1) in parallel at each address. As an example, the 6264 memory device contains 8192 ($a = 13$) addressable memory elements, where each element has 8 memory cells or bits; this is more usually described as 8K by 8, or 8 kilobytes. The 6264 therefore has 13 external address lines and 8 external data lines. Other popular arrangements are memories with elements of 4 bits wide and 1 bit wide.

The address bus, which forms part of the microprocessor system bus, is passed through an address decoder such that each address value which appears on the bus will uniquely activate a memory array word select line, given that the memory device itself has already been selected by an active (low) chip select (CS/) signal from external address decoding logic (see section 10.1.1). The word select line enables a complete row of memory cells. If data is being written to the memory element from the external data bus, the write signal in the control bus of the microprocessor will become active during the time that data is present

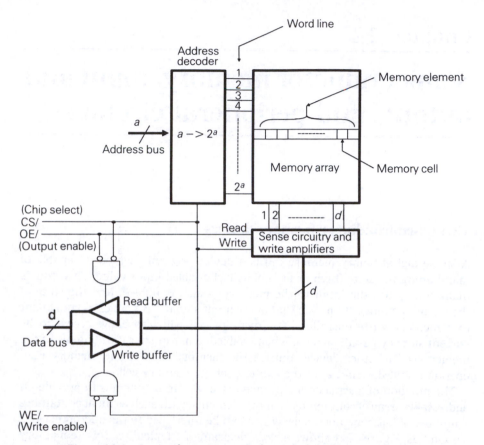

Figure 12.1   Internal organization of a typical read/write memory.

on the external data bus, as was shown in Fig. 10.3b. Hence the write buffer will pass the data from the external data bus into the memory array and write it to the memory element at the address on the address bus. Similarly, during a CPU read cycle, if the chip is selected, data stored at the memory element corresponding to the selected word line will be detected by the sense amplifiers and placed on the external data bus through the enabled read buffer.

There are many types of semiconductor memory which essentially have the same structure as described above. Their differences come from the design of the circuits used to make the single memory cells which go to make up the array.

12.1.1   READ ONLY MEMORY

As its name suggests, a microprocessor system may only read data from read only memory (ROM); ROM devices therefore provide a form of non-volatile

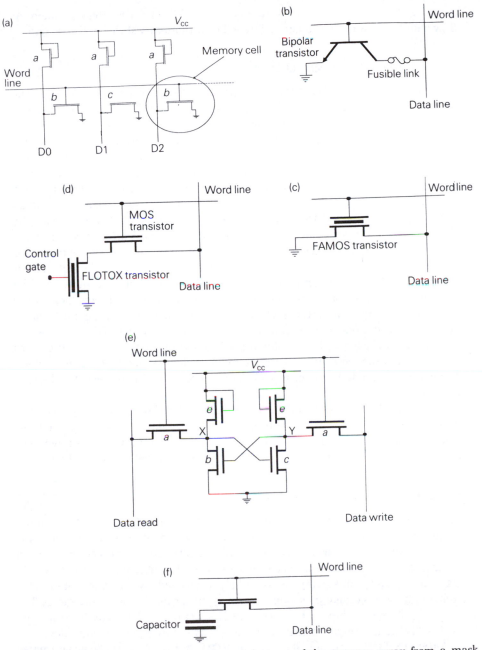

Figure 12.2   ROM and RAM structures: (a) part of the memory array from a mask programmable ROM (b) bipolar transistor ROM memory cell with a fusible link (c) EPROM memory cell with a FAMOS transistor (d) EEPROM memory cell (e) a static RAM bit cell (f) a dynamic RAM bit cell.

memory. They are important to the operation of the microprocessor system as they allow programs and constant data to be permanently stored in the system.

(a)   THE MASK PROGRAMMABLE ROM

The contents of a masked ROM are determined during its manufacture. The structure of part of a masked ROM memory is shown in Fig. 12.2a. The devices marked *a* and *b* are MOS transistors, which are described in section 13.3.2, whose operation can be considered as a simple switch. For example, the transistor at the intersection of the word line and the D2 data line operates as follows. When the word line is enabled and therefore at a high logic level, the transistor will be switched on and effectively connect the data line to 0 V. The transistor at the top of the data line has been configured such that its electrical characteristics are similar to a resistor, which simply limits the current drawn from the power line V_{cc} when the memory cell transistor is turned on. Hence, in places in the memory array where the transistor is not connected, as in the transistor marked *c*, or when the word line is not enabled, the output of the data line is a high logic level. The contents of the ROM are therefore determined by the presence or absence of connections to the transistors at the intersection of the word lines and the data lines. Each transistor corresponds to a memory cell, or bit cell because it also corresponds to a single bit of memory. The ROM can be fabricated with each cell transistor initially making a connection to the word line, and then the required bit pattern can be implemented by selectively etching away the connections to those transistors which are to represent a logic 1 memory bit.

A masked ROM represents the most economic form of memory available because its bit cell is very simple to implement. In addition its contents can be determined at the time of its manufacture, so that no extra functions have to be added to facilitate programming after it has been fabricated and packaged.

(b)   THE PROGRAMMABLE ROM (PROM)

This type of ROM may be programmed with information after fabrication. Programming can take place only once, after which the device has the same characteristics as a masked ROM. To allow the device to be programmed, each bit transistor will contain a fusible link in its main current path as shown in Fig. 12.2b. These links can be selectively destroyed by the application of a sufficiently high voltage pulse, and therefore program the bit as a logic 1. The transistor shown is a bipolar type, described in Section 13.3, because the majority of PROMs made using this structure are fabricated in bipolar technology.

Although PROMs are more flexible than ROMs in that they can be programmed by the user, the additional flexibility comes at extra cost because of the extra circuitry required to program the part once it has been fabricated and packaged. The user must also be equipped with a PROM programmer capable of accepting the required bit pattern and programming it into the PROM.

(c) THE UV ERASABLE PROM (UV-EPROM)

To allow erasure and reprogrammability of a ROM, a special type of MOS transistor is used in the bit cell. The transistor is fabricated using FAMOS technology, which allows it to be switched on and off by the presence or absence of electrical charge which is trapped inside the structure of the transistor. The charge can be introduced into the memory array in a similar way that a PROM is programmed, and to avoid damage to the remainder of the MOS circuitry, a series of programming pulses may be applied until sufficient charge has been forced into the selected transistors. This programs the transistor in the bit cell to turn on when it is enabled by the word line, shown in Fig. 12.2c. Similarly, absence of charge would cause the transistor to remain switched off when enabled by the word line.

Owing to the structure of the transistor, the charge remains trapped inside it until the memory is erased. Erasure of the memory involves removal of the charge within the transistor, which may be achieved by irradiating the device with ultraviolet (UV) light, a process which normally lasts 15 to 20 minutes. UV erasable devices may easily be identified by a small quartz window in the package which exposes the chip itself.

UV-EPROMs give the user an additional level of flexibility in that the device may be reused, and therefore is convenient for use during the development stage of a microprocessor based system. A wider variety of high density UV-EPROMs exist than PROMs, and so the UV-EPROM has become the most popular read only memory device for use in microprocessor systems. Typically, UV-EPROMs are more expensive owing to the special packaging required to allow UV erase, but may be reprogrammed many times.

(d)   THE ELECTRICALLY ERASABLE PROM (EEPROM)

The bit cell of an EEPROM (E²PROM is shown in Fig. 12.2d. It employs two MOS transistors, one normal transistor and the other a modified version of the FAMOS transistor used in the UV-EPROM bit cell, called a FLOTOX transistor. The FLOTOX transistor has a control gate added to it, and is responsible for adding and removing charge from the transistor, which programs the bit cell as either a logic 0 or a logic 1. Its operation is such that if a positive programming voltage is applied to the control gate, it causes a negative charge to build up inside the transistor which causes it to turn off, and the bit cell will be read as a logic 1 when it is enabled by the word line. The positive programming voltage therefore causes the bit cell to be programmed as a logic 1. The charge remains trapped inside the transistor until it is reprogrammed by a negative programming voltage applied to the control gate. This forces a positive charge into the transistor which effectively stores a logic 0 in the bit cell.

The erasure and reprogramming of a byte of FEPROM may each take tens of milliseconds, which is much too slow for the EEPROM to be considered as a read/write memory. In addition, EEPROMs have a life expectancy of 10^4

erasure/reprogramming cycles. Typical applications of the EEPROM are therefore for storing system configuration values which may be changed during the operation of the system, and which must be stored while it is powered down. The EEPROM bit cell is more complex and larger than that of the UV-EPROM and requires more processing steps in its fabrication.

12.1.2   READ/WRITE MEMORIES

Read/write memories, also known as random access memories or RAMs, allow data to be both written to and read from them. They differ from the ROM devices in that they are only capable of storing information while they have power applied to them. Read/write memories are therefore a type of volatile memory, and they are used mainly for the storage of variable data in a microprocessor system.

(a)   STATIC RAM (SRAM)

The bit cell in an SRAM is based on a bistable circuit, so called because its output is stable in one of two states and can be changed according to the logic value of a valid input signal. Once the input signal is no longer valid, the output of the bistable remains in this state until changed by a subsequent valid input. The bistable can be considered a static circuit as it will retain its state indefinitely without the need for any further external actions, provided power is applied.

The SRAM bit cell is shown in Fig. 12.2e. It is shown constructed from the MOS transistors used previously to describe the operation of a ROM device. The transistors marked e are configured to behave as resistors, which limit the current flowing through transistors b or c when either is switched on. Two data lines are connected to the bit cell. One is used to write data to it and the other is used to read data from it. Both data lines are connected to the bit cell by transistors, marked a and d, which are turned on by an active word line.

The bistable itself consists of transistors b, c and e. Consider what happens when a logic 1 is written to the bit cell. The first action is that the word line becomes active while a logic 1 is applied to the data write line. Transistor d is enabled by the word line and so the logic 1 appears at point Y in the bistable, and causes transistor b to switch on and draw current through e. Transistor e acts as a resistor, and so the current passing through it will cause the voltage at point X to drop to a logic 0 and hence switch off transistor c. This maintains the logic 1 at point Y. When the word line is disabled, it turns off transistors d and a, but the bistable remains in its stable state with point Y at logic 1, keeping transistor b switched on. This in turn causes a logic 0 at point X, which keeps transistor c switched off and point Y at a logic 1, thereby completing a stable loop. A similar cycle of events occurs when a logic 0 is written to the bit cell.

The bit cell is read from the data read line when the word line becomes active. However, from the previous discussion it can be seen that the inverted value of the data is read, such that if a logic 1 is stored in the bit cell then a logic 0 will be read from it. As this is always the case, a simple inversion of the data in either the read or write data lines makes the polarity of the data input equal to its output.

The bit cell of an SRAM is considerably larger than in a ROM part, which means that the SRAM bit cells cannot be so closely packed within a given area of silicon. This drawback has led to the popularity of dynamic RAM, whose bit cell is very simple, but whose operation requires the addition of external logic.

(b)   DYNAMIC RAM (DRAM)

The data storage device in the DRAM bit cell, shown in Fig. 12.2f, is a capacitor capable of storing charge delivered to it through the access transistor. A single data line is used both to write data into the bit cell and to read data from it. When a logic 1 is written to the bit cell, it charges up the capacitor until the voltage across it is also a logic 1. However, in reading the bit cell, some of the charge stored on the capacitor will escape and cause its voltage to drop. Even if the bit cell is not read, charge will still tend to leak out of the capacitor, and unless it is periodically refreshed the bit cell will eventually lose its data. The need to continually refresh the contents of the memory gives it its name of dynamic RAM.

Refreshing the memory contents is performed on groups of bit cells by reading data and writing it back into the same cell. This usually employs circuitry external to the memory chip to control the operation of the refresh cycles which are interleaved with normal microprocessor read and write cycles. The inconvenience that may be caused by the addition of the refresh control logic is usually justified in systems that require large amounts of read/write memory owing to the higher packing density achieved with DRAMs compared with SRAMs. This means that beyond a certain limit, more memory can be implemented in a given area using DRAMs than SRAMs. More modern DRAM chips have their refresh control logic on-chip, making them appear to the circuit designer as an SRAM. These devices are also known as pseudo-SRAMs.

12.2   Input and output devices

Input and output (I/O) devices provide the means by which a microprocessor system can convey information between itself and the outside world. As we have seen, the microprocessor system processes information in the form of digital data, and the purpose of the I/O device is primarily to convert information from the outside world, such as analogue signals, into a digital form which is compatible with the microprocessor system. In addition, such a device may be equipped with secondary features, such as interrupt capabilities, which improve its functionality and the way it is interfaced to the microprocessor system bus.

In general, I/O devices contain two types of register. The first type is typically referred to as a control or status register; through it the program can control the mode of operation of the I/O device and indicate to the CPU its status. The second type of register provides the data path to enable the microprocessor system to read or write information to the outside world.

The specific purpose of each type of register will be dependent on the function of the I/O device. In the remainder of this section, three types of popular I/O device are used as typical examples.

12.2.1   PARALLEL I/O

The purpose of a parallel I/O device is to provide a set of data lines that will allow devices external to the microprocessor system which can be controlled by a binary state signal, or devices that output a binary signal, to be interfaced to the microprocessor system. For example, devices such as switches or lamps could be interfaced by connecting each lamp or switch to an individual data line of the device. The interface to the outside world is called a port. In the case of the parallel I/O device, each port consists of eight digital data lines which may be configured either as inputs or as outputs. In addition to interfacing devices such as lamps and switches, a parallel port may be used to interface a digital device which does not have a microprocessor type bus interface.

(a)   SIMPLE I/O USING LATCHES

The simplest form of parallel I/O device can be made from a general purpose tristate latch device, such as the 74LS373 shown in Fig. 12.3. The operation of the latch is similar to that of register or memory element. Whenever the enable input En is high and the outputs have been enabled by making the OE/ input low, the value of the D inputs are reflected on the Q outputs. When the latch is disabled by making En low, the value of the D inputs at the time will be maintained, or latched, on the Q outputs. The output can be disabled, or tristate, at any time by making the OE/ input high. Therefore, a high going pulse on the En input will latch data in the device, and its output can be independently controlled by the OE/ input.

Figure 12.3a shows how the latch can operate as an output port to control a series of lamps. The D inputs are connected to the microprocessor system data bus, and the latch output is permanently enabled so that data written to the latch will always control the devices to which it is attached, in this case the lamps. The enable input is controlled by a combination of the write signal WR/ from the microprocessor system bus and a chip select signal CS/ from the address decoding logic, which is designed such that it becomes active whenever the address bus contains the address chosen to correspond to the latch. The logic gate operates such that whenever both CS/ and WR/ are active low (refer to Fig. 10.3b, which shows the write cycle timing of the CPU), which indicates

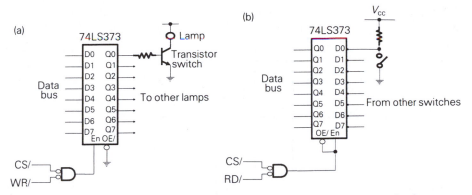

Figure 12.3 Simple (a) output and (b) input ports using a tristate latch.

that the latch is being written to by the CPU, the value which is written on to the data bus by the CPU will appear at the latch output. As the WR/ signal pulses low during the write cycle, this is translated into a high going pulse by the logic gate and has the effect of latching the data into the latch.

The transistors which drive the lamps are necessary to provide sufficient voltage and current levels to turn the lamps on and off. The transistor operates as a switch such that when a high logic level is output from the latch, the transistor turns on and causes current to flow through the lamp, thereby making it light up. Writing a data value 0x55 (01010101 in binary) to the address which corresponds to the latch would turn on alternate lamps, with the lamp corresponding to the data line D0 being illuminated.

Figure 12.3b shows the same latch configured as an input port for eight switches. The switch is interfaced to the latch using only a resistor, such that when the switch is open a high logic level is applied to the D latch input. The Q latch output is now connected to the data bus of the microprocessor system, and the latch must transfer the logic levels on its D inputs to the data bus whenever it is read by the CPU. This is achieved by controlling both the OE/ and En function of the latch with the signal which indicates that the latch is being read by the CPU. When the latch is not being read, the output of the logic gate will be high and the latch will be enabled so that internally it will reflect the status of the switches. However, as the OE/ and En inputs are connected together, when the latch is enabled its output is disabled and so nothing is output on to the CPU data bus. When the CPU reads the address corresponding to the latch, both CS/ and RD/ becomes active (see Fig. 10.3a for the CPU read cycle timings) and the logic gate produces a low signal which disables the latch function, thereby freezing the status of the switches within the latch, and enables the output, so this value is transferred on to the microprocessor data bus and is read by the CPU. For the switches in the position shown, the CPU would read the value 0xAA (10101010 in binary).

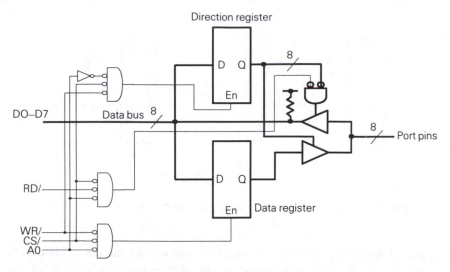

Figure 12.4   A bidirectional I/O port.

(b) PARALLEL I/O CHIPS

The simple I/O latches described in the previous section provide dedicated input or output functions. A more flexible solution may be provided by a general purpose parallel I/O device which may also incorporate functions that are responsible for configuring its mode of operation, such as input or output or a mixture of both, and controlling the flow of information across the port. First we look at the features typically found on a general purpose parallel I/O device.

Figure 12.4 shows the basic features of a bidirectional I/O port, where individual lines may be programmed as either an input or an output. The circuit is representative of a single I/O line, and would be replicated eight times to produce an 8 bit I/O port. The port has two registers: a data register as before, and a direction register whose contents configure each line to be either an input or output. Each register corresponds to a different address, usually arranged as two consecutive addresses by enabling either register with the status of the A0 address line.

The data register now consists of a latch which is used to output data, and two tristate buffers. The direction register is a simple latch whose output is used to select one of the two tristate buffers in the port data interface such that when a logic 1 is written to the direction register the port line is an output, and when a logic 0 is written the port line becomes an input. In addition, the input tristate buffer goes directly to the CPU data bus and is enabled both by the status of the direction register and by the CS/, RD/ and A0 lines. When the CPU reads from the data register, if the port line has been configured as an input, then the buffer will be enabled and data on the port pin will be transferred to the microprocessor data bus. This arrangement allows the data register to be able to either input data from, or output data to, the port.

The direction register would normally be written to during an initialization routine at the beginning of a program in order to configure each port line as an input or an output according to the type of device physically connected to the port line. Writing data to the data register will produce an output only on those port lines configured as outputs by the direction register. Similarly, reading data from the data register will only read valid data from those lines configured as inputs. The status of the lines configured as outputs during a read of the data register will be a logic 1 owing to the pull-up resistor which imposes a high level on the data line when the output of the input buffer is tristate.

12.2.2 INTERRUPT SUPPORT

Interrupts, as described in Chapter 11, are a means of gaining the immediate attention of the CPU to process some task in response to an asynchronous event which occurs externally to the CPU and the microprocessor system. A high proportion of interrupts in a system is involved in data transfer between the outside world and the microprocessor system, and of this proportion the majority will normally be used to input data to the microprocessor system.

For many applications, the provision of interrupt generation on a parallel port may not always be of use, as the port itself will not normally be used to indicate that an event has occurred to which the CPU should respond. Instead, the port would be used to implement the response to the event, such as reading new data. This distinction can be illustrated by considering how a sampled data controller would be implemented with a microprocessor system. In Fig. 12.5 a pulse generator is used to cause an interrupt every T seconds. In response to the interrupt, the CPU reads the parallel port A, calculates a new value of control signal, and outputs this value to port B. The pulse generator therefore determines the event, which is the sampling instant, and the parallel port is

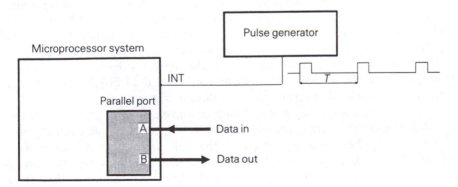

Figure 12.5 A sampled data system.

used to read new data, the presence of which is signalled to the CPU by the event as an interrupt signal.

12.2.3   DATA TRANSFER USING HANDSHAKING

Apart from providing an interface to devices such as switches and lamps, the parallel port may also be used to provide asynchronous data transfer between two microprocessor systems. To ensure that the data transfer is reliable, by which we mean there is some guarantee that one system has read the data before the other device attempts to send it new data, control signals are used to indicate the events 'new data ready' and 'acknowledge that new data has been read'. The operation of these control signals between the two systems is called handshaking and is frequently incorporated as part of the functions of the parallel port device. In addition, handshaking may also be combined with interrupt generation by the port device to allow it to effficiently process the handshaking signals.

Figure 12.6a shows how two microprocessor systems may be interfaced using parallel ports and handshaking lines to provide data transfer between the systems. Each port has a set of handshaking lines consisting of data ready (DRdy) and data acknowledge (DAck) control lines. The handshaking protocol which uses these lines must indicate and acknowledge the data transferred over the port data lines. Port 1 on each system is configured as an output port and port 2 as an input port. Consider what happens when system A sends a data byte to system B. First, system A writes data to the data register in port 1 which activates its DoutRdy/ control line, as shown in the timing diagram of Fig. 12.6b. This control line is connected to the DinRdy/ control input on port 2 of system B, and either would be configured to generate an interrupt to the CPU when its input was activated, or would set a flag in a status register which could be polled by the CPU. When system B recognizes that data is ready, it reads its port 2 data register and indicates to system A that it has done so by activating the DinAck/ control line. Similarly, in system A, the active DinAck/ control line would either cause an interrupt to the system A CPU, or set a flag in the port 1 status register which could then be polled by the CPU. Once the CPU of system A has seen the active acknowledgement signal, it resets its DoutRdy/ signal as it now knows that the data has been read by system B, and when system B sees that the ready handshake signal has been reset, it knows that it has seen the acknowledgement and so resets the DinAck/ control line. System A is then ready to transmit more data to system B.

Much of the handshaking protocol may be implemented in hardware by the parallel port device, thereby relieving the CPU of a lot of simple control tasks. For example, when data is written to the data register in a port it may automatically activate the DoutRdy/ line, and when the data register is read by the CPU it may automatically activate the DinAck/ line. Also, it is

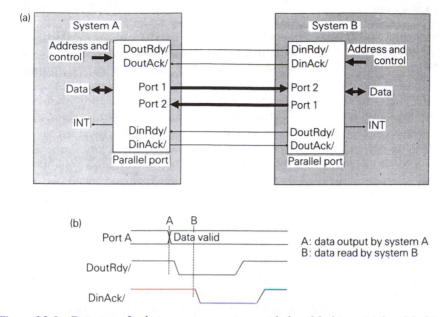

Figure 12.6 Data transfer between two system with handshaking: (a) handshaking signals (b) handshake timing.

comparatively easy to arrange that when DoutAck/ input becomes active, it automatically resets the DoutRdy/ line, which in turn would reset the DinAck/ output. When the cycle is complete, a flag can be set in an internal status register of the port to indicate that the port is no longer busy.

An identical arrangement could be used to transfer data from system B to system A with an identical set of handshake control lines. However, a disadvantage of having two separate port connections for the transfer of data between the two systems is that this uses many external connections when the volume of data transferred may not be very high. An alternative would be to use a single data port as a bidirectional port with the same handshaking lines to control the flow of data in either direction. However, additional control lines are required to determine which system is going to send and which to receive data, and also to arbitrate in situations where both systems decide to send data simultaneously.

Handshaking is an example of a data transfer method which takes place asynchronously between the two systems. No clock signal is involved which determines when signals must change, and so the rate at which data transfer can take place is limited only by the logic delays in the control circuits, the speed with which the CPU can respond to the interrupt and read the data, and the propagation delays over the connecting cable between the two systems.

12.2.4   SERIAL I/O

Serial input and output devices are used as a means of transferring data between microprocessor systems, or between components within a system, over single data line connections. Instead of transferring data in a parallel format, for example a byte at a time, each byte is sent one bit at a time with a clock signal which allows the bits to be reassembled at a receiver into a complete byte. This has the advantage that only three wires are needed in a cable to send information one way, namely a data line, a clock line and a ground line. On the other hand, parallel transfer of a data byte would require at least nine lines, one for each data bit and a ground connection, and in addition any necessary handshaking lines. The cost of the interconnecting cable is therefore one advantage of serial data transfer, especially over distances greater than several metres.

However, unlike the parallel I/O port which can be used to interface simple sensors and actuators to a microprocessor system, there are no primary sensing or actuation techniques which result in devices that may conveniently interface to a serial digital port. (Shaft encoders may produce a serial digital output, but the information is encoded in the time or frequency domain rather than the amplitude sequence of a number of consecutive pulses.) Therefore, to achieve serial data transfer between systems or devices, it is necessary that each system or device has a serial data port, and that the way the information is transferred, for example the most significant bit first or least significant bit first, is the same for all ports.

(a)   SIMPLE SYNCHRONOUS SERIAL DATA TRANSFER

Figure 12.7a shows how simple serial data transfer can be achieved with devices called shift registers, which form the basis of all serial data communications. As its name suggests, a shift register is a register whose contents are shifted by one bit position when it receives a clock pulse. As each data bit in the register is shifted out with successive clock pulses, new data is shifted in from the input of the shift register. By connecting the output of one shift register to the input of another as shown in Fig. 12.7a, and using a common clock to shift data out of one shift register and into the other, data can be transferred serially between the two devices.

In order to transfer data between shift register A and shift register B, shift register A is first loaded with a data byte by the CPU. (The CPU bus signals have been omitted in the diagram for clarity.) Control circuitry within the serial port A then generates eight clock signals which progressively shift out data from one shift register and into the other. When port B has received eight clock signals, it can either interrupt the CPU to indicate that it has received new information, or set a flag in a status register as before.

The data transfer occurs synchronously with respect to the clock, and so the data transfer rate will be dependent on the clock frequency used. The operation

(a)   System A   System B

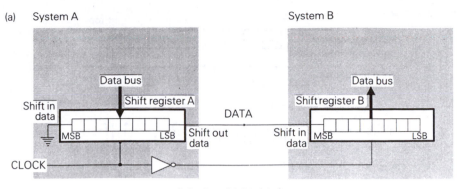

a. A simple serial data interface.

(b)

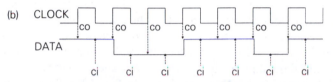

Figure 12.7   A serial data interface using shift registers: (a) interface (b) timing.

of shift registers is such that the shifting operation takes place on a specific edge of the clock signal. For example, if data was shifted out of port A and into port B on the rising edge of each clock pulse, then data would be clocked into port B as it was changing due to the new data being output from port A. The solution to this synchronization problem is to input data to the port B shift register on a different clock edge to that used to clock data out of the port A shift register, which allows time for the data to become stable on the data line before it is read by port B. This is the reason for placing an inverter in the clock line between the two ports in Fig. 12.7a. The timing of the data transfer is shown in Fig. 12.7b.

This type of simple synchronous data transfer can be expanded to provide bidirectional data transfer, and data transfer between more than two devices. It is commonly used between devices within the same system, and can provide a more simple and cost effective replacement for the parallel system bus described in Chapter 10, although its operation would be slower (dictated by the speed of the clock) and each device would need to have a similar serial interface.

(b)   ASYNCHRONOUS SERIAL DATA TRANSFER

An asynchronous serial data transfer technique is commonly used between systems for low speed communication. The asynchronous transmission and reception of serial data do not rely on a common clock between the two systems. Instead, each system will have its own independent clock running at the same nominal frequency. However, it can be seen that for even a small difference in

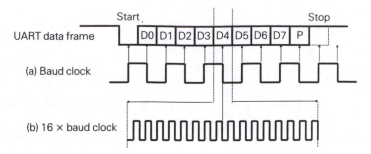

Figure 12.8   An asynchronous serial data frame and baud clocks.

frequencies between the two clocks, the corresponding clock pulses soon become out of step with one another, which would cause data to be received incorrectly. To avoid this, each clock is synchronized at the start of transmission of every byte by the transmission of a start bit. This is then followed by the serial data bits, an optional parity bit for error detection, and a stop bit which resets the logic level on the data line ready to accept another start bit. The frame for asynchronous transfer of serial data is shown in Fig. 12.8. The circuitry used to implement the asynchronous transmission and reception of bytes of data is typically contained in a single chip called a universal asynchronous receiver and transmitter (UART).

(c) THE UART

A UART will contain all the necessary registers and timing circuitry to provide independent asynchronous transmission and reception of data bytes. The ability to transmit and receive different data at the same time is referred to as full duplex operation. A simplified block diagram of a UART is shown in Fig. 12.9.

Both transmit and receive channels will normally have separate transmit and receive buffer registers, TxBuf and RxBuf (although these may have the same address), as well as their individual transmit and receive shift registers, TxSR and RxSR. The purpose of TxBuf and RxBuf is to provide a buffer between the CPU and the shift registers so that, in the case of transmission, data to transmit can be written to the UART while transmission of the previous byte is still in progress, and for reception, received data does not have to be read immediately before it is overwritten as new data is received. The CPU may therefore only access the buffer registers.

The rate at which data is transmitted and received is called the baud rate. The UART transmits and receives one bit of data per baud clock, and the baud rate is therefore specified as the number of bits per second. During the reception of data, account must be taken of the fact that the clock in the transmitting UART is asynchronous to that in the receiving UART. To overcome this problem the incoming byte contains a start bit which is used to synchronize the receiver

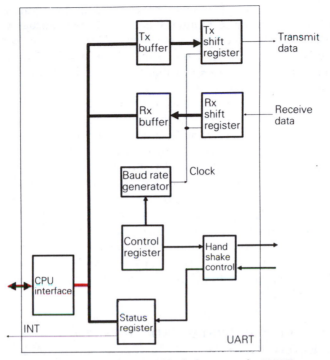

Figure 12.9   A simplified block diagram of a UART.

clock with the incoming data. The data line is then sampled in the middle of a bit transmission using a signal derived from the baud clock, as in Fig. 12.8a. A common way of implementing this condition is for the receiver to use a clock frequency which is a multiple of the baud rate, and to provide an internal divider to create a narrow pulse signal, shown in Fig. 12.8b, which is used by the UART to sample the data line. A typical multiply/divide number is 16.

Control registers allow the various functions of the UART to be selected. Typically it is possible to select the number of data bits to be transmitted, the number of stop bits that are transmitted, and the baud rate multiply/divide ratio, and to enable or disable the parity.

Parity is a simple error detection mechanism which operates by making the number of data bits and the parity bit which are equal to a logic 1, an even number for even parity, and an odd number for odd parity. Even parity for the data byte 0x07 would set the parity bit to 1 in order to make the total number of 1s an even number. Should a single data bit become inverted owing to electrical noise in the transmission cable, the parity calculated by the receiver will not agree with the parity bit in the received frame. A parity error flag would then be set in a status register in the UART which would indicate to the CPU that data in the receive register had been corrupted. The CPU could than take corrective measures.

The purpose of stop bits is to ensure that the data line returns to a logic 1 before the beginning of the next frame, which is indicated by a logic 0 start bit.

As well as controlling the internal oprations of the UART, the control register may allow the CPU to control handshake lines which control the flow of information between two systems, similar to those described for the opration of the parallel port in section 12.2.3.

The UART status register indicates to the CPU the state of the UART. Two important status flags indicate that the receiver buffer is full, and therefore that the CPU should read the receive buffer, and that the transmitter buffer is empty and is ready to accept another data byte from the CPU for transmission. The data transfer between most UARTs and a CPU can be achieved using an interrupt signal to indicate an active receiver buffer full or transmitter buffer empty flag. In response to this interrupt, the CPU may either read the received data byte from the receiver buffer register, or place new data in the transmit buffer register for transmission. In addition, the status register will be used by the UART to indicate a number of error conditions: for example, if a parity error occurred during the reception of the last frame, or if an overrun error occurred where the UART received more data before the CPU had read the previous data from the receive buffer register.

12.2.5   ANALOGUE TO DIGITAL AND DIGITAL TO ANALOGUE CONVERTERS

The purpose of analogue to digital (A/D) and digital to analogue (D/A) converters is to provide an interface between devices which produce or are driven by analogue signals, and digital systems such as the microprocessor system. The principles of the main conversion techniques were described in Chapter 8, and this section illustrates the use of these components in a typical microprocessor based system.

Figure 12.10 shows a block diagram of a circuit which provides a microprocessor system with eight analogue input channels and a single analogue output channel. It has been assumed that the A/D and D/A converters have 8 bit microprocessor bus interfaces, although if they did not it would be an easy task to provide interfaces to them from parallel I/O devices.

The analogue signal to be converted is selected by means of an eight-to-one analogue multiplexer which is controlled by the output from a digital latch. The latch is memory mapped and operates as a simple output port as described in section 12.2.1. Only three output lines from the latch are necessary to select the required input signal, as the multiplexer accepts a binary number as its input and has its own internal decoding to select the required channel. If the low order bits from the latch are connected to the multiplexer, then writing the number 0x05 to the latch would select the fifth analogue channel.

The selected analogue signal is then filtered in order to limit its highest frequency component to half the sampling rate of the interface circuit, which

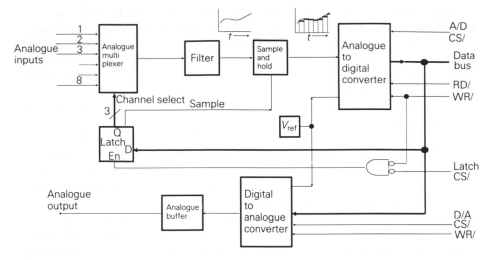

Figure 12.10   An eight-channel analogue to digital interface and single-channel digital to analogue interface.

is determined by the maximum rate at which the A/D converter and sample and hold circuit is able to make conversions. This is necessary in order to satisfy the sampling theorem for sampled signals introduced in section 8.6.2, and can be considered as the minimum sampling frequency at which theoretically all the information contained in the signal is captured and represented by its sampled equivalent.

The operation of the sample and hold circuit is in principle similar to the latch described in section 12.2.1, except the signal which is latched is analogue instead of digital. It is important to include this circuit because any A/D converter will take a finite time to convert an analogue signal to its digital equivalent, and during this conversion time the analogue input should be maintained at a steady value so as not to cause errors in the conversion. This is particularly important when the input signal is changing rapidly. The sample function of the sample and hold circuit is enabled by a digital signal from the latch, during which time the output of the sample and hold tracks its input signal. When the sample and hold is disabled, its output is frozen at the value of its input signal at the instant it was disabled. The signal may then be converted by the A/D converter.

The A/D converter will contain a data register from which the converted value is read by the CPU, and a control/status register which will contain a control bit to start a conversion, and a status bit indicating that a conversion is complete and that the data register can be read by the CPU. Alternatively, some A/D converters have a hardware input which may be used to start the conversion process. Where the conversion time of an A/D converter is of the same order as the execution time of a CPU instruction it may not be necessary for the converter to generate an interrupt on completion of a conversion, as the CPU will not have to wait any significant period for the result to be ready. This may be the case for flash A/D converters and fast successive approximation converters. However, integrating converters generally take much more time to make a conversion, typically of the order of milliseconds rather than microseconds, and may generate the interrupt to the CPU to indicate that new data is available. This allows the CPU to process other functions instead of waiting for the A/D converter to complete its conversion.

The D/A converter provides the system with a single analogue output channel. Unlike the analogue inputs, it is not practical to generate multiplexed outputs from a single D/A converter using the outputs from sample and hold circuits because they tend to drift over long periods. It is therefore necessary to have one D/A converter for each analogue output channel required.

Typically, a single D/A converter will have only a data register to hold the digital value which it converts to its equivalent analogue output. Many D/As have an output which is a current signal as opposed to a voltage signal because their conversion principle is based on a resistor ladder which sums the current components from each binary digit at the output of the D/A (see Fig. 8.35). The output of the D/A in Fig. 12.10 is therefore buffered to produce a voltage output. The same value of voltage reference is used for both the A/D and the D/A converters so that the analogue input channel range is the same as the output channel range.

12.3   Peripheral circuits

Microprocessor peripheral circuits add specific functions such as timers and interrupt controllers to the microprocessor system. They are distinct from the basic functions required to make a microprocessor system, that is the CPU, memory and I/O, in that they supplement the existing I/O functions of the microprocessor system.

12.3.1   PROGRAMMABLE COUNTER/TIMERS

Many of the activities of an embedded microprocessor system require timing and counting functions. For example, real-time processing by a microprocessor system may require it to synchronize its activities to a stable time reference. The measurement of frequency or periods is also common, as is the need to generate pulse trains and timing waveforms. The programmable counter timer therefore acts both as an internal peripheral to the system and as an I/O device. There are numerous modes in which a programmable counter timer may operate, and these may vary between different devices. This section describes the general features of programmable counter/timers and how they may be used to generate waveforms and measure time and frequency.

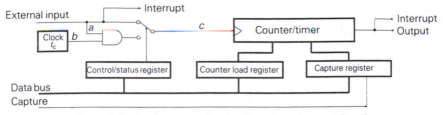

Figure 12.11   An example of a timer/counter peripheral.

The programmable counter/timer may be configured as a timer or a counter. As a timer, it will be driven by a reference clock signal generated within the microprocessor system, and used to time the interval between events or generate specific time intervals. As a counter, it is driven by a signal external to the microprocessor system, such as a series of pulses which are to be counted. A simplified block diagram of a typical counter/timer is shown in Fig. 12.11. The counter is a simple binary counter which counts up (or down) until it overflows (or underflows), at which point the state of the output toggles such that if it was a 1 it changes to 0, and if it was a 0 it changes to a 1. As the output changes state, it may also generate an interrupt to the CPU. Counting can be started and stopped under the control of the CPU by bits in the control/status register. Note from Fig. 12.11 that to enable the internal clock, it must be selected by the control register and enabled by making the external input a logic 1. This is a useful feature when the timer is used to make period measurements, described later.

When operating in timer mode, the counter/timer may be programmed to generate various logic pulse waveforms. This is achieved by loading the timer with different values, which are held in the counter load register, each time the timer output changes state. In order to generate accurate waveforms, the timer is automatically loaded with the value in the counter load register whenever a timer underflow occurs, assuming that the timer counts down from its loaded value to zero. This also generates an interrupt so that the CPU can then reload the counter load register with the value for the next period of the waveform. If the clock period to the timer is t_c, then the counter load register must be loaded with a value d to generate a period T where $d = T/t_c$. This is shown in Fig. 12.12a. A square wave could easily be generated by keeping the same value of the counter load register for every state of the timer output.

The timer mode can also be used for period measurement. The pulse whose period is to be measured is input through the external input and used to enable the internal clock to the timer. For the duration that the pulse is at a logic 1, the timer will count, and if the internal clock is set at a convenient frequency, for example 1 MHz, then each count will represent a microsecond. When the input pulse returns to 0, it disables the clock and so freezes the contents of the timer, and also triggers an interrupt to indicate to the CPU that it must read the timer, the value of which will represent the number of clock cycles that

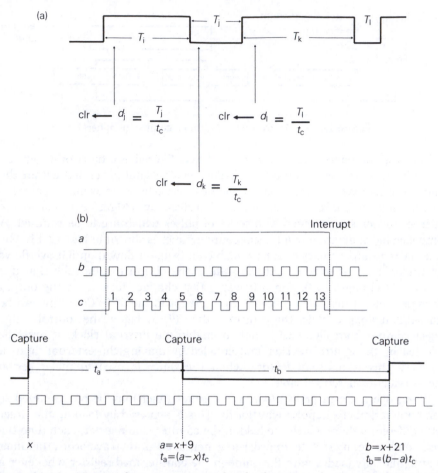

Figure 12.12 (a) Waveform generation (clr = load 'counter load register') (b) pulse measurement (c) waveform measurement with the help of a capture register.

elapsed during the pulse. The timer can then be reset to 0 ready for the next pulse. This method of period measurement is shown in Fig. 12.12b, where the waveforms are related to points in the counter/timer diagram of Fig. 12.11. If the value in the timer was initially 0, then it would count 13 pulses.

A drawback of this method of period measurement is that the clock to the timer is disabled by the input pulse, and so it is not possible to measure the time for which it is high and the time for which it is low. A more flexible method of period measurement is produced when the timer is run continuously, and the condition of the input pulse is used to trigger a capture register which is loaded with the instantaneous value of the timer at the instant that the condition is met. Triggering the capture register will also cause an interrupt to allow the CPU to read the value in the capture register. The condition on which the capture register is triggered should be programmable via a register so that both a rising edge pulse and a falling edge pulse may cause it to capture the timer contents. Period measurements between two successive triggers can be made by subtracting the two values read from the capture register, as illustrated in Fig. 12.12c.

Frequency measurement operates on the same principle as the gated period measurement, except the gate signal is now generated internally, and the signal whose frequency is to be measured is input as an external clock to the counter. The internal gate signal is chosen to be some convenient low frequency clock such as 2 Hz, whose half period, which is the time for which the gate is active, is then 1 second. The gate signal must also generate an interrupt on its falling edge to indicate that the CPU must read the counter whose value is then the frequency in hertz of the input signal. In general, the frequency of the input signal is the counter value multiplied by half the frequency of the gate signal, and judicious choice of the gate signal frequency as a modulo two (binary) number reduces the multiplication to a simple binary shift operation.

12.3.2   DIRECT MEMORY ACCESS

Some I/O related operations transfer large amounts of data to and from memory. However, occupying the CPU to move the contents of large blocks of data between memory and I/O devices is wasteful of the time that it could otherwise spend executing other more useful instructions. To move blocks of data, the CPU will have to repeatedly fetch and execute load and store instructions during which it may spend more time fetching instructions than it will moving the data. The CPU is therefore a bottleneck in that it restricts the fast transfer of data between memory and I/O, because for each transfer the data must pass through the CPU under the control of the program that the CPU is executing. A solution to this problem is to provide a data path directly between memory and I/O which is under the control of a dedicated device called a direct memory access (DMA) controller, as shown in Fig. 12.13. As its name suggests, a DMA controller provides I/O devices with direct read and write access to memory without the data passing through the CPU. However, because the DMA controller and the CPU make use of the same system bus, they must both implement a means by which they can share the bus without disturbing their operation.

(a)   CONTROL OF BUS MASTERSHIP

The operation of the microprocessor system bus, as described in Chapter 10, relies on the CPU to generate the necessary address and control signals in order to execute the program stored in memory. The CPU is therefore the bus master, while other devices, such as memory and I/O devices which respond to the control of the CPU, are the bus slaves. The DMA controller is also a bus master because it may control the reading and writing of data between memory and

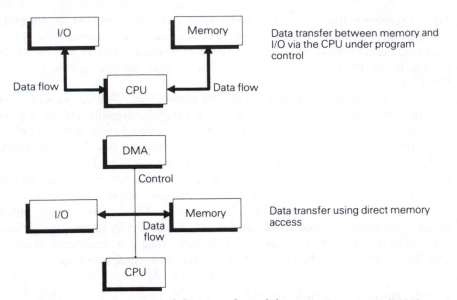

Figure 12.13   Programmed data transfer and direct memory access (DMA).

I/O. There may only be a single bus master at any time; therefore the CPU and the DMA controller must arbitrate the bus mastership so as not to cause a bus conflict, which would occur when more than one device attempts to become the bus master.

Figure 12.14 shows the interconnection of a CPU with a DMA controller which is to be used to transfer data between memory and I/O. Under normal operation the CPU is the primary bus master, and it is the responsibility of the DMA controller to request bus mastership from the CPU whenever the DMA device is requested by the I/O to transfer data to and from memory. To achieve this, two handshake lines are provided between the DMA controller and CPU, which operate in a similar way as described for the parallel port. For example, consider the I/O device to be a communications controller which has

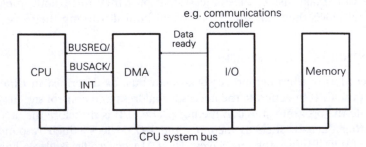

Figure 12.14   Interconnection of components for direct memory access.

just received 10 bytes of data representing a message which it has stored in internal registers, but which must now be transferred to memory so that the CPU can process the received message. The I/O device indicates to the DMA controller that it must transfer data to memory by the data ready line. In turn, the DMA controller signals to the CPU that it wishes to gain control of the system bus by activating the BUSREQ/ control line. The CPU then finishes fetching and executing the instruction currently in progress, halts any further operation, tristates its address and control bus outputs (the data bus should already be tristate), and then acknowledges to the DMA controller that is can now take control of the system bus and become the bus master by activating the BUSACK/ control line. The DMA controller then directly transfers the 10 bytes of information between the communications controller and memory by performing read followed by write cycles. During this time, the BUSREQ/ signal remains active, as does the BUSACK/. When the DMA controller has completed the transfer of the tenth byte, it tristates its address and control lines and releases the BUSREQ/ signal, which indicates to the CPU that it can now regain bus mastership. Having done so, it releases the BUSACK/ control line and continues execution from where it was stopped. In cases where it is necessary for the DMA controller to inform the CPU that DMA has taken place, the DMA controller may also then interrupt the CPU.

(b)   THE DMA CONTROLLER

To perform direct memory access, the controller will contain registers which define the source address, the destination address, and the number of bytes to be transferred. As data is transferred from the source address to the destination address, the address registers will be updated to point to the next memory locations.

DMA data transfer can operate in one of a number of different modes; the three most popular are byte mode, burst mode and cycle stealing mode. In byte mode, for each byte that is transferred under DMA control, the DMA controller must go through the procedure of gaining bus mastership and then relinquishing it. However, in burst mode several bytes are transferred at a time while the DMA controller has control over the system bus. With the operation of DMA as described, the DMA achieves data transfer at the expense of CPU bus cycles such that, as the amount of DMA increases, the occupancy of the system bus by the CPU decreases, thereby slowing down the average rate at which it normally executes instructions.

Some CPUs have specific cycles during which the CPU performs only internal operations and does not make use of the system bus. The cycle stealing DMA mode takes advantage of this by using these cycles to perform DMA transparently to the operation of the CPU, and therefore does not impact the CPU bus occupancy or throughput. Cycle stealing DMA, however, is not possible with many of the new CPUs available today because the CPU itself makes use of potentially idle bus cycles to fetch more instructions or data into the CPU in order to improve its throughput.

12.3.3   INTERRUPT CONTROLLERS

In a typical embedded microprocessor system there may be numerous sources of interrupt from devices within the system itself, such as I/O or peripheral functions, and from devices external to the microprocessor system, such as switches. To accommodate large numbers of interrupts in a system that typically has a CPU with only a single general purpose interrupt, a separate interrupt controller may be used to manage the interrupts and the way in which they are given access to the CPU. Figure 12.15 shows how the interrupt controller is connected to the CPU. (The action of the CPU to an interrupt signal is discussed in more detail in Chapter 11.)

The interrupt controller must ensure that the interrupt requests that arrive at its input are serviced one at a time by the CPU. The registers inside the interrupt controller will assign a number of different attributes to each physical interrupt request line, will define the conditions on which the interrupt is triggered, and will also help the controller and CPU to manage multiple overlapping interrupt requests. Some typical attributes controlled by these registers are as follows:

Interrupt enable   Each interrupt request input may be individually enabled and disabled. In addition, the controller may provide a global enable/disable function for all interrupt inputs, and may be used to disable interrupts while the interrupt controller is initialized or reprogrammed, or to protect the execution of an interrupt service routine from being interrupted.

Interrupt trigger conditions   These attributes define the conditions under which an interrupt request input will cause a CPU interrupt. Typically, a CPU interrupt may be caused by a specific level of an interrupt request, or by a rising or falling edge of the interrupt request as it changes state.

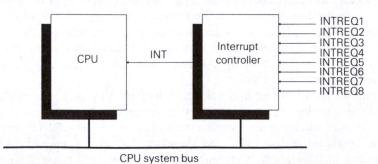

Figure 12.15   The interrupt controller.

Interrupt priority   The interrupt priority level determines which interrupt request is serviced first when two or more interrupt requests occur simultaneously. The priority may be fixed in the interrupt controller by hardware, so that the priority of interrupt request input zero will be higher than that of interrupt request input one, and so on.

Interrupts pending   Although interrupts may be disabled during an interrupt service routine, interrupt requests may still occur. These are recognized by the interrupt controller but held in a pending state until the CPU is ready to process more interrupts. The fact that more interrupt requests have occurred may be indicated by the status of an interrupt pending register in the interrupt controller.

12.4   Coprocessors

Coprocessors are used in a microprocessor system as an extension to the processing function of the CPU. A floating point mathematical processor is an example of a coprocessor, and would be used to perform fast floating point calculations where the speed at which the CPU could perform the same calculations using its standard instruction set was not sufficient. This additional processing power is often required in applications such as robotics control which make very intensive 'number crunching' demands on the CPU.

The operation of a coprocessor is achieved at a more intimate level with the CPU than that of other I/O and peripheral devices. A coprocessor is designed to operate with a specific CPU, and will itself have a unique instruction set which supplements that of the CPU. The CPU is responsible for fetching instructions and data from memory, and this action is continuously monitored by the coprocessor. If the coprocessor detects that the CPU is fetching a coprocessor instruction, it captures the instruction from the bus while the CPU ignores it. When the coprocessor has read the necessary opcode and operands, it can then start to execute the instruction. Should it need to further access the system memory, it can do so by making use of a bus request and acknowledgement procedure similar to that described for DMA.

It is important to note that during execution of an instruction by the coprocessor, the CPU also continues to fetch and execute instructions. The execution of instructions may therefore take place in parallel; this requires the CPU and coprocessor to be coordinated to make sure that one processor does not get out of step with the other in terms of the execution of instructions which require a specific ordering. For example, the CPU may have to wait for a result to be produced by the coprocessor before it can continue execution of an instruction. Also, if the coprocessor is busy executing an instruction when the CPU reads another coprocessor instruction, the CPU must suspend its operation until the coprocessor has finished. Many synchronization schemes exist for different CPUs and their coprocessors, and these may be based on hardware mechanisms, such as a BUSY status signal from the coprocessor, and software mechanisms, such as a WAIT instruction which would force the CPU to wait for a result from the coprocessor.

12.5   Microprocessor types

The variety of different microprocessors available today is huge. Over the years, the microprocessor has evolved from a general purpose device to a more application oriented product. Table 12.1 shows a variety of applications together with the type of microprocessor that may be used today, and some general requirements imposed upon it by the application.

The applications in Table 12.1 are roughly ordered according to the complexity of the task being performed; as the tasks become more complex, so the performance of the microprocessor system is required to increase. Although microprocessors are now designed for specific application areas as well as having improved architectures for increased performance, it is always necessary to match the performance of the microprocessor system with the specific application requirements in order to produce the most cost effective solution.

The many types of microprocessor available to the designer today can be placed into one of three broad categories: microprocessors, microcontrollers, and digital signal processors.

Any microprocessor together with the necessary memory, I/O and peripheral components is capable of performing control functions. However, some

Table 12.1   Comparison of microprocessor applications

Application	Microprocessor	Requirement
Washing machine	4 or 8 bit microcontroller	Cheap; immune to electrical noise
Intelligent transducer	8 bit microcontroller	Analogue I/O; low power consumption; communications
Autofocus camera	8 bit microcontroller	High level of integration
Engine management unit	8 or 16 bit microprocessor or microcontroller	Analogue I/O; signal processing; immune to electric noise; high current output
Speech processing	16 or 32 bit digital signal processor	Fast multiply and accumulate operation
Flight control system	16 or 32 bit microprocessor	High reliability; communications
Robot controller	16 or 32 microprocessor	High speed calculations; communications

microprocessors may be more suited to embedded real-time control applications for a number of reasons. For example, the instruction set of a microprocessor may more closely match the functions it is required to perform than other devices. More specifically, control applications frequently need to process information as bits, and so specific instructions which perform operations on individual bits of data help to speed the execution of the program as well as making the program the easier to write. A microprocessor for real-time applications should also have a fast interrupt response, in order to minimize the time between an interrupt request occurring and the point at which the CPU starts to execute the interrupt service routine. Another important real-time feature is the time taken by the microprocessor to perform a context switch; this occurs when the CPU switches between the execution of two tasks, and is the time taken by the CPU to save its registers, and if necessary reload them with the status of the new task to be executed. Interrupts and context switching were discussed in Chapter 11.

12.5.1   MICROCONTROLLERS

The microcontroller is a highly integrated microprocessor chip containing a CPU and an amount of I/O, peripherals and memory, and whose application is oriented towards embedded control. When the microcontroller contains both RAM and program ROM it is a complete microprocessor system on a single chip, and is known as a single chip microcontroller.

The general I/O functions found on a typical microcontroller include parallel input and output ports, serial input and output for communications with other devices or a terminal, and timer/counters to allow frequency and period measurements to be made and to generate fixed or variable pulse waveforms. The microcontroller will have a number of interrupt sources, both from internal peripherals and I/O such as the timers, and from general purpose interrupt inputs, all of which will be managed by an on-chip interrupt controller. Microcontrollers designed to meet the requirements of more specific application areas would contain other functions such as analogue to digital converters, voltage comparators, and other frequently used interfacing circuitry such as display drivers and stepper motor drivers.

Although a high level of integration is desirable, there is an economic and practical limit to the size of a silicon chip, based on the yield of the number of working chips that can be produced from a silicon wafer, its power consumption and packaging constraints, and its cost. The philosophy behind the purpose of the single chip microprocessor is that it should be used for simple I/O intensive embedded control functions, and so the emphasis is towards providing a high level of integration for the CPU and the I/O peripherals. As a result, only a comparatively small amount of memory is integrated into the microcontroller, and provisions are normally made for this to be expanded with off-chip memory where necessary. Memory sizes of 512 bytes of on-chip read/write memory and 16 kilobytes of program memory are typical of larger microcontrollers.

The instruction set of a microcontroller is also designed to match its hardware capabilities. For example, in single chip devices where the memory size is limited, instructions will be designed to create dense code. Features such as bit addressing are also common, where instructions operate on individual bits within registers or memory instead of using separate instructions to read a byte, modify a bit with a logic operation, and then write the complete byte back to the same address.

Microcontrollers with 4, 8 and 16 bit architectures are available today. The following section briefly describes the features of a 16 bit microcontroller.

12.5.2 EXAMPLE: NATIONAL SEMICONDUCTOR HPC16083 MICROCONTROLLER

The HPC16083 microcontroller is a member of a series of microcontrollers, all of which have a common core consisting of a set of basic functions including the CPU, and to which are added amounts of memory and I/O functions in order to make the microcontroller suitable to a specific area of application. The HPC16083 is intended for general purpose embedded applications, but other microcontrollers from the same series are aimed at communications and at industrial and automotive control applications. The device aimed at communications applications is equipped with on-chip communication ports and a DMA controller which is responsible for transferring data between the memory and the communications ports. The device for industrial and automotive applications has an 8 channel 8 bit A/D converter.

A block diagram of the HPC16083 is shown in Fig. 12.16, where the elements within the dotted line are those features contained within the core. As can be seen, the core contains not only a 16 bit CPU, but also three timers with capture registers, interrupt control, a serial communications port called Microwire, clock circuitry, and watchdog logic, the purpose of which is explained shortly. The core also has two power saving modes in which the microcontroller is 'put to sleep' by stopping the execution of its program and disabling its various I/O functions and peripherals in order to significantly reduce its power consumption. The HPC can subsequently be 'woken up' by an interrupt signal. These power saving modes are important when the HPC is being used in battery powered devices, or in devices which are remotely controlled and powered, such as remote weather recording stations or telephones.

In many applications, especially control functions, it is necessary to provide failsafe functions. If the failure occurs in a piece of external equipment, the microcontroller should have the means of detecting it as a signal input to one of its ports or interrupt lines. In Chapter 11 the use of the NMI interrupt line was discussed as a means of implementing failsafe procedures in the case of a power cut. However, in harsh operating environments it is still probable that

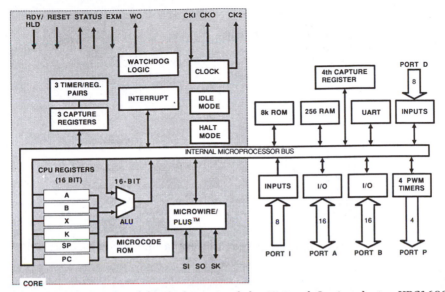

Figure 12.16 Functional block diagram of the National Semiconductor HPC16083 microcontroller (National Semiconductor; HPC is a trademark of National Semiconductor).

the normal operation of the microcontroller can be disturbed by electrical noise, with the likely result that the program gets lost or crashes; the only way to guarantee its recovery is to provide it with an interrupt or reset signal. However, if the microcontroller is embedded in some other equipment, say for example an engine management unit, it would be extremely inconvenient to only have a manual form of reset. Instead, a mechanism called a watchdog is frequently used on microcontrollers to automatically provide the reset or interrupt should such an event occur. The operation of the watchdog is quite simple; it is based on a time-out principle. A timer is loaded with a time-out value and proceeds to decrement to zero, or increment to overflow. If the program does not reload the timer before zero count or overflow, it generates an interrupt or reset signal which brings the CPU back under control. This means, however, that the program running on the microcontroller must, during the normal course of its execution, periodically reset the watchdog timer.

The instruction set and addressing modes of the HPC were chosen by its designers to produce dense code. In particular, the instructions which support conditional execution are based on a load and skip operation as opposed to a more typical test and jump operation. For an HPC skip instruction, this means that if the condition is satisfied the CPU will skip the following instruction. To illustrate this, Fig. 12.17 shows two simple routines to move a block of data from source to destination, the first with the HPC (Fig. 12.17a) and the second with some other CPU possessing similar registers (Fig. 12.17b). In both examples,

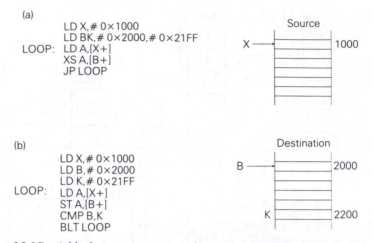

Figure 12.17   A block move comparison between (a) HPC and (b) another CPU.

register X is used to point to the source address and register B to the destination address. Both registers can be incremented automatically as part of an instruction, as indicated by the + sign. The K register is used as the limit of the block move. In Fig. 12.17a the XS is the exchange and skip instruction; this exchanges the contents of the accumulator, which was loaded by the previous instruction with the source data, with the memory location pointed to by the B register. The B register is then automatically incremented to point to the next destination location, and if this is equal to the contents of the K register then the block move has been performed and the HPC skips the jump instruction, thereby exiting the loop. If the value of the B register does not equal the K register, then the jump instruction is executed and the loop continues. In Fig. 12.17b the loop is formed instead by a comparison instruction CMP, followed by a conditional jump instruction which closes the loop. Each of the instructions inside the loop, in both cases, can be encoded as a single byte; this means that the HPC loop requires only 3 bytes of program code, whereas the other loop requires 4 bytes. Although this is a saving of only a single byte, it is significant when moving large blocks of data.

The HPC16083 is a single chip microcontroller. It has 8K bytes of on-chip read only memory and 256 of on-chip RAM. All I/O and peripheral features of the HPC, including the registers of the CPU, are memory mapped together with the on-chip RAM in the address range 0x0000 to 0x01FF. The 8K bytes of read only memory are memory mapped in the range 0xE000 to 0xFFFF and the interrupt and reset vectors are held in memory locations 0xFFF0 to 0xFFFF. The memory area between 0x0200 and 0xCFFF can be used to add memory external to the HPC16083, to give it a maximum memory size of 64K bytes. When external memory is added, the connections for some of the parallel I/O lines are used as the external system bus interface, and may no longer be used as general purpose I/O.

In addition to the HPC core, the HPC16083 has a number of I/O and peripheral functions. Amongst these are a UART, more timers, and parallel I/O ports.

12.5.3   DIGITAL SIGNAL PROCESSORS

Digital signal processors (DSPs) are a specialized form of microprocessor device optimized for processing digitized signals in realtime. The architecture of the DSP is tailored to match the recursive nature of digital signal processing algorithms, an example of which is the low pass filter of Fig. 8.30a. The ALU of the DSP is designed to perform typical calculations such as multiply and accumulate much faster than a normal microprocessor, and some devices may also provide floating point mathematical functions.

In many digital signal processing algorithms, an operation or instruction must be performed on each of a set of data values. It is soon realized that a single bus used for the transfer of both instructions and data creates a bottleneck in the system and prevents the DSP from achieving a high performance throughput and hence higher frequency real-time processing. Many DSP architectures therefore employ separate system buses for instructions and data, so that the fetch of an instruction may be overlapped with a data transfer. Mechanisms inside the DSP provide synchronization to ensure that the correct data is operated on by the instruction being executed. The arrangement of separate instruction and data buses is referred to as Harvard architecture.

DSPs can typically be used in one of three configurations. The first is a stand-alone configuration, where the DSP appears as would a microprocessor system or a single chip microcontroller. The second is a multiprocessor configuration where several DSPs in the same system run in parallel, and exchange data and synchronize their activities using memory which is shared between the devices. In the third configuration, the DSP acts as a peripheral to a general purpose CPU.

Chapter 13

Semi-custom devices, programmable logic and device technology

The evolution of microprocessor system components is dependent on the processes used to make semiconductors and the design tools with which they are designed. Until recently, this type of technology has only been available to the semiconductor manufacturer owing to its expense and sophistication. This meant that the semiconductor manufacturers could dictate the type and functionality of components they made to sell to their customers. Today the situation is somewhat different thanks to the automation of much of the design process through computer aided design (CAD). As a result, the customer now has the opportunity to define the functionality of an application specific integrated circuit (ASIC); designers can tailor such a device to meet their own specific requirements and achieve higher levels of integration within their systems. ASICs are semi-custom ICs, in that an amount of processing by the semiconductor manufacturer is still required in order to dedicate the chip to the function defined by the customer. This chapter describes two important techniques of implementing ASICs, referred to as gate arrays and standard cells.

ASICs represent the half-way house between a full custom design by a semiconductor manufacturer and a circuit built from discrete components. In terms of user definable functionality, programmable logic devices are a class of device between ASICs and a discrete solution. Programmable logic devices have evolved from user programmable memory devices, and have now reached a level of complexity comparable with small ASIC devices, with the advantage that no further processing steps are required. The section on programmable logic (section 13.2) describes a range of devices from simple PROMs to logic cell arrays.

Finally in this chapter, several of the important semiconductor technologies are described. The semiconductor technology refers to the types of transistors, and hence the type of process, used to fabricate the integrated circuit. Frequently it can be found that similar devices are available in different technologies, and it is important from a designer's viewpoint to understand the basic characteristics of the technology in order to choose the device which meets his needs in terms of performance and cost.

13.1   Application specific integrated circuits

The application specific integrated circuit meets the needs of the circuit designer who wishes to define the functions of his own integrated circuit without the costs and long design times associated with a full custom integrated circuit. The major incentive to design an ASIC for an application is its impact on the overall cost of a system. Many functions can be integrated into a single device, and therefore the number of components and the size of the circuit board can be reduced.

Previous to the availability of ASICs, both the design and the fabrication of an integrated circuit was the responsibility of the semiconductor manufacturer, with the result that the circuits they produced were general purpose in order to meet a wide variety of applications. With an ASIC, the application designer can define the function of the integrated circuit and therefore tailor its design to meet his requirements more closely.

The design and fabrication of an integrated circuit follows a well defined sequence of steps from circuit design through to its packaging and final testing. Each stage now employs a high level of computer aided design (CAD) and engineering (CAE) assistance, which has had the effect of progressively hiding the details and interdependencies between the fabrication process and the circuit design from the engineer. This allows the majority of design effort to be spent on circuit design.

Digital circuit design is well suited to top down design methodology because the interfaces between various elements may be easily formulated. It is typically found that a design will make frequent use of a common set of components, such as latches and gates, whose specific design details may be fixed, and whose performance characteristics are well understood. Such common components can be entered into a library on a CAD system, and from there can be used as building blocks to design more complex circuits in less time, with greater ease and with more reliability than would be the case if the designer had to design each circuit at the transistor level.

Today, this building block approach to the design of integrated circuits is such that complete CPUs, memory and I/O functions are available as building blocks, so that the designer can create a complete microprocessor system on an ASIC. The incentive to use ASICs is based primarily on a reduction of the overall system cost due to the high levels of integration that can be achieved. The two most popular types of ASIC which we review in this section are gate arrays and standard cell ASICs.

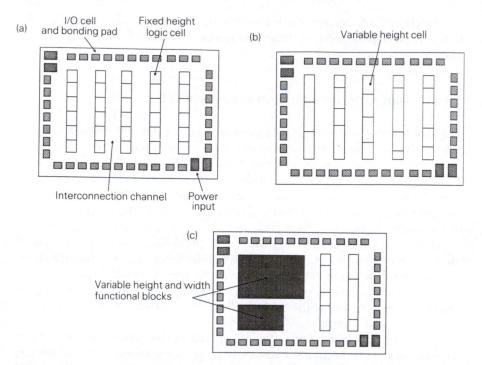

Figure 13.1 The layout of different types of ASIC: (a) typical gate array layout (b) typical standard all layout (c) ASIC with both standard cells and functional blocks.

13.1.1   GATE ARRAYS

The common denominator for all logic circuit designs is the logic gate, and any logic function can be created entirely of either logic NAND gates or logic NOR gates. The gate array is simply an array of logic gates which has been fabricated as far as the final step in the process, which defines the interconnection of the individual gates on the array. It is the responsibility of the circuit designer to define the interconnection between the gates, and therefore to define the function of the gate array. The advantage of this approach is that the fabrication on the uncommitted chip is almost complete, and the time to finish the manufacture of the final device when the design is complete will therefore be shorter than for a full custom device, which has to go through all stages of the fabrication process once the design is finished.

Figure 13.1a shows the layout of a typical gate array. The gates are arranged in fixed columns with spaces between to allow vertical wiring between the gates. On the periphery of the chip are the bonding pads which connect to the pins on the package of the integrated circuit.

To configure an ASIC for a particular application, the designer must first define the circuit which he wishes to implement. Entering the circuit diagram into the computer is a process called schematic capture and is normally done on a CAD workstation, or a computer with a graphics terminal. The process of schematic capture involves the designer in drawing the circuit on the graphics terminal; the CAD workstation makes a representation of this circuit which can then be used as an input to circuit simulation and layout programs. In producing the circuit, the designer may make use of library functions such as latches and multiplexers which exist as predefined groupings of gates with fixed interconnections. A typical library will implement many useful logic building blocks and is of great benefit to designers, as they no longer have to create their own set of commonly used functions, and also have the assurance that the library functions are known to work and have characteristics which are well documented.

In addition to designing the circuit to meet the basic functional requirements, designers must also be aware that once the device has been fabricated they will only have access to the input and output pins of the circuit. They must therefore design the circuit so that it can be tested and, if necessary, debugged via these pins. Typically, this can lead to a significant overhead in terms of the extra logic circuitry required to implement, testing, and may add to the number of I/O connections to the chip, a parameter to which the finished device is very cost sensitive.

Before fabrication, the designer will normally perform a computer simulation of the circuit to ensure that it functions correctly. The simulation relies on the designer of the circuit writing a series of test inputs, called test vectors, to the device. The simulation program then calculates the corresponding output signals of the circuit, taking into account the delays that the signals experience in passing through the logic gates. This helps the designer to identify problems, or potential problems, when the circuit is eventually fabricated on the gate array and run at its operating speed.

The next step is to choose a gate array with the required number of gates. The gates must then be allocated to specific circuit functions and interconnected according to the circuit design. This is usually done automatically by a place and route software package. A net list is produced by the computer from the circuit diagram and contains a list of all the components in the circuit together with their connections. The place and route software takes this information and maps it on to the gate array, allocating gates to specific functions of the circuit and interconnecting them. The space for wiring between the gates will be fixed, so that it may not always be possible to utilize all the gates on the array for a particular application; 70 to 95% utilization is typical with automated place and route, depending on the circuit complexity. Manual intervention in the place and route procedure can always be used to achieve higher utilization, or to optimize the layout for areas of circuit where the timing is critical.

The last step is to perform a final simulation of the circuit, which includes the different signal delays caused by the different lengths of interconnect between the individual gates of the array. The test vectors and output signals from the simulation can then be used as test patterns for a hardware tester to test the device after it has been configured with its metallization layer to identify good or bad parts.

13.1.2 STANDARD CELL AND FUNCTIONAL BLOCK ASICs

A standard cell ASIC is different from a gate array in that none of the integrated circuit is prefabricated, and therefore it has to be made as would a full custom integrated circuit. The name 'standard cell' is derived from the fact that the layout of the individual transistors in a standard component, such as a latch, can be fixed to define a standard cell, which can be placed together with other standard cells in a library and used as building blocks in the design of a more complex circuit or system. The concept is similar to the library described earlier for the gate array, except that the definition of the cell includes its layout on the silicon at the transistor level. Each standard cell is a functional building block with a set of defined input, output and power supply connnections. A design implemented with standard cells will implement only those gates required in the design, while a similar design on a gate array will frequently have gates that are not used. The standard cell approach therefore makes better use of the silicon area of the chip and, in comparison with the gate array, can accommodate approximately 30% more complex circuitry in the same area of silicon. However, the cost of a standard cell ASIC is higher than that of a gate array because it is fabricated as a normal full custom integrated circuit.

The layout of a typical standard cell design is shown in Fig. 13.1b. It is similar to that of a gate array in that the logic functions are arranged in columns with wiring channels between them. The layout shows that each cell has a fixed width and variable height, which allows cells to be stacked in columns. This arrangement works provided each standard cell is of a similar complexity and is defined with a fixed width and variable height. It is also convenient in that the cells can be defined with fixed I/O points which face the wiring channels, and may have fixed power supply connections on edges adjacent to each cell, allowing a power bus to be formed.

In the standard cell approach, any frequently used logic circuit may be defined as a standard cell. Cells which are much larger and more complex, for example a microprocessor or a block of memory, are themselves complex systems and will include circuitry that will allow the cell to be tested when it is part of the ASIC. Larger cells are often referred to as functional blocks, and are too large to have a fixed width and variable height format; instead they have variable height and variable width dimensions. An example of a layout of an ASIC containing both standard cells and functional blocks is shown in Fig. 13.1c.

The HPC microcontroller described in section 12.5.2 illustrates how ASICs have influenced the design of microcontrollers to take advantage of the flexibility that ASICs provide. Each of the family of HPC microcontrollers is partitioned into a core, shown surrounded by the dashed line in Fig. 12.16, and a number of peripheral functions which target the microcontroller at a specific application area. The core consists of the CPU and a number of important peripherals, and has also been implemented as a functional block. The HPC core may therefore be implemented on an ASIC together with other functional blocks and standard cells chosen by the application designer to tailor the microcontroller more closely to a specific application than a standard microcontroller.

The basic design process for standard cell ASICs follows the same procedures as outlined above for the design of a gate array, although a full custom process is required for their fabrication.

13.1.3 ANALOGUE ASICs

The idea of logic gate arrays and standard cells has been extended to the implementation of semi-custom linear integrated circuits. The analogue equivalent of a gate array is a prefabricated set of components such as transistors, resistors and capacitors, which are configured as a circuit by the final metallization layer of the chip. As in the gate array, the interconnection pattern may be determined through schematic capture of the circuit, and library functions such as operational amplifiers and comparators may also be available to the designer. More emphasis must be placed on the simulation of circuits to be designed with analogue ASICs, as the performance of the analogue circuit will be subject to larger deviations due to the process of its fabrication. More recently, the ability to mix both analogue and digital functions on the same chip, and to mix the different process techniques required, has given rise to mixed analogue and digital ASICs.

13.2 Programmable logic devices

Programmable logic devices (PLDs) are collections of logic gates whose interconnections are user programmable. They are typically used as replacements for small amounts of discrete random logic functions. The benefit to the user is that they can program these devices to their requirements in a matter of minutes, without the expense or time delays associated with the design of an ASIC However, they are less complex than ASIC devices, and provide the designer with an intermediate solution between a semi-custom ASIC and discrete logic.

The structure of a programmable logic device is similar to that of a programmable read only memory (PROM), discussed in Chapter 12. In fact the PROM is often used as a PLD, and has one of the three common structures found in PLDs.

13.2.1 THE PROGRAMMABLE READ ONLY MEMORY

The programmable read only memory consists of a series of AND gates whose inputs are derived from an array of the input signals, followed by a series of

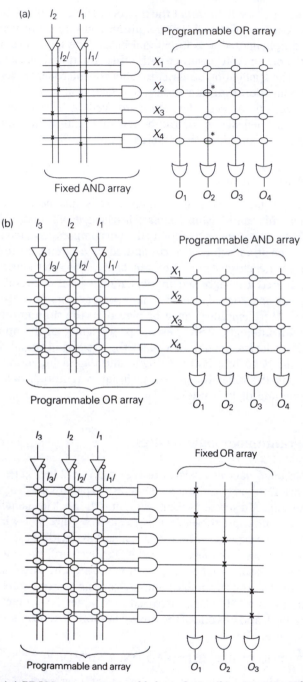

Figure 13.2   (a) A PROM as a programmable logic device (b) a programmable logic array (PLA) (c) a programmable array logic (PAL) structure.

OR gates whose inputs are derived from a programmable array of the outputs from the AND gates, as shown in Fig. 13.2a. The outputs from the OR gates are the outputs of the PROM.

For the PROM that was discussed in Chapter 12 as a memory element, the address inputs to the PROM correspond to the I inputs in Fig. 13.2a, and the data outputs correspond to the O outputs. As the PROM used as a PLD no longer has to interface to a microprocessor bus, there is no need for tristate output gates. The input AND array is the equivalent of the address decoder used in the memory PROM to generate the word line enable signals. Similarly, in the PLD PROM the output of each AND gate becomes active for each combination of the input signals, and is fixed by the crosses in the array indicating a connection between the vertical and horizontal lines. The programmable output OR array is equivalent to the bit cell array of the memory, and each programmable connection, indicated by a circle, will activate its output when the link between the crossing lines has not been fused and the output of the relevant AND gate is active. The single input OR gate does not exist in the circuit as a real gate, but is the natural result of the output line being active when either one or any other of the links in the vertical output line is in place. This is more commonly known as a wired-OR connection.

The AND followed by OR structure of the PROM generates Boolean expressions which can be most easily represented by a sum-of-products function. For example, if we consider all of the programmable links in the OR array to be broken apart from those marked with a *, then the output O_2 will be active whenever either X_2 or X_4 is active or they are both active together. This can be expressed as a Boolean summation by

$$O_2 = X_2 + X_4$$

where the summation indicates a logical OR function of the two operands X_2 and X_4. It can also be seen from the input AND array, which is fixed, that variables X_2 and X_4 are generated by AND combinations of the input signals such that $X_2 = I_1.I_2/$ and $X_4 = I_1.I_2$, where the . indicates the AND function, and $I_1.I_2$ is an example of a product term. The equation above then becomes

$$O_2 = I_1.I_2/ + I_1.I_2$$

which is a typical sum-of-products expression.

It can be seen that the fixed AND array automatically generates all the possible product terms of the input variables, and is the defined function of the address decoder for the PROM. For n input variables, the PROM will always generate 2^n product terms with 2^n AND gates. This proves to be a disadvantage in that the logic functions to be implemented will not normally make use of all these product terms. The unused product terms are therefore wasted. Also, the addition of an input variable has the effect of doubling the number of AND gates required to generate the product terms. If we were to add a variable I_3 to the PROM in Fig. 13.2a, we would also require an additional four AND gates to generate the eight possible product terms. This becomes impractical beyond a certain number of input variables, as the ratio of used to unused product terms becomes too small for the PROM to be cost effective.

13.2.2   THE PROGRAMMABLE LOGIC ARRAY

The programmable logic array (PLA) – sometimes referred to as the field programmable logic array (FPLA) – contains a programmable AND array as well as a programmable OR array. Thus only the product terms required by an expression need be implemented, resulting in an array whose size is primarily influenced by the number of product terms required rather than the number of input variables. Figure 13.2b shows the arrangement of a PLA with its programmable AND array. For n input variables, the PLA is designed to generate a number of product terms which varies between different types of PLA, and will always be less than 2^n.

Although the PLA is more flexible than the PROM in that it allows the product terms of an expression to be programmed individually, it is comparatively more expensive owing to the additional circuitry required to provide a programmable AND array. The PLA is therefore more suited to a logic function that has a few product terms with a large number of variables, while the PROM is more suited to functions with a large number of product terms having a smaller number of variables.

13.2.3   PROGRAMMABLE ARRAY LOGIC

The programmable array logic (PAL) chip is constructed from a programmable AND array to give flexibility in the programming of product terms, and a fixed output OR array to achieve a lower cost than the fully programmable PLA. The structure of the PAL is shown in Fig. 13.2c, where the fixed OR array can be seen to constrain the number of product terms available to an output variable. In the case shown in the figure, it can be seen that this is two product terms per output. PALs are therefore made available with a range of different output OR gate configurations which must be chosen to satisfy the logic function to be implemented.

13.2.4   PROGRAMMING AND REPROGRAMMING PROGRAMMABLE LOGIC DEVICES

Most PLDs are programmed with the aid of a computer program. This translates the required logic functions into a fuse pattern that can be loaded into a programming device capable of destroying the necessary fuse links in the PLD. A typical program will allow the input and output pins of the device to be defined in terms of input and output variables, and then the relationship defined between the input and output variables as a series of Boolean logic functions.

The program may also generate a series of test patterns or test vectors; these may be applied to the input pins of the PLD after it is programmed, and its output compared with that predicted by the program in order to verify that the device has been correctly programmed.

PROMs, PLAs and PALs use technology that allows the devices to be programmed only once. Embedded systems may often require a degree of reconfiguration and, while the microprocessor may provide this through software, it can only do so within the bounds set by its hardware. Hardware reconfiguration is generally more difficult to achieve, although some possibilities exist with reprogrammable PLDs.

PLDs emply the same techniques for reprogrammability as memory devices. Erasable PLDs, or EPLDs, allow the fuse pattern to be erased with exposure of the PLD to ultraviolet light, in the same way that EPROMs may be erased. As is the case with EPROMs the erase time is 15 to 20 minutes, and the erase and reprogramming cycle is not conveniently done while the PLD remains in-circuit. Consequently the EPLD does not provide on-line reconfigurable hardware, so that the system containing the EPLD must be stopped in order that it may be removed, erased and reprogrammed before it is reinstalled and the system set running again. The EPLD may however be used as a convenient prototyping tool during the development of systems, after which its function may be transferred to a less expensive PAL.

Electrically erasable PLDs (EEPLDs or E^{2}PLDs) are PLDs whose program pattern may be erased by the application of suitable electrical pulses to the device, in the same way that EEPROMs may be electrically erased. This gives it the ability to be reprogrammed in circuit during system operation, and allows on-line reconfiguration of the system. While these devices give the designer more flexibility than PLDs and EPLDs, they are generally more expensive and smaller in terms of their memory capacity.

More sophisticated PLDs enhance the basic AND/OR structure with additional circuitry to provide extra functionality. For example, latched inputs, outputs which look like registers, programmable output polarity, and internal feedback paths from outputs to the inputs, are all functions which may be found on the more advanced devices, and are very useful features in many applications.

Within the spectrum of user programmable and semi-custom logic devices, the PLD has a relatively low logic complexity in terms of the functions it may be programmed to perform, but has the advantage that it can be programmed in a matter of minutes and is then ready for use. EPLDs have the additional advantage that, during periods of system development, the devices may be reprogrammed. Gate arrays, on the other hand, offer a higher level of logic complexity and integration at the expense of taking several weeks to produce. A device which partially fills the gap between PLDs and gate arrays is the logic cell array (LCA). This device has an architecture which consists of three types of user programmable elements. These elements are I/O blocks, configurable logic blocks and interconnections, and are configured by a program which is stored in static read/write memory internal to the device. The physical layout of the logic blocks, I/O and interconnections within the LCA is similar to that of a gate array, and programming the device consists of entering a circuit schematic followed by logic simulation of the circuit and the generation of a configuration bit pattern on a workstation or personal computer with the necessary software. The configuration pattern is then loaded into the LCA. In an embedded system, the configuration pattern must be stored in an external EPROM and loaded into the LCA when the system is initialized. The LCA is in effect a reprogrammable gate array, and offers the designer the logic complexity of a gate array combined with the convenience of reprogrammability.

13.3   Semiconductor technologies

There are many different ways in which a transistor can be made from a piece of silicon, and each produces a device with specific characteristics that make it suitable for particular applications. For example, a bipolar transistor is a rugged device which is well suited to switching large currents on and off, and is typically found in applications such as driver circuits for electric motors. If a battery powered circuit was being designed, the CMOS transistors would be used because they consume relatively small amounts of power when they are not switching or in a static state. The physical structure and material composition of a device, and its associated characteristics, are referred to as the technology of the device. This section reviews the characteristics of some of the important semiconductor technologies used in VLSI devices such as microprocessors and their associated circuits.

13.3.1   SOME IMPORTANT CHARACTERISTICS

In comparing various device technologies, we can use a set of common characteristics which may influence the choice of particular device technology for an application.

(a)   SPEED

The speed of a technology is usually measured by the delay of a logic signal through a simple gate, and the maximum frequency logic signal which can be used to clock the gate.

(b)   PACKING DENSITY

The packing density refers to the number of devices which may be fabricated within a given area of silicon. Technologies with a high packing density have the capability of producing very highly integrated and complex circuitry.

(c) COST

The cost of a particular technology is a function of a number of variables. These include the cost of the materials, the complexity of the process by which the devices are fabricated, and the maturity of the technology, which may determine the yield of good devices obtained, all of which contribute towards the eventual cost of an integrated circuit.

(d) POWER CONSUMPTION

The power consumption of a technology is an important system consideration as it will influence the system power supply requirements. Large power supplies are not only costly but also physically large and heavy. Devices which consume large amounts of power also need to dissipate large amounts of heat, and this again imposes further cost on the system in that it must provide cooling for these devices where necessary. A further consequence of a high power, high speed technology is that devices may tend to radiate large amounts of electrical noise, causing interference to other devices and systems.

(e) NOISE IMMUNITY

The noise immunity of a technology will determine how susceptible a device is to electrical noise. The most common source of noise is that generated by the devices themselves, and is caused when transistors switch on and off. This noise often travels by the power supply connections to other devices within a system. Similarly, electrical noise generated external to the system may find its way into a system through the power supply or through its interface connections. Other common sources of interference can occur from electromagnetic and radio frequency emitters, and from nuclear radiation. Should any of these cause problems in the operation of a system, then the solution is either to change the technology of the device to one with higher noise immunity (usually a more expensive technology), or to provide additional protection in the form of shielding sensitive components, isolation of the I/O lines and improved regulation for the power supply.

The family tree of the most popular VLSI technologies is shown in Fig. 13.3. The range of technologies is split into two groups consisting of bipolar technologies and metal oxide semiconductor (MOS) technologies. MOS technologies were the first used for VLSI designs, and at present are still the most popular for the fabrication of devices such as microprocessors and memories. Bipolar technologies are more typically found in analogue integrated circuits, although the various forms of transistor transistor logic (TTL) have been the predominant technology for small and medium integrated circuits such as logic gates, latches and multiplexers.

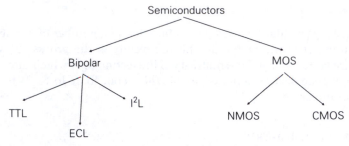

Figure 13.3   Semiconductor technologies.

13.3.2   MOS TECHNOLOGIES

MOS transitors have simple structures which make them ideal for VLSI device fabrication. A key feature which allows high packing densities to be achieved is that each transistor requires no isolation from adjacent transistors, as is the case with bipolar technologies. In addition, all of the necessary circuit elements, such as resistors and small capacitors, can be made from a basic MOS transistor structure, whereas bipolar circuits require these components to be created as different components on integrated circuits.

MOS technologies are generally slower than their bipolar counterparts, consume less current and have good noise immunity, but they require protection against the build-up of static charges or large current pulses which can destroy the MOS transistor. This protection is normally included as part of the device.

(a)   nMOS/pMOS

The nMOS transistor is one of two basic types of MOS transistor. The n means n-channel, and refers to the type of conducting channel induced in the transistor when it is switched on. The other type of device is the pMOS transistor, which has a p-type channel. pMOS was the original MOS technology, and was used to fabricate the first microprocessor chip. However, nMOS technology has taken its place owing to its faster speed.

The input of a logic gate will have a switching threshold which is the voltage level at which the output of the gate changes state. The switching voltage level of a MOS gate is determined by many factors, including the physical dimensions of the device and the doping concentrations of the different types of silicon which constitute the device. In practice, the switching levels of the devices which interface to external circuits are chosen to be TTL compatible (see section 13.3.3).

(b)   CMOS

CMOS employs a combination of both nMOS and pMOS transistors. The C in CMOS means complementary, in that the switching properties of the two transistors are such that when one of them is switched on, the other is switched off. A consequence of this operation is that when a CMOS circuit is in a static condition, such that none of the gates is in the process of switching, the circuit consumes virtually no current. Current is only consumed in a CMOS gate when it switches from one state to another, and the magnitude of the current is then proportional to the frequency at which it switches state. This is ideal for low power circuitry as the current consumption of the device can be traded off against its speed of operation. The maximum speed of CMOS, however, is slower than nMOS because each CMOS gate contains a pMOS transistor which is slower than the nMOS type.

The switching voltage level of a CMOS gate is chosen to be half that of the power supply voltage in order to give the CMOS circuit a symmetrical noise immunity. This means that it would take the same amount of noise voltage to cause the gate to switch from a logic 0 to a logic 1 as it would to switch from 1 to 0. CMOS circuitry therefore has a good immunity to electrical noise, and is usually preferred for systems which must operate in particularly noisy environments, such as in factory and automobile applications. CMOS circuits also operate from a wide range of power supply voltages, from 3 to 18 volts, and their low power consumption make them ideal for use in battery powered systems.

One of the disadvantages of CMOS gates is that they have a lower packing density than their equivalent nMOS gates because the pMOS and nMOS devices require electrical isolation from one another. Also they are not suitable for interface circuits, which require high current outputs.

13.3.3   BIPOLAR TECHNOLOGIES

Bipolar technologies in general have applications in high power and high speed circuits. Most bipolar technologies have suffered in the past in that they required a relatively large chip area to create a gate compared with the equivalent MOS gate, and so have not been suitable for highly integrated circuit applications. This was due to the isolation required between transistors, the more processing steps required to fabricate a bipolar device, and the need to also fabricate other components, such as resistors, to complete the circuit. However, the development of integrated injection logic has overcome many of these problems and may provide a bipolar VLSI technology which is able to compete with MOS technologies.

(a)   TRANSISTOR TRANSISTOR LOGIC

Transistor transistor logic (TTL) and its many variations, such as low power Schottky TTL, have been used to provide the circuit designer with a family of small and medium logic integrated circuits consisting of groups of simple gates, latches, multiplexers, shift registers etc. The 74 series of integrated circuits is used in microprocessor systems mainly as the discrete logic which interfaces the various VLSI functions to the system bus. For example, the address decoder described in Chapter 10 was designed using a 74138 '1 of 8 decoder'. Large scale integration devices, such as memories and simple microprocessors, have been produced in TTL in order to achieve high speed devices, at the expense of low density and high power consumption.

Several forms of TTL exist today. Schottky TTL produces devices with an improvement in speed (maximum clocking frequency) and gate delay over standard TTL, but at the expense of higher power. Low power Schottky TTL was therefore developed; it consumes much less power than standard TTL or Schottky TTL, but has a higher speed and similar gate delay as standard TTL. Advanced low power Schottky TTL provides even more improvement in the speed/power ratio, as do the newer TTL families such as FAST and FACT. A similar range of logic functions is also available in CMOS, called the 4000 series logic family. A high speed version of CMOS has recently been developed and used to produce a family of circuits which are pin compatible to the TTL 74 series, offering gates with delays comparable with low power Schottky TTL yet with the power consumption of CMOS.

The power supply voltage for TTL circuits is 5 volts. The input voltage switching level of a TTL gate is typically 1.3 volts but may vary between 0.8 and 2.0 volts, which means that the noise immunity of a logic 0 input to the gate is quite poor. A typical low level output voltage from a TTL gate is 0.2 volts; for the worst case switching level a voltage spike of only $0.8 - 0.2$ volts (with sufficient duration) would be required to cause the output of the gate to change state. These voltage levels are significantly different from CMOS levels that some consideration is required whenever CMOS circuits are interfaced to TTL circuits.

(b)   EMITTER COUPLED LOGIC

Emitter coupled logic (ECL) is the major technology used in the design of large computer systems. The technology can produce gates with sub-nanosecond gate delays because the transistors used as the switching elements within the gates are not driven to their limit, or saturation, as is the case with TTL technology. ECL logic levels are different from the nominal standard 0 and 1 logic levels of 0 and 5 volts, and typically use a value of $- 0.75$ volts for logic 1 and -1.55 volts for logic 0, together with $- 5.2$ volts power supply. Additional interfacing circuitry is required to translate these logic levels should the ECL circuit be used together with logic devices of a different technology.

The increased speed of ECL is gained at the expense of increased power consumption, which may require special packaging of the integrated circuit and design of the printed circuit boards for a high heat removal rate using specialized cooling systems. High speed ECL applications are consequently very costly.

A variant of ECL technology known as current mode logic (CML) is used to produce ECL circuits whose speed and power consumption may be optimized. Within the gate, the logic states are represented as currents instead of voltages. One of the logic states is represented by zero current, and the other can be chosen by trading the speed of the gate against its power consumption to meet the requirements of the application. Very low power circuitry can therefore be designed with CML provided low speed, and hence long gate delays, can be tolerated.

(c)   INTEGRATED INJECTION LOGIC

Integrated injection logic (IIL or I^2L) achieves higher packing densities than other bipolar technologies because the connection of the transistors within the gate allows parts of them to be merged within the same area of silicon. In other words, the same structure in the silicon is used as a part of a number of transistors within the gate. In addition, the I^2L gate requires no internal resistors which would normally occupy large areas of silicon.

As with CML, the action of the transistors within the gate is to switch current on and off between the various transistors, and variation of the current supplied to the gate, called the injector current, allows the speed and hence the power dissipation of the gate to be adjusted.

Typical logic levels within an I^2L gate are 0.7 and 0.1 volt for logic 1 and 0 respectively. As is the case with ECL, additional circuitry is required to interface I^2L to other logic technologies. A wide range of power supply voltages may also be used with I^2L technology.

Chapter 14

The development of microprocessor systems

14.1 The system specification

During the development of a microprocessor based system, the designer will make use of hardware and software tools that enable him to implement the design of the system. The microprocessor system will be originally conceived from a functional requirement, for example the need to control a robot arm or to monitor the performance of a machine. This requirement provides the basis on which a specification of the microprocessor system can be made, and which should define the following system features.

(a) INPUT/OUTPUT (I/O)

The number and type of I/O interfaces required by the microprocessor system can be determined from the sensors and actuators needed to monitor and control specific functions, and the type of operator interface required. Communications with other systems to provide remote configuration and control will define the type of communications I/O required, and this must be chosen such that its physical interconnection and the type of protocols used to convey information between the systems are compatible.

(b) CENTRAL PROCESSOR UNIT (CPU)

The number and complexity of functions to be performed will indicate the performance and hence the type of microprocessor suitable for the application. Performance is the most critical factor to be considered here, and is the most difficult to assess. The selection of a microprocessor on crude measurements, such as the number of instructions per second that it is capable of executing, is normally not sufficient for a specific application. A more objective means of assessment is benchmarking, where a section of program which is representative of the operations to be performed is run on the microprocessor, or a simulator, to determine its speed of execution. Constraints such as environment, size and power must also be considered.

(c) MEMORY

The size and speed of system memory are important parameters. The number of functions to be performed by the program that is to be executed by the microprocessor system will indicate the size of memory required. In practice it is not possible to know accurately what the memory requirements will be before the program has been written, and so it is usual to design the system with more memory than necessary. This is frequently found to be an advantage in that it allows future additions to be made to the program with no hardware modifications.

Having produced a specification for the microprocessor system, the designer can detail the hardware and software. This activity produces circuit diagrams of prototype hardware for the microprocessor system, and a specification for the software that is to be written for the application.

14.2   The development environment

With these designs, the microprocessor system enters its development phase. Figure 14.1 shows a typical development environment for a microprocessor based system; this is the set of tools with which the designer can verify the hardware design, write and test software, and finally bring together both hardware and software to test the complete system.

The prototype hardware is referred to as the target system because it acts as the focus for both hardware and software development work. The software for the target system is written on a computer system referred to as the host, as during the development of the system it hosts the software development tools used to develop the application programs for the target system. The software development tools allow the designer to write the program and generate the equivalent machine code instructions which can then be loaded and run on the target system. When eventually the program functions correctly, it may be programmed into a ROM device and permanently installed in the finished target system.

Today, a popular host computer system is the general purpose personal computer (PC). With the addition of software development tools for the target

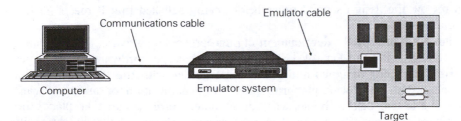

Figure 14.1 Development equipment for an embedded microprocessor system.

microprocessor, the PC creates a cost effective single user development environment for microprocessor software. In larger corporations, and for large projects involving several designers, it may be preferable to use a multi-user minicomputer as a host so that several designers may have simultaneous access to the software tools and program files. However, this possibility depends on the availability on a particular host system of the software development tools for the target microprocessor. Some semiconductor manufacturers produce their own host computer systems together with software and hardware development tools, although the popularity of this approach has decreased due to the low cost and general purpose usage of the personal computer.

The interface between the host and target systems is a hardware emulator for the target microprocessor. This emulator is called an in-circuit emulator (ICE). It also has the ability to control the execution of the application program in the target system. Typically the emulator is a stand-alone unit, as shown in Fig. 14.1, and interfaces to the host system by a simple communications link, for example a serial RS232 interface, by which the host can download the machine code application programs to the emulator, and control and monitor its execution. Alternatively, when the host development system is a personal computer, the emulator may take the form of an add-in card to which the host can directly load programs using the internal bus of the personal computer instead of the much slower RS232 interface.

To the target system, the emulator appears as would the real microprocessor. The emulator achieves this by having an emulation pod which plugs into the socket normally occupied by the microprocessor in the target system. The emulation pod contains the target microprocessor with circuitry which allows the emulator to control the execution of its programs.

14.3   The development cycle

The steps taken by a design engineer to develop the hardware and software for the target system follow a well defined cycle typical of any product development. In general terms the basic cycle is as shown in Fig. 14.2. It is very rare that the first pass through this design cycle yields a fully functional target system which meets the required design specifications. Instead, the designer may need to review the design several times before being satisfied that it meets all the requirements.

For the more specific development of microprocessor software, the cycle may be represented as in Fig. 14.3. This assumes that the hardware of the target system has been designed and built, and together with the ICE provides the designer with a hardware platform on which the application, or target, software may be tested. Before the software development process can take place, the software itself must be designed to meet certain criteria. Obviously, the main criterion is that it should perform the operations required by its functional

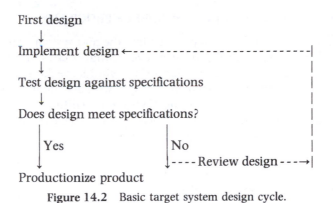

First design
↓
Implement design ←-------------------------|
↓ |
Test design against specifications |
↓ |
Does design meet specifications? |

|Yes |No |
↓   ↓----Review design----→|
Productionize product

Figure 14.2   Basic target system design cycle.

specification. In addition, secondary features such as ease of maintenance, testability and portability (use of the same software on different target machines) are now considered as being of equal importance owing to the size and complexity of some systems, and the investment in terms of manpower required to develop the software. Software engineering is now recognized as a discipline in its own right, and is beyond the scope of this book. Readers interested in this area are referred to *Software Engineering* by Ian Sommerville (Addison Wesley, 1985).

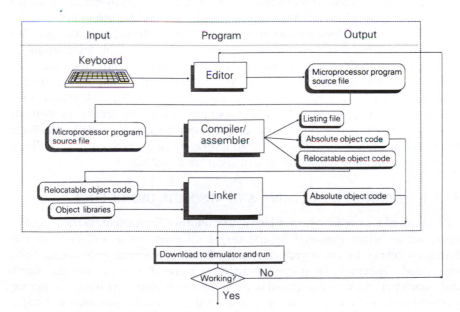

Figure 14.3   Development cycle of microprocessor software.

14.3.1   THE EDITOR: ENTERING THE SOURCE PROGRAM

The designer enters the application program into the host system using an editor. The editor is similar to a word processor, except that it does not provide text formatting facilities. The program then exists within the host as a text file consisting of lines of program statements, as one would write out the program on a sheet of paper. The contents of the file are referred to as the source code of the program because they act as the origin from which the machine code for the target microprocessor is finally produced.

14.3.2   COMPILATION AND ASSEMBLY:
THE GENERATION OF OBJECT CODE

The next step is to convert the source code program into machine code instructions which are understood by the target microprocessor. This is done by a language compiler, or by an assembler, depending on which language has been chosen to implement the target program. The assembly language of the target microprocessor is the most simple form of language that can be used, and is the mnemonic representation of its machine code instructions. For example, the assembly language statement

ADD A, #0x10

when passed through the assembler would generate the correct machine code instruction for ADD A followed by the operand 0x10. Hence, with assembly language, there is a one-to-one relationship between the statements in the source code and the machine code that is produced from it by the assembler.

Alternatively, the designer may choose to implement the program using a high level language such as C or Pascal. The statements in a high level language are more meaningful to the reader, and are independent of any instruction set specific to a particular microprocessor. The purpose of the compiler therefore is to translate the high level language statements into machine code instructions for the target microprocessor. Typically, a single source code program statement will cause the language compiler to generate many machine code instructions owing to the greater complexity of high level language statements. This allows the designer to more easily implement complex functions.

14.3.3   LINKING: RELOCATABLE AND ABSOLUTE OBJECT CODE

The output of a compiler or assembler will be another file containing the machine code, which when executed by the target microprocessor will perform the functions defined by the source code program. This machine code is called the object code. However, for the microprocessor to be able to execute the object code, each byte must be associated with an address which is within the program memory area of the target microprocessor system. Instructions such as jumps, and operations on variable data memory, must also refer explicitly to addresses defined by the memory map of the system. If the object code file contains these addresses, then it is called absolute object code and can be immediately transferred to the target system via the emulator and executed.

Large programs are frequently developed as a collection of smaller and more manageable programs. It is the function of the linker to take the individual object modules produced by the assembler and combine them into a single absolute object code module which can be executed on the target hardware. Each of the smaller programs may call subroutines defined in other programs, and also may use variables defined in other programs. Consequently, at the time a program module is assembled, it may not have knowledge of these address values. The assembler therefore produces relocatable object code which contains no fixed addresses. The linker must therefore assign the absolute addresses to the program and data variables which correspond to program and data memory areas of the target hardware on which the program is to be executed.

There are many benefits to be gained from this modular approach to the implementation of an application program. Some of these benefits are as follows:

1. Modularity can ease the definition of the organization of a complex system from its specification. This definition often leads to a more structured implementation of the system, with each task having its own specification including well defined inputs and outputs in terms of the information it receives from, and transmits to, other tasks.
2. The detailed specification of each task allows a number of programmers to work on implementing the application program at one time. The specification of the task will allow it to be developed independently of other tasks, and will define what functions must be implemented, what information is used as input to the task and produced as an output, and what variables are to be used.
3. Modularity also makes debugging the system easier as it forces the interdependence between tasks to be minimized. It also helps the software to be maintained with updates. Documenting the software and its operation is also made significantly easier.

14.3.4   OBJECT CODE LIBRARIES

An application program may make frequent use of a particular routine, in which case the routine may be placed into a library of commonly used routines which are made available for use in any subsequent program. For example, a routine may be written to perform the floating point multiplication of two numbers. If the routine is written such that it can be used to multiply any two floating point numbers, then it may be reused as a general purpose function. A library consists of the relocatable object code of such routines, and those which are used by the application program are linked into the absolute object code by the linker.

14.3.5   EMULATION AND DEBUGGING

The absolute code can be loaded into the emulator and executed to determine if it performs the functions as specified in its design. Finding the cause of operational errors is often a time consuming exercise which can be made easier with the use of software and hardware debugging tools.

An ICE typically provides two features, breakpoints and single stepping, which allow the execution of the program to be controlled. A breakpoint defines a condition which allows the emulator to execute the program in the target system until the condition is satisfied, at which point execution of the program is broken and control is returned to the host system. During times when the emulator is not executing a program, it will allow the user to read and alter the contents of the internal registers of the microprocessor, and the memory of the target system. The most common way of using a breakpoint facility is to set a breakpoint condition on a program address to determine if certain instructions are executed. During execution of a program, an emulator may also provide a trace facility which makes a record of the most recent instructions to be executed, and may be used in conjunction with a breakpoint to find out how the program execution arrived at the breakpoint condition. The benefit of breakpoints when debugging a real-time application is that the microprocessor system runs at full speed, which allows it to respond to real-time events such as interrupts. This frequently means, however, that the embedded microprocessor system must be debugged *in situ* together with the necessary sensors, actuators and other systems which generate the real-time conditions under which it must operate.

The single step feature of the emulator allows the user to step one instruction at a time through a program, and between each step to have access to micro-processor registers and memory. This is useful for the line-by-line verification of a section of program which is not real-time dependent.

Once the causes of errors in the operation of the microprocessor system have been identified, and the required modifications to the program source code have been made, the development cycle in Fig. 14.3 is repeated until the system meets the requirements of its specification.

14.4   Assemblers, linkers and assembly language

The mnemonic form of a microprocessor's instruction set is referred to as its assembly language. The assembler is a program which converts an assembly language program into machine code instructions which can be executed by the microprocessor. The mnemonic itself is a simple abbreviation which indicates the action to be performed by the instruction.

As well as translating mnemonic instructions into machine code, the assembler provides the programmer with a number of features which make the task of programming easier. For example, reference points within the program, such as data storage locations or points to which jump instructions must go,

```
 1                         ;Program to add the contents of array1 and array2
 2                         ;together and place the result back into array1.
 3                         ;
 4 0000                    .sect DATA,RAM16        ;Define the type of memory.
 5 0000         count:     .dsb 1                  ;Allocate storage for count.
 6 0002         array1:    .dsw 10                 ;Allocate storage for array1.
 7 0016         array2:    .dsw 10                 ;Allocate storage for array2.
 8 0000                    .endsect                ;No more variables to be defined.
 9
10 0000                    .sect PROG,ROM16        ;Define the type of memory.
11 0000         start:                             ;'start' marks the beginning of the program.
12 0000 9300               ld x,#0                  ;Initialise the index register.
13 0002 830A00008B   R     ld count.B,#10          ;Initialise the loop counter.
14 0007         loop:                              ;'loop' marks the beginning of addition loop.
15 0007 A60002CEA8   R     ld a,array1[x].W        ;Load acc. with value in array1.
16 000C A60016CEF8   R     add a,array2[x].W       ;Add to acc. value in array2.
17 0011 A60002CEAB   R     st a,array1[x].W        ;Store acc. back in array1.
18 0016 A9CE               inc x                   ;Increment index by two to point to the next
19 0018 A9CE               inc x                   ;value in the array.
20 001A B600008A    R      decsz count             ;Decrement loop counter and if zero skip
21                                                 ;next instruction.
22 001E 77                 jmp loop                ;Do loop again.
23
24                         ; program will continue onwards from here when the two arrays have been added.
25
26 0000                    .endsect
27 0000                    .end start
```

Figure 14.4   Assembler listing of the array adding program.

may be given user defined labels instead of addresses, which give the program a more meaningful structure and make it significantly easier to read and understand. The labels are then assigned absolute addresses when the object code modules are linked together by the linker.

A typical assembly language program consists of two types of instructions. The first type are the microprocessor mnemonic instructions which cause the generation of their equivalent machine code instructions. The second type are called assembler directives and control the operation of the assembler. For example, assembler directives can be used to allocate areas of read/write memory as data storage locations, define constant values as labels which have more meaning within the context of the program, and define the beginning and end points of the program. Some typical characteristics of assemblers are illustrated in the following example.

14.4.1   AN ASSEMBLER PROGRAM EXAMPLE

In section 11.5.5 a small section of an assembler program was used to illustrate how the indexed addressing mode of a microprocessor could be used in a routine that adds two arrays of numbers, array1 and array2. This routine has been implemented using the instruction set of the HPC microcontroller, which was described in section 12.5.2, and assembled using the HPC assembler. Figure 14.4 shows the listing produced by the assembler. The figure contains the original lines of source program in the middle and on the right of the listing, and the corresponding line number, relocatable address and machine instructions on the left. The main features shown by the listing are as follows:

(a)   LABELS

So that the assembler can distinguish between assembler mnemonics, assembler directives, user defined labels, and comments, each is recognized by having a different format. User defined labels are always followed immediately by a colon. The label 'start' has been used to identify the beginning of the program, and the label 'count' has been used to refer to the variable which holds the count of the number of times the program loop, defined by the label 'loop', is to be executed.

When large programs are written as a combination of several smaller program modules, instructions in one program module must be able to refer to labels in another module. The assembler achieves this by declaring variables either as public, in which case the label may be used in other program modules, or as external, in which case the variable has been defined in another program module and is being used in the current module.

(b)   ASSEMBLER DIRECTIVES

Assembler directives are preceded by a full stop, as illustrated by the '.dsb 1' directive which allocates a single byte to the variable 'count'. The address assigned to the variable 'count' is treated by the assembler as a relocatable address, and it is the responsibility of the linker to give the variable 'count' an absolute address when the relocatable object code is linked. Similarly, the '.dsw 10' directive assigns ten words of memory to each of the arrays, and the address of the labels 'array1' or 'array2' then refers to the first element in the corresponding array. Each element in the array is a word value, and the index pointer in the program must therefore be incremented twice on each addition to point to the next word in the array.

The assembler also allows the programmer to assign a specific absolute address to a label, as is required in the case of I/O related registers which have a fixed address in memory. For example, if an I/O data register was memory mapped at address 0x8000, then the HPC assembler would allow the name 'port' to be assigned to this address using the statement 'port = 0x8000'.

Other assembler directives in Fig. 14.4 are the '.sect' and '.endsect' statements, which help to partition the program into sections in which the statements are referred to a particular type of memory. This segmentation is useful to the linker when several programs are combined together to fit within the memory

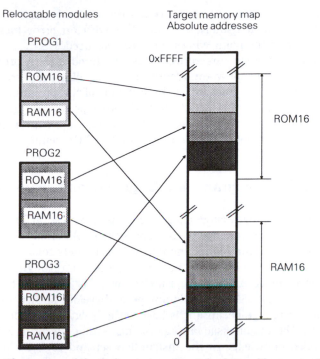

Figure 14.5 The action of the linker is to assign absolute addresses to the program modules and to locate them in areas of the target memory map where memory exists.

implemented in the target microprocessor system. For example, the target system will have a memory map containing elements of ROM and read/write memory or RAM. The statement '.sect DATA,RAM16' in Fig. 14.4 indicates that when the assembler subsequently reserves memory using the '.dsb' directive, it does so in a section of memory which has been called DATA by the programmer, and is of a type RAM16 or 16 bit read/write memory. The absolute address range of the RAM16 memory is defined by the programmer when the relocatable object module is linked with object modules from other programs, and the linker consecutively allocates addresses to all the variables from different object modules having the same memory type. This is illustrated in Fig. 14.5, which shows how three program modules are combined so that the program code for each is allocated addresses in the ROM16 range, and the read/write variables are allocated addresses in the RAM16 area.

(c)   ASSEMBLER MNEMONICS

The assembler instructions are recognized as normal text. However, the instruction set will have a syntax which must be obeyed if the assembler is to recognize what the instructions are. When the assembler cannot understand the source program, it usually indicates this with an error message in the assembler listing at the point where the error occurred.

The documention of a source program is very important. A program which is well documented will allow any programmer, including the person who wrote it, to understand or be reminded of the function of the program, how it may be used, and how it can be integrated with other program modules. Lines of documentation can normally be included anywhere in an assembler program following a special symbol, which in the case of the HPC assembler is a semicolon.

14.5   High level programming languages and compilers

A programming language defines a logical means of expression for actions which are to be executed by the microprocessor system. Assembly language is an example of a programming language which is intimately related to the specific hardware architecture of a particular microprocessor through its instruction set. As a consequence, each microprocessor will have its own instruction set and its own assembler. Assembly language is therefore known as a low level programming language because it is based strictly on the low level operation of the device. The major disadvantage of this is that the programmer has to have a detailed knowledge of the instruction set and registers of the micro-processor they are programming, and must always have these details in mind when writing a program. Alternatively, high level languages, such as C, mask these details from programmers by allowing them to express programs in a way which makes no reference to the registers or instruction set of the micro-processor. High level languages are therefore intended to be machine independent, and are constrained only by the programming rules of the language.

The high level language program is, like the assembler program, a source program which must be translated into the machine code instructions of the microprocessor before it can be executed. This is achieved by a compiler for the high level language being used. The output of the compiler is usually relocatable object code, which must then be linked in the way described above to produce machine code instructions which may be executed by the microprocessor.

Each line of high level language source code may produce several machine code instructions when passed through the compiler. If a high level language program is compared with a well written assembly language program that performs the same function, then the machine code generated by the compiler is always larger than that of the assembler. The difference in the size of machine code generated by a compiler and an assembler is the compiler overhead and is very dependent on how the compiler itself has been implemented. Assembly language programming is still recognized as being the most efficient way to produce the smallest and fastest machine code.

The major benefits of using a high level language are therefore that it:

1. Improves the productivity of the programmer because each line of high level source code produces several lines of machine code;
2. Makes the programmer less likely to produce errors because he no longer has to concern himself with the specific instruction set and register details of the microprocessor;
3. Allows more complex data manipulations to be more easily implemented, as a high level language will support a range of data types and data structures which can be directly manipulated by high level language statements;
4. Makes the program more portable between different types of micro-processor, given that a compiler and a linker are available to generate the object code for the desired microprocessor;
5. Makes the source program more readable in that the statements in a high level language program can be based on English words and on phrases whose meanings more adequately reflect the function of the program.

The disadvantages of a high level language compiler are that:

1. It generates more object code than an equivalent program written in assembly language;
2. As a consequence, the compiled object code will generally run more slowly than the assembly code program.

One of the most popular high level languages used for programming micro-processors is the C language. Figure 14.6 shows the C source code to implement the addition of the two arrays of numbers as described in section 14.4.1, and illustrates some of the features of a high level language. For completeness, the values of the arrays are initiallized with data and it is not assumed that they are filled with data by another subroutine.

At a first glance, the statements which make up the program bear a much closer resemblance to the function it performs than the assembly language

```
main()
{
        int count;
        static int array1[10] = {4,7,5,23,65,9,7,74,52,3};
        static int array2[10] = {7,12,9,30,44,2,1,86,34,7};

        count = 10;
        while ( count-- )
        {
                array1[count] = array1[count] + array2[count];
        }
}
```

Figure 14.6   C program to add two arrays of numbers.

program did. The function of statements such as 'count = 10' are obvious to readers, and help them to grasp the function of the program more quickly.

All C programs start with the label 'main' and have the following format:


```
main ()
{
        declaration of variables
                :
        program statements
                :
}
```

Variables which are made use of within the program must be declared so that the necessary memory storage can be allocated by the compiler. Variables can be of different types. In Fig. 14.6 the variable 'count' has been declared as an integer variable, and any attempt within the program to assign it a non-integer value, such as 13.56, would cause the compiler to generate an error when the source program is compiled. It is necessary to distinguish between the different types of variable because their processing by the microprocessor will be different, and hence the compiler will need to generate different object code for similar operations on different data types. For example, the machine code which is generated by the compiler to implement 'count = count − 1' when 'count' is an integer will be different form that generated if 'count' were a floating point number.

The arrays are declared as having ten integer elements, and each element may be referenced by an index so that the first value of array1 is 'array1[0]' and the last is 'array1[9]'. The index itself can be a variable, and in Fig. 14.6 the value of the variable 'count' is used to index each array as well as to determine the condition for further execution of the 'while' block. The statement 'while (count − −)' causes the variable 'count' to be evaluated as a Boolean variable, the result of which will be false if the value of 'count' is zero or true if 'count' has a value other than zero. If the evaluation is true, the addition of the elements within the arrays indexed by 'count' occurs. The process is then repeated until 'count' has a value of zero. The '− −' causes 'count' to be decremented immediately after the evaluation so that the index value is correct for the arrays. A statement such as 'while (count + +)' would cause 'count' to be incremented before evaluation.

In the declaration of the two arrays, the term 'static' is used to indicate to the compiler the type of integer to be used. More formally, the term 'static' is referred to as a 'storage class identifier' and determines how functions, or subroutines, access those variables.

As the arrays are initialized, the code that is generated to run on the micro-processor will include a routine to load the specified values into the arrays in memory before the main program loop is executed.

Examination of the program statements shows that not only does the syntax of the C language differ significantly from assembler, but the structure of the program is also different. It can be seen that there are no jump or conditional jump commands in the program, which, unless the programmer is extremely careful, can lead to code whose path of execution is scattered confusingly about the program. Instead, the C language confines relevant code to a block which follows a conditional statement, as for example in

```
while (condition true)
{
    execute this section of code
}
```

This block structuring of a program is a typical feature of high level languages, and means that the consequential actions of a conditional statement may easily be related to the conditional statement itself. This helps to give the reader a quicker and easier understanding of the source code.

The use of high level languages in programming microprocessor based systems has now become established practice. In addition to the benefits of such programming languages, the designer of embedded real-time microprocessor based systems can also make use of executive programs which provide him with an infrastructure which has been designed specifically for operation in real-time systems. The counterpart of this executive in a general purpose computer is its operating system, which controls the loading and running of programs from commands entered into the computer by the user. Similarly, the executive controls the execution of programs, called tasks, from the occurrence of real-time events which cause interrupts to the microprocessor. The remainder of this chapter briefly describes the facilities and operation of a real-time multitasking executive, illustrated by a simple example.

14.6   The real-time multitasking executive

An executive is a program which provides a set of services to an application program which must perform a number of concurrent functions in real time. The term 'concurrent' means that the functions appear to be executed in parallel, although in reality they are executed sequentially by the microprocessor whose speed gives the impression of parallel execution.

The application program is broken down into a number of smaller programs known as tasks. Each task will perform a specific function, and will cooperate with other tasks to execute the overall function required of the program.

The executive integrates the tasks into a single real-time application program. The executive provides the infrastructure that coordinates the order in which the tasks are executed, and provides services that allow the tasks to cooperate with one another.

A real-time multitasking executive used in an embedded microprocessor based environment will typically provide the following services:

task scheduling
intertask communication and synchronization
timing
memory management.

The basic principles on which the operation of these services is based are described in the following sections, followed by a simple example which illustrates their application.

14.6.1   TASKS AND TASK SCHEDULING

A task is a program whose execution is controlled by the executive. A task is typically written as a continuous loop, and will have a point at which it receives input data from other tasks, or from input devices via interrupt routines, and, following any processing, will output data either to other tasks or to output devices. The task will then wait to receive further input data. As there may be several tasks that comprise the complete system, when a particular task is waiting to receive input and therefore not performing any useful function, the executive will start to execute another task which is ready to be executed. Each task can therefore be in a blocked state waiting for input, or be in a ready state. However, because there may be more than one task in a ready state, the executive must employ a scheduling mechanism to make sure that the tasks that are ready to be executed gain access to the CPU and are executed sequentially in a particular order based upon their relative priority. Consequently, each task can be in one of three states: executing, ready and blocked.

Figure 14.7 shows these three states together with the possible state transitions that may occur between them. When a task is being executed by the CPU it runs as would any normal program. The task will make use of functions, or subroutines, that are part of the executive to perform intertask

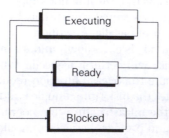

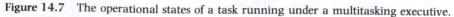

Figure 14.7   The operational states of a task running under a multitasking executive.

communication, memory management or timing functions. As part of these functions, the executive can also review the status of the currently executing task and other tasks that are ready to execute or that are blocked.

If a task is blocked, it is typically waiting for some condition to occur to unblock it, such as it receiving data. Because the intertask communications are performed as functions within the executive, when a task receives all the necessary information required to unblock it, the executive can change the status of the task from blocked to ready. If there is more than one ready task waiting to be executed, the executive must schedule the ready tasks to reflect the importance of their execution so that the most important task is the next task to be executed. If the execution of this task is more important than the currently executing task, then the executive may also perform a task switch, where the highest priority ready task starts execution and the previous executing task joins the other ready tasks.

For example, consider a microprocessor system controlling a power plant in which an error condition occurred. This event would be signalled to the microprocessor via an interrupt input, and the interrupt service routine would then send information to the error handling task identifying the type of error that had occurred. Under normal operation of the system the error handling task would be blocked, but the receipt of data from the interrupt service routine would cause the executive to change to a ready state. The error handling task should be assigned the highest priority, such that when the executive reschedules the ready tasks, it will be the next task to execute.

In order to perform task rescheduling and task switching functions, the executive must know some specific information about the tasks. For example in order to perform a task switch and start executing another task, the executive must know the start address of the task. Also, if the executive is to perform rescheduling of the ready tasks, it must know the relative priority of each ready task. Information such as this is contained in an area of memory called a task control block (TCB). Each task will have a TCB which is created by the executive during its initialization and placed in a queue called the ready queue. The ready queue corresponds to the collection of those tasks which are ready to execute, and during rescheduling it will be reordered to place the highest priority ready task at the top of the queue.

Task rescheduling and task switching are functions which are not normally visible to the programmer, but are instead implicit in the various executive functions which are available to the programmer. Not all executive functions will cause a task reschedule or task switch to occur. Typically, a task will execute until it must wait for further input data, or until another task is made ready which has a higher priority than the currently executing task. Hence, executive functions which cause the task to wait for input data or synchronization events will cause it to become blocked, at which point the executive must also reschedule the ready queue and perform a task switch. However, functions such as memory allocation and deallocation are not expected to change the operational status of the task, and so the executive would not be required to reschedule the tasks as part of these functions.

14.6.2   INTERTASK COMMUNICATIONS AND SYNCHRONIZATION

(a)   SEMAPHORES

Cooperating tasks which execute sequentially must have a means of synchronizing their operation. A mechanism implemented by executives to achieve this is the semaphore.

The implementation of semaphores differs between executives, but their fundamental usage is as flags to indicate to a task that specific events have occurred in another task. For example, the results of calculations performed in one task, say task1, may be used in further calculations in another task, say task2. Consequently, task2 cannot be executed until task1 has generated its results, which could be indicated to task2 by means of a semaphore. Task2 would therefore be in a blocked state until task1 sets the semaphore. Figure 14.8 shows the basic operation of task1 and task2. Here a, b, c, d and e are program variables and SEM1 and SEM2 are the names given to semaphores which are initially in an idle state (i.e. not set). If task2 attempts to execute before task1, then task2 must be forced to wait until task1 has calculated the value c before it can read the value c. Task2 calls the executive function 'wait semaphore (SEM1)' which, because the semaphore has not yet been set, causes it to cease execution and be placed in the blocked state. A task switch therefore takes place and so task1 starts to execute. Task1 then performs the calculation $c = a + b$, stores the result in an area of memory which can be accessed by both tasks, and then sets the semaphore SEM1 using the executive function 'set semaphore (SEM1)'. The consequence of this is that task2 is no longer in a blocked state, and the executive will reinstate is TCB on to the ready queue and reschedule it.

At this point, the executive may or may not perform a task switch depending on the relative priority of the two tasks. If the priority of task2 is less than task1, then no task switch occurs; task1 will continue to run until it executes

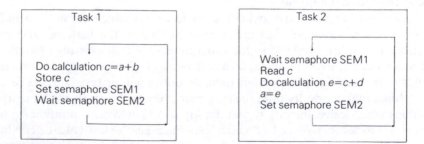

Figure 14.8   Two tasks that use semaphores to synchronize their execution.

the 'wait semaphore (SEM2)' function, therefore forcing itself to be put in a blocked state awaiting further execution, and causes task2 to start execution. However, if task2 has a higher priority than task1, then the executive does perform a task switch on the 'set semaphore' function, and task2 starts to execute immediately.

As task2 executes, it reads the value c calculated by task1, performs the calculation $e = c + d$, stores the result back into a, and sets the semaphore SEM2 which causes task1 to become ready to execute again. The execution of task2 then loops back to its beginning again and waits for semaphore SEM1. The cycle then repeats itself continuously, performing the calculations as described above.

(b) MAILBOXES

In the above example, semaphores were used to signal the fact that tasks had reached certain points in their execution. Data was passed between the two tasks simply by task1 writing the variable c to an area of memory that could be subsequently read by task2. Having stored the variable, task1 then used the semaphore SEM1 to indicate this event to task2.

Passing data between tasks is a common requirement, and an executive will normally implement this function through a mechanism called a mailbox. As its name suggests, a mailbox is an area of memory under the control of the executive into which tasks can post mail messages or data, and from which tasks can subsequently read the information. As is the case for semaphores, tasks which wait for data to arrive at an empty mailbox are put into a blocked state, and the action of a task that sends data to a mailbox which is being waited on by a blocked task will cause the blocked task to become ready.

Task1 and task2 in the previous example can be rewritten to make use of mailboxes as shown in Fig. 14.9. Instead of semaphores, task1 uses mailbox1 to pass the variable c to task2, and task2 uses mailbox2 to send the variable e back to task1.

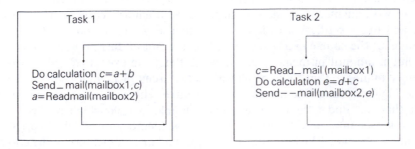

Figure 14.9   Two tasks that use mailboxes to pass information to each other.

Mailboxes are often implemented in an executive as a queue, so that the mailbox may hold more than one mail message at a time. Typically the message queue will be operated on a first in, first out principle, and the maximum number of mail messages will be limited in order to control the size of the queue.

14.6.3   TIMING

Most executives make use of a real-time clock to provide timing functions for the application tasks, and for functions which are internal to the executive.

Timing functions are driven by a hardware timer within the target system, which is initialized by the executive to continuously generate interrupts with a fixed period. The hardware timer will interface to the executive through an interrupt service routine, which may then call a specific executive function which notifies it that a timer interrupt has occurred, and therefore that a specific period has elapsed. The executive may create a number of software timers from memory registers that it can initialize with count values, and may update each active counter when it is informed that a timer interrupt has occurred. If the executive finds that any of the active timers have timed-out, it can arrange to set a semaphore associated with the timer in order to put a blocked task into the ready state. Tasks can therefore make use of several software timers under the control of the executive, each of which is synchronized to a single hardware timer.

A common use of timers is to provide a time-out recovery function to act in a similar way to the watchdog described in section 12.5.2. When a software timer expires, it will cause a specified blocked task to become ready and then execute.

A special use of the timer by the executive is to implement a form of task scheduling called time slicing. This method is used to execute each ready task having the same priority for a fixed time. The tasks are executed one after the other in round robin fashion.

14.6.4   MEMORY MANAGER

An embedded microprocessor system will be designed to have a certain amount of read/write memory. In addition to the fixed memory requirements of each task and the executive, a task will need to use varied amounts of memory throughout the course of its execution. Because not all tasks require to use the maximum amount of memory at the same time, the executive will employ a memory manager that dynamically allocates areas of memory for use upon request by each task. The memory is a resource which must be shared by the tasks that run under the executive. For example, a task which receives data from a sensor may need to create an array in memory in which it can store the time series data. The task must first request a block of memory of the correct size from the executive, which will then allocate the memory for use exclusively by the task. Once the task has processed the data and no longer requires the memory, it can return the memory block for use by other tasks. Should no memory be available, the task must wait until sufficient memory is returned by other tasks before continuing.

The organization of the memory by the memory manager will vary between different executives. A common approach is to create a memory pool, where the pool is typically divided into a number of fixed size blocks in order to prevent it from becoming fragmented when variable sized blocks are allocated and subsequently returned. However, fixed size memory blocks can lead to uneconomical usage of memory, as for example when a block of 50 bytes is used to store only a single byte character. To overcome this, many different memory pools may be defined, each of which can contain different fixed size blocks which correspond in general to the storage requirements of the tasks.

An executive must be flexible in its use for different applications, and for different configurations of target hardware systems that have different memory maps and different hardware components such as UARTs, timers and interrupt controllers. The executive typically consists of a core which implements the above features, and a configuration program in which all application dependent features are defined. This configuration program will be run during the initialization of the system and will typically perform the following functions: the creation and initialization of mailboxes and semaphores for the application tasks; the creation of a task control block for each task which is then placed on the ready queue; the organization of memory pools; the creation of any software timers; and the initialization of any hardware components. The executive will then start to execute the highest priority task on the ready queue.

14.6.5   AN APPLICATION EXAMPLE

The following description is a simple example of how an executive may be used in an edbedded real-time application.

(a)   SYSTEM HARDWARE

A microprocessor system, shown in Fig. 14.10, is to be used to control a conveyor belt. The major task of the microprocessor system is to control the speed of the conveyor belt. The motor speed is controlled from the D/A port of the micro-processor system via the motor drive unit, which provides the necessary signal conversion to drive the motor. The speed of the motor is measured with a shaft encoder which is attached to the opposite end of the conveyor and provides a stream of pulses whose frequency indicates the speed of the conveyor belt. Frequency measurement is performed using the gated counter technique as described in section 12.3.1.

The speed control of the conveyor belt is a closed loop operation, whereby the microprocessor continually monitors the speed of the conveyor belt and

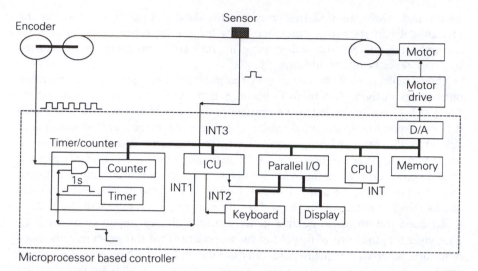

Figure 14.10   A microprocessor controlled conveyor belt system.

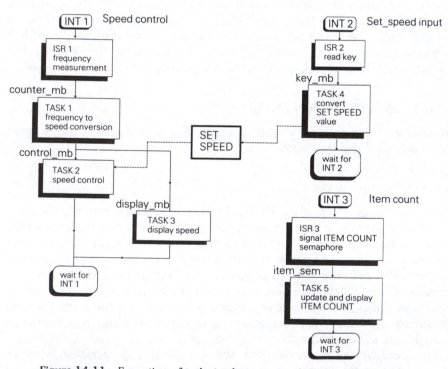

Figure 14.11   Execution of tasks in the conveyor belt control example.

adjusts the output to the D/A converter accordingly in order to maintain the desired speed. The actual speed, and the desired speed which can be input from the keypad, are output to the display. Both the keypad and the display are interfaced to the microprocessor using a parallel I/O port, and depression of a key is indicated by means of an interrupt to the CPU via an interrupt control unit (ICU).

The items that travel on the conveyor system are counted using an optical detector positioned in the middle of the conveyor belt. Whenever an item passes in front of the detector it causes a single pulse to be sent to an interrupt input of the ICU, which in turn causes a CPU interrupt and consequently a software counter to be updated and displayed.

(b) SOFTWARE

The system performs a number of tasks which must cooperate in order to achieve the overall function required of the system. A top down approach may be used to split the system operation into a number of convenient tasks. At the highest level the system must control the speed of the conveyor belt and count the number of items that it conveys. The speed control of the conveyor may be decomposed into a number of tasks as follows:

speed control of conveyor belt
    (measure the output frequency from the shaft encoder: this produces a pulse
       train whose frequency is directly proportional to shaft speed
    convert this frequency to a speed
    calculate speed control output
    display speed)

In order to calculate the speed control output, the task will compare the measured speed to a desired set speed which can be changed at any time by the operator. Therefore a further task must be defined for speed control, that is to enter a new set speed value from the keypad:

set speed input
    (read key
    enter new set speed value)

The task of counting the number of items can be simply described as:

item count
    (update and display count)

Figure 14.11 shows the relationship between the tasks and the mailboxes used to transfer information between them. The speed control loop is activated by an interrupt from the timer on INT1. The interrupt service routine ISR1 then reads the counter value, COUNTER, which contains the value of the shaft encoder output frequency. The ISR will reset the counter, pass the value

COUNTER to the counter mailbox, counter mb, and then return from the ISR. Task 1 will then become ready because of the data placed in the counter_mb, and start to execute. It reads the contents of the counter_mb and converts it to a SPEED value suitable for processing by the speed control task. Task 1 places this value in both the control_mb and the display_mb which causes both task 2 and task 3 to become ready. In terms of real-time control, the execution of speed control will have more importance than the display of the speed, and so task 2 will be assigned a higher priority than task 3 during the initialization of the system. Task 2 will therefore execute next, and in doing so will make use of the variable SET SPEED in order to calculate the control output for the motor drive. It is not necessary to pass this variable through a mailbox as this data is not responsible for activating a task. Instead, its current value is used each time task 2 runs, and must only be updated by task 4 whenever a new value is entered at the keypad. When task 2 has finished, task 3 will start to execute and display the speed value. (In real applications, task 1, task 2 and task 3 may be combined into a single task because their execution will always occur sequentially. They have been described separately in this example to illustrate the possible interaction between the tasks.) When task 3 has finished, all tasks in the speed control loop are in a blocked state waiting to be triggered by further input to their respective mailboxes. The control loop is then started again by the occurrence of another timer interrupt on INT1.

Data entry from the keypad is driven by an interrupt service routine in response to INT2. Although the operator may decide to enter a new set speed value at any time, its update will not be as time critical as the execution of the speed control loop, and so the priority of task 4 will be set to a lower value than those tasks in the main speed control loop. ISR2 will be executed each time a key is depressed, and the KEY values queued up in the key_mb mailbox. The executive will not allow task 4 to execute until the main speed control loop has finished because of its lower priority. When task 4 does execute, it reads the key values from the mailbox and stores them in a BUFFER until it detects that the ENTER key has been pressed. It then converts the value stored in BUFFER to a set speed value and stores this value in an area of memory where it may also be accessed by task 2.

Task 5, which counts items on the conveyor belt, is also initiated by an interrupt, INT3. Only the fact that an interrupt has occurred need be signalled to task 5 as this is sufficient indication that an item has passed the detector. ISR3 therefore makes use of a semaphore instead of a mailbox to trigger task 5, and when it is executed, updates software counter, ITEM COUNT, and displays its value.

Figure 14.12 shows how the individual tasks and ISRs may be implemented. In addition, it shows the functions that would be performed by the configuration program executed during the initialization of the system. When the TCBs are created, they are immediately placed in the ready queue. Therefore, when the configuration program completes its execution, the executive will start to execute

ISR1 (Frequency measurement)

```
disable timer interrupt
stop timer
sendmail(counter_mb, COUNTER)
zero counter
return from ISR1
```

TASK1 (Speed conversion)

```
COUNTER = readmail(counter_mb)
convert COUNTER to SPEED
sendmail(control_mb, SPEED)
sendmail(display_mb, SPEED)
```

TASK2 (Speed control)

```
SPEED = readmail(control_mb)
compare SPEED and SET_SPEED value
calculate new CONTROL value
output CONTROL value to motor
start timer
enable timer interrupt
```

TASK3 (SPEED display)

```
SPEED = readmail(display_mb)
convert SPEED for display
display SPEED
```

ISR2 (Read key)

```
disable interrupts
read key
KEY = key value
sendmail(key_mb,KEY)
enable interrupts
return from ISR2
```

TASK4 (Enter set speed value)

```
KEY = readmail(key_mb)
if KEY = ENTER then do
     convert BUFFER
     update SETVALUE
else do
     display KEY
     store KEY in BUFFER
```

ISR3 (item count)

```
disable interrupts
set_semaphore(item_sem)
enable interrupts
return from ISR3
```

TASK5 (Update and display item count)

```
wait_semaphore(item_sem)
increment ITEM COUNT
display ITEM COUNT
```

CONFIGURATION PROGRAM

```
create mailboxes:    counter_mb
                     control_mb
                     display_mb
                     key_mb
create semaphore:    item_sem
create TCBs:         TASK1, priority 10
                     TASK2, priority 10
                     TASK3, priority 10
                     TASK4, priority 5
                     TASK5, priority 5
initialise CONTROL = 0
output CONTROL to motor drive
initialise SPEED = 0
display SPEED
initialise SET_SPEED = 0
display SET_SPEED
initialise BUFFER
initialise timer
enable interrupts
```

Figure 14.12 Algorithm for tasks and interrupt service routines in the conveyor belt control example.

the task at the top of the ready queue. However, the system is initialized into an idle state, so the first action performed by each task is to either read a mailbox or wait for a semaphore. As each task executes it is immediately placed in a blocked state waiting for information. The speed control loop is then triggered by an interrupt from the timer, and the 'item count' function is triggered by an interrupt received from the optical decoder.

Chapter 15

Communications

Digital communications have become, and will continue to be in the future, an important component of control system technology. In this chapter some of the fundamentals of a digital communications system are introduced, with particular concentration on the computer networking techniques used in large control systems.

A digital communications system provides information transfer between two or more remote digital processors comprising a distributed control system. Concentration thus far has been on single processor systems exercising comparatively simple control functions. Control systems exist today which are considerably bigger in both physical size and complexity. Instead of using a more powerful single processor system to implement the control system, the control functions are divided between a number of more simple processors; these, together with a communications system that allows the individual processor systems to communicate with one another, comprise the distributed control system.

As an example, consider an automated manufacturing cell. This may consist of a number of machines, such as drills and lathes, integrated with a robot arm to manipulate the workpiece and service the machines with different tools, and a conveyor system to supply the cell with work material. One approach to the control of such a manufacturing cell would be to provide it with a single processor system to control all the machines within the cell. Such a system can be seen to have a number of drawbacks:

1. The microprocessor system hardware is expensive because the overall processing requirements are high, forcing the use of an expensive high performance microprocessor.
2. The software is complex because of the high degree of multitasking required.
3. The single processor system has an adverse influence on the reliability of the system. A failure in processor operation causes the entire machine cell to stop operating.
4. In general, maintenance costs are higher because the hardware and software are more complex.

An alternative approach would be to provide each machine in the cell with its own local control system and to give a supervisory system the responsibility for coordinating the functions of the local controllers. The local controllers would be connected to the supervisory controller by a digital communications system which allows the supervisory controller to send commands to each of the local controllers and to monitor their status. The advantages of this approach are:

1. Each of the local controllers requires less performance than the single processor solution because they are not required to implement so many functions. The hardware may therefore be implemented by lower cost microprocessors.
2. The software in each local controller is not as complex as in the single processor case and so is easier to write and maintain.
3. A failure in any of the local controllers has an effect only on that part of the manufacturing cell which it controls. However, it can still be the case that a failure in a critical area may cause operation of the cell to cease. System reliability may be improved by the use of fault tolerant techniques, such as dual redundancy at critical areas of the system, but the cost of the implementation of these techniques has to be weighed against the cost of failure of the normal system.

The first approach described is often referred to as a centralized processing system. The second, where the controllers are distributed throughout the system to points where they are close to the locality of their control, is known as a distributed system. Because many controllers are based on microprocessor technology, the term 'distributed' also implies distributed intelligence or processing capability under programmed control. Hence, a distributed control system may consist of a number of microprocessor based controllers and computers which must be capable of communicating with one another in order to exchange command and control information. The communications system is therefore a vital part of a distributed system and its performance and characteristics will have a direct influence on the overall performance of the control system.

The remainder of this chapter considers the requirements of the communication systems, how they are specified and some of the services they provide to the control system. Finally, some examples of communication systems currently used in control systems are presented.

15.1   Control and communication system hierarchies

A large or complex control system will contain a hierarchy of control functions. For example, a general view of the control hierarchy of a manufacturing system is shown in Fig. 15.1. At the top of the hierarchy are those systems which

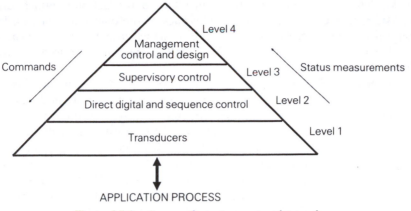

Figure 15.1   A manufacturing system hierarchy.

support the global management functions of the manufacturing process, for example work scheduling programs, inventory control and performance monitoring routines. These provide command inputs to the systems at the level below and receive status reports concerning the operation of these systems, from which they can generate information on the performance of the process as a whole and, where necessary, indicate how its performance may be improved. This level also interfaces to other engineering systems, such as computer aided design.

The level below, the supervisory control level, is concerned with the high level control functions which are responsible for the global control of the system. In the context of the totally automated manufacturing system, the design systems will produce information which will instruct the individual machines, or manufacturing cells, on how to make the required component. In addition, information will be generated on what material is to be used, where it can be found, how it can be delivered, in what order and where the final product is to be stored. Supervisory controllers will ensure that the necessary control information is delivered to the machines and that the progress of the product through its various manufacturing stages is controlled and monitored.

The next level down in the hierarchy, direct digital and sequence control, is concerned with the low level control functions of the system. Direct digital control refers to the digital control of analogue variables, such as temperature or position, whereas sequence control exercises control over logic variables which are either on or off. In either case, the control may involve a transducer which senses the variable being controlled – for example, temperature or position – and this information is used by the controller to adjust its output, via some kind of actuator, in order to make the controlled variable equal to a command value. This type of control is referred to as closed loop because the flow of information, from the controller, to the actuator which drives the system, to the transducer and finally back to the controller, forms a loop. Alternatively, the control may be open loop, in which case the controlled variable is no longer sensed and the controller simply outputs a value to an actuator which drives the system.

At the bottom of the hierarchy, the transducers provide the interface between the control system and the application process which it controls. Both analogue and digital transducers are common (Chapters 2 to 6), and the trend towards and increasing popularity of intelligent transducers –transducers with integrated microprocessor systems use to enhance their performance and functionality – has the advantage that the intelligence becomes distributed to the interface between the control system and the process, machine or system which it controls. This allows many of the signal processing functions described in Chapter 8 to be implemented within the transducer where they are most effective.

From this somewhat simplistic description of the manufacturing system hierarchy, it can be seen that the order of the levels in the hierarchy defines a command structure running from the top to the bottom and a flow of applications information from bottom to top. In a distributed system, digital communications allow this information to be efficiently transmitted between systems at the same level within the hierarchy as well as between systems at different levels.

The nature of the functions performed at the different levels in the hierarchy gives rise to correspondingly different characteristics for the data that is to be communicated between the computers and controllers. The information transfer at the highest level, level 4, would typically consist of occasional file transfers between computers and some terminal activity where an operator is connected via the communications system to a computer. The characteristic of the data transfer across the communications system is therefore one of infrequent transmission of large amounts of data – file transfers – and the more frequent transfer of small groups of characters associated with terminal access. Both types of data transfer can be considered to occur randomly in time and, where these types of data are not involved in any real-time process, the transport of the data across the communications system is not required to be deterministic in the sense that the maximum delay experienced by the data in its transfer between two machines does not have to be guaranteed.

At level 3 the characteristics of the data transfers would be similar, but with the exception that the data is now loosely real-time dependent as it is to be used in supervisory control functions. There are varying degrees of real-time dependency, as outlined in Chapter 9. The term 'loosely' in this context implies that although the data is used for real-time control functions, its time dependency is not a demanding performance requirement of the system; operation outside the desired time-out limits causes a degradation in system performance and not a catastrophic failure.

Progressing down the hierarchy, the trend is for the data transfers to become more frequent and to be increasingly real-time dependent. In addition, the majority of data transfers at the lower levels will consist of small amounts of data representing measurements, command or status information and may be transmitted periodically owing to the sampled nature of digital control systems.

It is clear that the requirements for data transfer between different levels of the hierarchy place differing demands on the communications system. In addition, consideration of the cost of the communication system in comparison with the cost of the equipment it interconnects imposes a different set of requirements. For example, the cost of a communication system for a number of minicomputers is considerably higher than that which is reasonable for a communications system to interconnect a number of small logic controllers. In response to this, a hierarchy of communications systems is provided which is capable of meeting the differing requirements of cost and performance at each level.

Because much of the equipment used to implement control systems is based on microprocessor and computer technology, the communication systems techniques that are employed in distributed control systems are based on those used in distributed computer systems. In particular, local area networks are frequently used as the communications system in distributed control systems.

15.2   Local area networks

Local area networks (LANs) allow a group of computers spread over a local geographic area to be interconnected in order to share each other's information and resources. Many of the standards associated with LANs define 'local' such that the length of the LAN cable is no greater than 1 kilometre, and in practice they are used mainly within a building or factory or in and between a group of buildings on the same site.

Physically, the LAN consists of a cable and an interface for each computer or system connected to the cable. The transmission of data over the cable is in a serial format, with data rates typically of the order of 10 megabits per second (10 Mbps). The serial to parallel conversion of the data is performed by the interface.

The transmitted data will contain two types of information. The first of these is the information which one computer wishes to send to another. Referred to as application data, its nature is dependent on the type of application in which the computers are involved. For example, this type of information may be data files, program files, terminal data or perhaps real-time data. The second type of information is referred to as protocol data and is that used by the interface to control the transfer of the application data between the computers. The protocol is the set of rules by which the interface controls the access of the computer to the LAN and the flow of information across the LAN.

A significant reason for the popularity of LANs is the standardization of their protocols and the concept of open systems interconnection (OSI), whereby LANs may be constructed and made operational with the minimum of effort from different equipment purchased from different vendors, but with each interface conforming to the same standard protocol or set of protocols.

15.2.1 STANDARDS AND THE COMMUNICATION SYSTEMS REFERENCE MODEL

The LAN protocol standards are therefore extremely important in that they provide a comprehensive definition of the way in which a protocol should operate. As mentioned previously, it is impossible to define a single protocol which meets the requirements of all applications. Instead, in order to provide a standard interface architecture, an OSI interface will consist of several different protocols, each of which has a distinct set of functions to perform and may be chosen from a set of associated standards in a way that best meets the needs of the application.

The organization of these various protocols can be conveniently represented as a stack in which the standard protocols are layered on top of each other, as shown in Fig. 15.2. This is known as the seven-layer reference model and was developed by the International Standards Organization (ISO) as a general representation of a communications system to be used as a framework for the definition of compatible standard protocols to achieve open systems inter-connection (OSI). The objective is to provide a mechanism which will allow a variety of different systems from different manufacturers to be interconnected with one another for the purpose of communications.

The main idea behind the layering is that each layer adds to the functions or services provided by the lower layers. Therefore the uppermost or application layer in Fig. 15.2 is capable of providing the full set of communications functions to the system to which it is connected. In addition, the independence of each layer is maintained by defining the interface between layers such that this interface is independent of the way each layer is implemented. As a consequence of this, each layer hides the details of its own operation and of the lower layers from the upper layers.

As a simple example, assume that a layer provides the single service of error detection and recovery of data which has been received incorrectly or corrupted by noise during its transmission over the network. The bad data will be passed to this layer from some layer below. On receipt of the data the layer will proceed to execute an error detection algorithm on the data. If no error is found, then the data is passed on immediately to the layer above. However, if an error is found, the layer initiates the sending of a request to the transmitter of the bad data for it to resend the data. Only when the layer detects that the data transmitted is error free does it pass it on to the layer above. Consequently, this higher layer may assume that the information it receives is error free and so

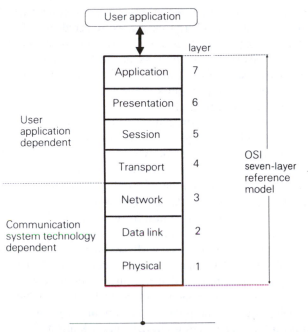

Figure 15.2   The OSI seven-layer reference model of a communications system.

need not concern itself with providing further error detection services. Similarly, at the applications layer, for example, an application program may request the transmission of a file to another system and then assume that it has been successfully received unless it is otherwise notified, indicating some reason why the file was not transmitted.

In the following description of the seven layers, a general communications system is assumed, after which a return will be made to the more specific topic of LANs used within mechatronic systems. The seven-layer reference model recommends the functions and services to be provided within each layer.

Physical layer   This layer is the lowest layer of the model and is concerned with all aspects of the physical interconnection of the communications interface to the cable. For example, the standards will define the mechanical aspects such as the type of connector to be used, the electrical levels of signals transmitted and received via the cable, and functional and procedural aspects such as the type of handshaking to be used. Interestingly, the reference model does not extend down to the physical medium itself and so does not include a specification of the type of cable to be used.

Data link layer   This layer is concerned with the functional and procedural means of data transfer over the network. The layer therefore defines several important aspects, such as the way the serial data, consisting of the protocol data and application data, is organized (also referred to as framing), and the detection and possibly correction of errors that occur at the physical layer.

Network layer   This layer has the task of masking from the transport layer all of the characteristics of the transfer medium, for example whether it is a telephone network, a satellite link or a LAN. In the case where several networks are interconnected, the network layer ensures that the data is routed correctly in order for the data to arrive at its destination.

Transport layer   This layer provides a transparent data transfer service between systems connected to the network. It builds on those services provided by the network layer to give a reliable and efficient service to the higher protocol layers. The boundary between the transport and network layers marks the division between those services which are dependent on the communications technology used and those services which are dependent on the user of communications system.

Session layer   This layer is involved with establishing the interactions between two user applications on different systems connected to each other by the network. Anyone who has used a mainframe or minicomputer will know that the first thing a user does is to log on in order to gain access to the sevices provided by the computer. In a similar way, two user applications must first establish a connection via the network and then interact with one another in some defined way.

Presentation layer   The main function of the presentation layer is to provide an independence to the user application from differences in the representation of data. Hence differences in the way that computers talk to one another (the syntax) and in the actual data can be resolved.

Application layer   This layer provides a set of services which can be directly called by application programs. A common set of services called common application service elements (CASE) is defined for general usage as well as a number of application specific services such as file transfers and terminal operation.

The first three layers of the seven-layer reference model mask the specific details of operation of the communications system technology from the higher layers. Therefore the transport layer (layer 4), for example, is unaware of the type of LAN employed, whether optical fibre or copper cable is being used, or if errors have occurred during the transmission of data. Layers 1 to 3 of the reference model are therefore dependent on the communications systems technology, while the operation of the upper layers 4 to 7 depends on the application program running on the system connected to the network (see Fig. 15.2).

15.2.2 LAN STANDARDS

The three main types of LAN available today are carrier sense multiple access with collision detection (CSMA/CD: more commonly known as Ethernet), token

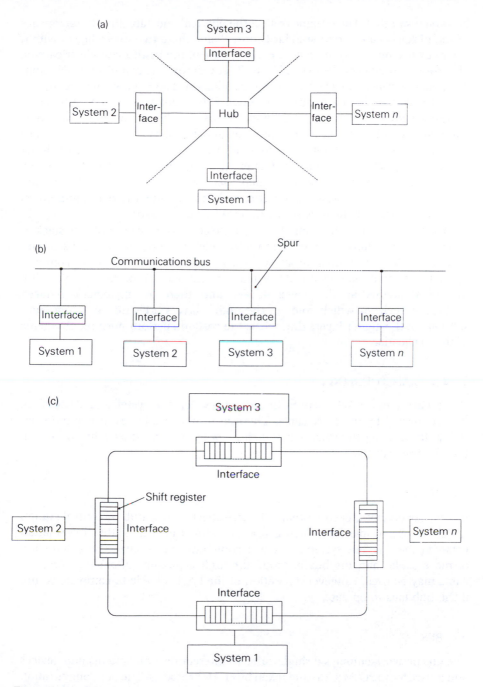

Figure 15.3 (a) A star topology communications system (b) a bus topology communications system (c) a ring topology communications system.

bus and token ring. These names reflect the physical and data link characteristics of each LAN because their standards only define these two lower layers within the reference model, as for example in the case of the Institute of Electrical and Electronics Engineers (USA) IEEE 802 standards. A complete LAN communications system may be built up by the addition of higher layer protocols.

A benefit of having hardware and software that conforms to standards is that it has a high potential for acceptance by users who know that compatible equipment can be bought from a number of different vendors. A consequence of this has been that semiconductor companies have been willing to invest in the manufacture of chips that support these standard protocols. Today, it is usual to find chips that implement part or all of the physical and data link layer protocols for the LANs mentioned above, while the remaining layers are typically implemented as software on a microprocessor.

In reality, the implementation of a complete seven-layer protocol stack is expensive and requires a lot of memory in which to store the protocol software together with a fast microprocessor with which to execute it. A common approach in many low cost LAN systems is to make use of chips conforming to the standards at the lower layers and then to implement reduced protocol – stacks in which some of the higher layers are null – or proprietary software at the higher layers that meets the cost and performance requirements of the application.

15.2.3   LAN TOPOLOGY

The topology of a LAN refers to the way in which the systems or devices are interconnected by the LAN cable. There are three topologies in common use today, these being the star, bus and ring configurations shown in Fig. 15.3a, b and c respectively.

(a)   STAR

In the star network, each system is interconnected by a hub or central station which is responsible for relaying and routeing the transmitted information between the different systems. Such a configuration requires the installation of more cable than the bus or ring, although a mixture of different types of cable may be used. However, operation of the LAN is liable to complete failure if the hub fails to operate.

(b)   BUS

The bus network employs a single cable to interconnect all systems and devices and is used for the CSMA/CD and token bus LAN protocols. The bus configuration is sometimes referred to as a multidrop bus on the grounds that a common way of connecting to the bus cable is through a spur dropped from the main cable to the interface. The bus is therefore a passive medium and the failure of any system or device to operate should not prevent other devices or systems from communicating.

(c) RING

The ring network, used in the token ring LAN, creates a loop by connecting each system or device to its neighbour. The interface is responsible for relaying the information around the ring and so is part of the physical structure of the ring. Operation of the ring therefore relies on the correct operation of each interface. As the interface forms part of the ring it is an example of an active medium and requires the use of failsafe measures to bypass faulty interfaces in order to keep the ring intact.

15.2.4 LAN FRAME STRUCTURE AND MEDIUM ACCESS TECHNIQUES

Bus and ring topologies require systems to share the medium – or cable – in order to communicate. The LAN protocols therefore include a medium access control (MAC) protocol which defines the way in which each system gains access to the medium.

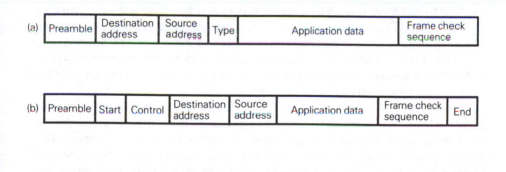

Figure 15.4 Frame formats for (a) CSMA/CD (b) token bus (c) token ring LANs, with (d) token ring token.

Operation of the CSMA/CD, token bus and token ring MAC protocols can be described as the sequence of events that occurs when system 1 in Fig. 13.3b or c wishes to send data to system 2. The data is serial and is formatted into a frame in a way which is specific to the requirements of the protocol. The frames of information for the three types of protocol are arranged as shown in Fig. 15.4. For the CSMA/CD and token bus protocols the frame starts with a bit pattern called a preamble which, when transmitted on to the bus, allows the receivers in the interfaces to lock on to and synchronize their receiver clocks with the transmitter clock in the system 1 interface. This is made possible by making the preamble a series of alternate 1s and 0s such that their frequency is equivalent to the data rate or clock frequency.

Each system or interface is uniquely identified by an address, and the frames contain both the destination address – the address of the receiving system – and the source address – the address of the system transmitting the information. Fields within the frame such as the access, type and control fields contain information concerning the control and status of the MAC protocol. A frame check sequence allows the majority of errors that occur during the transmission of the data to be detected. In the token bus and token ring frames, additional fields delimit the start and end of the frame. The application data field contains the application data which is transmitted between the two systems.

(a)   CSMA/CD

The CSMA/CD MAC is used to access a bus network. When system 1 attempts to transmit data to system 2 over the bus, it first monitors the bus to determine if any transmissions are currently in progress. If no activity is detected, the interface starts to transmit its data onto the bus. However, if another system is detected as already transmitting, then system 1 waits until it detects no further activity and then starts to transmit data.

Despite this 'listen before transmit' principle, it is still possible that two or more systems may start to transmit at the same time, causing a collision of their respective transmitted data on the bus, resulting in its corruption. The starts of transmissions by the systems do not have to be exactly simultaneous in order to cause a collision, and in fact this is determined by the propagation time of a signal along the complete length of bus. This arises from the fact that, as it takes a finite time for an electrical signal to travel along a length of cable, there will be a time delay between the instant at which system 1 starts to transmit data and the instant at which this transmission is detected by another system. The greater the distance separating the two systems, the larger will be the associated time delay, and if another system, say system n, attempts to transmit during this time, this will cause a collision of their respective data transmissions. The worst case condition occurs when system 1 and system n are at either ends of the bus cable. The time delay, or collision window, is then equal to the propagation delay of a signal travelling the full length of the bus and back again.

The reason the collision window is twice the propagation delay of the bus length arises from the way in which a transmitting system recognizes a collision. Each transmitting system receives data from the bus as it transmits, thus effectively receiving its own transmissions. In the absence of collisions this received data should be the same as that which was transmitted. However, when a collision does occur, data first travels from system 1 along the bus until it meets a transmission from system n, at which point the two transmissions collide. The resulting signal must then propagate back to the respective transmitting systems before the collision is detected. For the worst case condition stated previously, the signal must travel up and back along the entire bus length before the collision can be detected.

Should a collision occur, the transmitting systems continue to transmit for long enough to ensure that all other systems on the network know that a collision has occurred. All transmission is then stopped and each transmitting system waits for a different period, called the back-off period, before it attempts to transmit again. The back-off period is a random time interval and is chosen such that each system will attempt to retransmit at a different time, thereby avoiding further collisions by the same systems.

In the bus configuration, each system is effectively connected to every other system and so a transmission from one system will be received by all. The receiver section of CSMA/CD interface will typically make use of an address filter to ensure that it only continues to receive frames with a destination address that corresponds to that of its network address while ignoring all others. There is one exception for an address number defined as a broadcast address, which is used when a frame is intended to be received by all systems on the bus. Other functions within the data link layer protocol provide error detection and correction for the received frames.

(b)   TOKEN BUS

The token bus MAC protocol relies on the existence of a special frame, called the token, which is transmitted from one system to another in a round-robin fashion. The system receiving the token assumes the right to transmit a single frame to another system before it passes the token on to the next system. If the system holding the token has no information to send, it simply passes it to the next interface.

Although the systems are interconnected by a bus, their access to it is ordered as a logical ring, and each system must have a knowledge of the address of its successor in the ring in order to transmit the token onwards around the ring. If, for some reason, the token is not received by the successor, the interface that is passing the token on will try again until it is certain that the successor has failed and will then find the next successor system, to which it transmits the token.

The token bus interface will maintain the address of both its successor and its predecessor and will update these addresses whenever an interface is found to have failed or to have been removed from the bus. The token itself is, as shown in Fig. 15.4b, simply a frame containing no data but with a specific value in the control field that identifies it as a token. As in the CSMA/CD MAC, higher levels of protocol are used to correct for errors and control the flow of information between communicating systems.

Unlike CSMA/CD, the token bus does not allow the possibility of collisions because of the orderly access of each system enforced by the token passing technique. In addition, the time taken for a system to gain access to the bus is bounded by a limit which can easily be determined from the operational parameters of the token bus and which can therefore be designed to meet the real-time requirements of the data to be conveyed by the communications system. Consequently, the token bus is said to have a deterministic access delay. In comparison, CSMA/CD suffers a disadvantage in that its access time for a system cannot be guaranteed because of the random back-off delays caused by collisions. CSMA/CD LANs therefore have a non-deterministic access delay and are considered unsuitable to carry application data which is real-time dependent, unless the resulting degradation in system performance can be tolerated.

(c) TOKEN RING

The token ring MAC is very similar in principle to the token bus MAC in that a token frame is used to control the access of each system connected to the ring by passing the token in the same direction around the ring to each system in turn (Fig. 15.4c). However, instead of each system being physically connected to all other systems as is the case with a bus, the ring allows each interface to interconnect to only two other interfaces, and so the successor and predecessor systems are uniquely chosen by their physical arrangement within the ring. The ring token, shown in Fig. 15.4d, therefore does not need to contain the source and destination addresses as is the case for the token bus frame.

The token ring interface may only contain a few bits of storage for the information that passes around the ring. This storage is typically a shift register and, as each bit is received by the interface, it is clocked through the shift register at the data rate of the LAN. As information passes through the shift register, it can be monitored by the interface which can therefore perform functions such as address and token recognition. If the interface is receiving data from the ring, it strips the information bit by bit from the ring, making sure that it is not passed on to the following interface. An interface that is transmitting information will insert data onto the ring via the shift register in a similar manner once it has acquired the token.

The token ring MAC relies on the circulation of a single token around the ring. The token may be in one of two states, busy or free, as indicated by the state of one of the bits contained in the access field of the token frame. A system which has requested its interface to transmit data on to the ring must first wait until a free token starts to pass through its interface. When a free token has been recognized, the interface waits until the busy/free bit is in its shift register and then modifies this bit to indicate that the ring is busy. Once the remainder of the token has passed through the interface, it starts to transmit its data and, when finished, it restores a free token on to the ring.

A problem associated with the ring network is that the combined length of the shift registers for all interfaces that make up the ring must be sufficient to accommodate the token frame, because the free token must be capable of circulating freely around the ring. This problem places a limit on the minimum number of interfaces required to create an operational ring. This problem may be overcome by the inclusion of a special interface with a shift register long enough to contain the entire token.

Operation of the ring fails when a token is lost or when two or more tokens are present on the ring. The detection and correction of these conditions may be performed centrally in a special monitor interface – which may also contain the long shift register – or may be distributed as a function to be performed in each interface.

Like the token bus, the token ring has a deterministic access delay under normal operating conditions and so may be used to transport data with a real-time dependency.

15.3   A communications system hierarchy for industrial automation applications

As previously mentioned, the communications requirements of equipment at different levels in the manufacturing system hierarchy can only be met by the provision of a hierarchy of communications systems whose performance, cost and functional characteristics match those of the equipment at the different layers which it interconnects. This hierarchy of communications systems is well illustrated by three standard communications systems currently available for use in distributed manufacturing systems. These are the manufacturing automation protocol (MAP), MAP/EPA and Fieldbus. Their relationship with each other is illustrated in Fig. 15.5, which shows how these three networks may be interconnected to meet the communications requirements of a manufacturing system.

15.3.1   THE MANUFACTURING AUTOMATION PROTOCOL

The manufacturing automation protocol (MAP) is an example of a communications system based on the ISO seven-layer model, where the choice of protocols at the different layers reflects the application of the system in a manufacturing environment. The driving force behind the development of this standard has been the American automobile manufacturer General Motors (GM), who in the early 1980s realized that their manufacturing activities had to be

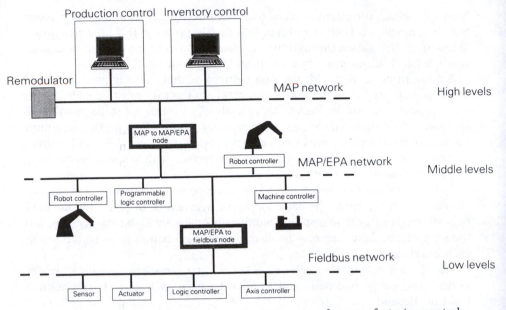

Figure 15.5   A hierarchical communications system for manufacturing control.

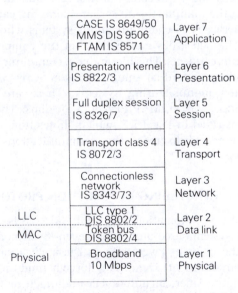

Figure 15.6   The MAP seven-layer protocol stack.

further automated in order for the company to remain competitive. This posed a major problem for them in that the equipment for their manufacturing systems was typically supplied by many different vendors, who at that time made use of a variety of non-standard communication protocols. A major drain in terms of cost and on the resources of GM engineers, therefore, was the effort required to successfully interconnect these pieces of equipment using proprietary protocol converters. This inspired them to propose a standard communications system for factory automation applications, in the hope that the standard would be adopted by the manufacturers and users of such equipment and eventually lead to lower installation and maintenance costs for automated manufacturing systems.

The seven layers of MAP are shown in Fig. 15.6. At the time of its initial specification, MAP consisted of several established OSI protocols, mainly at the lower layers, and a number of interim protocols specified by GM to fill in the gaps where no standards existed or were being developed. This was the case for most of the upper layer protocols.

The physical and data link layers of MAP use the ISO/IEEE broadband token bus LAN standard because of the need to have a deterministic medium access control method (section 15.2.4) to meet the possible real-time requirements of the data conveyed by the LAN. The broadband technique allows the bus cable to be used for services in addition to the MAP communications system, as well as providing multiple MAP channels. This is achieved by frequency multiplexing each service, including the MAP signals, into its own specific frequency channel in much the same way that signals from different radio stations are frequency multiplexed in order to share the 'air waves'. In this way, up to 30 individual channels may be carried by a single cable.

A MAP system requires the use of two channels, of which one connects the receivers of all interfaces and the other interconnects the transmitters. A head-end remodulator situated at the end of the cable then translates the signal in the transmitter channel into the receiver channel.

The broadband technology is derived from cable television and has been in use for many years. Consequently the status of the techniques and components is mature. The reasoning behind the use of broadband signalling at the physical layer is that, once a broadband trunk cable is installed, it can then carry all the necessary services required as well as MAP, thus eliminating the need to install further cables.

The data link layer of the IEEE standards is split into two sublayers called the media access control (MAC) and the logical link control (LLC). The combination of the physical layer and the MAC sublayer defines the LAN technology used, for example, broadband token ring, Ethernet on twisted pair cable, token ring on optical fibre, and so on. As already described, MAP uses the broadband token bus technology, the IEEE standard for which is contained in a single document, IEEE 802.4. The standard for the LLC layer is contained in a separate document, IEEE 802.2. The reason for partitioning in the IEEE standards compared with the OSI reference model is that the different MAC techniques used specifically by a LAN interface lie partly in the physical layer and partly in the data link layer in terms of the functions they perform.

The logical link control (LLC) layer implements functions such as framing, error checking of received frames and flow control of the transmission of frames over the network. The IEEE 802.2 standard defines two modes of operation. The first of these, type 1, is used by MAP and is a simple mode of operation where no flow control is used to regulate the flow of information between two communicating systems. The second mode, type 2, requires the LLC layer to first initialize a logical connection between the transmitting and receiving systems before transferring data with acknowledgements. These are used as a form of handshaking in order to control the flow of information between the two systems. Although MAP uses LLC type 1, it is a connection oriented communications system, where the connection service is negotiated at layer 4 of the protocol, the transport layer, instead of the LLC.

Connection oriented protocols are typically used for the transfer of large files of information between systems. Because the amount of information being transferred is comparatively large, the systems must first establish a connection which indicates that they can provide services to transfer the information over an extended duration and implement flow control to ensure that the transmitting system does not overrun the rate at which the receiving system can receive the information. In contrast, connectionless oriented protocols are typically used to transfer small amounts of information in a system where the logical connection between systems remains fixed, and so the process of establishing a connection would be an overhead and thus prevent the system from making a fast communication to other systems as a result of external events which may require a real-time response. Because MAP uses connection oriented protocols and a full seven-layer protocol stack which requires considerable processing time, it is not practical for use in real-time applications. Instead, its main function is as a communications backbone providing the communications services required at the higher levels of the manufacturing hierarchy.

The remaining layer protocols are chosen from existing standard protocols and those which have either become or are in the process of becoming standards. MAP was designed to meet the higher level communications requirements in manufacturing automation, and hence the definition of the application layer is critical as it provides the interface between the user applications program and the MAP communications system.

A range of different services may be provided by the application layer. When complete, MAP will provide common application service elements (CASE), specific application service elements (SASE), network management and a directory service. SASE includes:

1. File transfer;
2. Access and management protocol (FTAM);

3. Virtual terminal protocol (VTP), which makes a terminal connected to one system look as though it is connected to another system on the MAP network;
4. Job transfer and manipulation protocol (JTM);
5. Manufacturing message handling system (MMS);
6. Electronic mail system.

Only those standards which currently exist or are nearing completion are shown in Fig. 15.6.

It should be realized that the standards mentioned in this and subsequent sections undergo frequent revision, especially the new protocols for those layers which are more applications dependent. Some protocols, particularly those at the lower levels, are more established and have reached a degree of stability which has allowed them to be implemented as semiconductor devices. For example, Ethernet (CSMA/CD), token bus and token ring chip sets are currently available from many semiconductor manufacturers.

15.3.2   THE ENHANCED PERFORMANCE ARCHITECTURE (EPA) MAP

A full MAP communications system is unsuitable for use in applications at lower levels of the manufacturing systems hierarchy because of its relatively high cost and inadequate real-time performance. The high cost of a MAP interface is due to the fact that it uses a comparatively expensive broadband technology and implements a full seven-layer protocol stack which requires significant processing power and memory. In addition, the use of a connection oriented protocol, together with the processing time for the seven layers, results in a communications system which is generally too slow and unwieldy for use in real-time applications.

The MAP/EPA specification addresses these problems through the use of a reduced protocol stack in which protocol layers between the data link layer and application layer are eliminated. The remaining layers of MAP/EPA are shown in Fig. 15.7 and consist only of a physical layer, a data link layer and an application layer.

At the physical layer, the broadband technology of MAP is replaced by a much simpler and cheaper carrier band technique. Instead of the possibility of many channels, only a single channel is available on which the digital states (1 or 0) of the transmitted data are represented by two different frequencies transmitted with a duration equal to that of their equivalent digital states. This is more commonly known as frequency shift keying (FSK).

The data link layer MAC is the same as that of the normal MAP system. However, the LLC provides a connectionless acknowledged service instead of the more simple connectionless (type 1) service used in MAP. The difference is that the correct reception of data by the receiving system must now be acknowledged at the data link layer before the application data can proceed

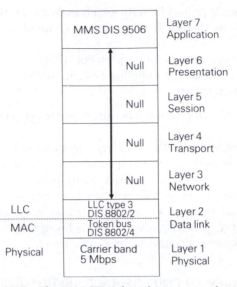

Figure 15.7   The MAP/EPA three-layer protocol stack.

to the application layer in the MAP/EPA interface. This type of service was originally defined for a communications system called PROWAY which was the forerunner of the MAP/EPA specification. Although not originally in the IEEE 802.2 LLC specification used by MAP, this is being revised to include this configuration as a type 3 service.

At the application layer, the protocol for MAP/EPA includes the manufacturing message services to support communications between factory devices. In a full seven-layer system such as MAP, the application layer would interface to the presentation layer. However, because this and other lower layers are missing from the MAP/EPA system, it is necessary to include other functions in the application layer to create a functional interface between the data link layer and the application layer.

Although MAP/EPA achieves an overall lower implementation cost and quicker response times by reducing the protocol stack to just three layers and by using a simpler physical layer, the drawback of the system is that it no longer conforms to the seven-layer OSI reference model and therefore cannot be considered as open. The interface between a MAP network and a MAP/EPA network will be a hierarchical connection facilitated by an interface called a MAP to MAP/EPA node. In the hierarchical organization of communications networks, the node must interpret requests by a system on the network for information from a system on the other network, and execute the request such that one network appears to the other as a single system and interface. This is in contrast to interworking between networks of similar functional capability, where the interface between the two networks makes one network appear as an extension of the other systems on either network can communicate directly with each other.

15.3.3 FIELDBUS

Fieldbus is the name given to a communications system standard for applications at the lowest levels in the manufacturing systems hierarchy. The Fieldbus standard is still in its early stages of development and is intended to provide a flexible communications system for devices such as sensors, actuators and low level controllers. In process control terminology these devices are commonly referred to as field mounted devices, as they are mounted on parts of the process equipment or machinery 'in the field' and as a result may also be exposed to extreme operating conditions. However, the Fieldbus standard is intended to cover a broad application area in both manufacturing and process control.

Before looking at the requirements of a Fieldbus network, it is necessary to describe how the transducer level of a large control system would currently be implemented. In a manufacturing system, the interface between a microprocessor based control system and its analogue transducers will take the form of analogue conditioning circuitry which interfaces to the analogue to digital (A/D) and digital to analogue (D/A) converters of the microprocessor system (section 12.2.5 and Chapter 8). Each transducer will be connected by its own cable to the control system over a distance typically in the range 1 to 100 m. Where electrical power is required at the transducer this is typically provided by a local supply. Digital transducers perform similar functions but interface directly to the microprocessor system through timers or parallel ports.

The situation is similar in process control systems, with the exception that the majority of transducers are powered centrally from an intrinsically safe power supply which prevents the occurrence of sparks in hazardous environments. Instead of using two cables for each transducer, one for power and the other for the signal, both functions are combined so that power and signal can be delivered to and from the transducer by a single cable, saving on cable and its installation and maintenance costs. The most popular method currently in use is the 4 to 20 mA signalling convention in which the signal to and from the transducer is scaled with an offset so that a zero measurement or command signal causes the transducer to draw 4 mA and a full scale (100%) measurement or command signal causes it to take 20 mA. In this way, at least 4 mA of current is always supplied to the transducer at a voltage which maintains the power below the intrinsically safe level.

The area covered by a process control system such as a petrochemical plant or oil refinery can be of the order of several square kilometres, and the cabling between hundreds of transducers and their controllers can total several hundred kilometres. Therefore two requirements of a Fieldbus communications system which is to be used in existing as well as new installations is that it must maintain compatibility with intrinsic safety requirements and with the star

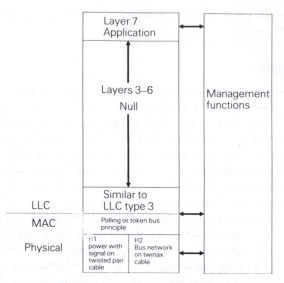

Figure 15.8   The Fieldbus architecture.

topology cabling of existing control systems, most of which use twisted pair copper cable.

A three-layer protocol stack, similar to that of MAP/EPA, is proposed for Fieldbus, with the addition of management functions to ease the problems of interfacing the data link layer to the application layer and to provide global functions for network and interface management. The system architecture is shown in Fig. 15.8.

At the physical layer, two types of interconnection are proposed. The first, called H1, has a cable length of up to 1.9 km and supports the supply of intrinsically safe power to up to five transducers which are clustered at the end of the cable. The transmission specifications of H1 will allow the use of twisted pair cable so that existing cable installations can be utilized. The second type of interconnection, H2, has a shorter length of 750 m but is to be used in a bus configuration to support around 32 devices. H2 will not therefore support power feed and so the connected devices must be powered locally. However, H2 will support faster information transfer rate and response times than H1 and will require the use of twinax cable.

The type of data link layer to be used by Fieldbus has not yet been decided. One possibility is the use of the token bus principle, which is attractive because the chips that perform this function are already available. Alternatively, a more common form of access method known as polling may be chosen. Polling is based on the idea that one of the devices on the network is the master while the others are slaves, and it is the responsibility of the master to send or request information from each slave in turn. This has the advantage that the interfaces for the slaves can be made comparatively simple, as most of the complexity will be required in the master which controls the communications on the network. Its disadvantage is that the Fieldbus communications system is vulnerable to a failure of the master device, although this situation can be improved by the use of a dual redundant master interface at the expense of increased system cost. It is expected that the LLC services of Fieldbus will be similar to those of the MAP/EPA (LLC type 3).

The main application of Fieldbus networks will be in low level monitoring and control for manufacturing and process control systems. A sensor will provide a measured variable to the control system and an actuator will respond to a given demand variable. The application layer must make the Fieldbus network invisible to the application program in each transducer or device and so must map the relationship between the different variables and the network address of the devices to which they belong. Support for a number of different data types such as text and floating point numbers will also be required in the application layer, as well as management features such as configuration – which may require configuration data to be downloaded to the device over the Fieldbus network – and maintenance functions such as fault diagnosis and repair. At the time of writing, a standard for the Fieldbus application layer has yet to be developed.

Part Three

Motion Control

Chapter 16

Drives and actuators

The simple definition of a mechatronic system (Fig. 16.1) indicates that the objective of such a system is to produce a physical effect such as a change of pressure or a mechanical movement. The effect itself will be quantified in terms of performance, for example the load capacity, speed, resolution and repeatability of a position controlled motor and toothed belt drive. The quality of the output of the system as a whole may however be judged by criteria which are more readily perceived by the user, such as the drawing quality and print legibility of a plotter. The drives and actuators which provide the means of conversion from the statement of intention, expressed as an electrical input, to the end effect will be discussed from an applications oriented point of view.

The term 'prime mover' is properly reserved for those devices or systems which convert chemical energy in fuel, or naturally occurring sources of power such as wind and water, into mechanical movement and thence commonly into electrical power. Steam turbines and diesel engines are examples of such devices.

Most industrial applications will therefore be at least one stage removed from primary energy conversion, with electrical power being the most commonly available service and the electric motor the usual means of providing a conversion into mechanical work. Also of great importance are further or intermediate stages of conversion such as hydraulic or pneumatic systems which may have a significant influence on the overall efficiency of a machine or plant. For example it may appear in a particular application that an air cylinder offers an effective means of positioning a load, and this may well be the case. However, in designing the whole system it would be prudent to take into account the losses accruing in each stage from the electrical drive to the air compressor, the compression process, the compressed air transmission system and finally the manner in which the cylinder itself is controlled. Effectiveness and efficiency may be properly assessed on quite different criteria, and efficiency of energy conversion may well be of less importance than operational effectiveness – for example speed or positioning accuracy – especially if the operation of the system is intermittent. These considerations appear frequently in the chapters to follow.

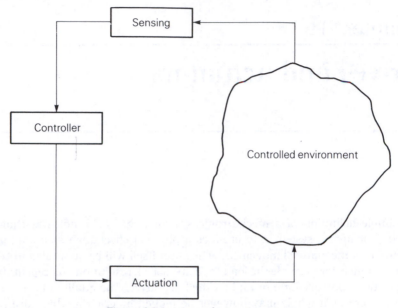

Figure 16.1   A mechatronic system.

The scope of this part of the book is shown in Fig. 16.2. At first sight the field of available services energy conversion devices and their attendant motion converters (or mechanical power transmission systems) is bewildering in its complexity, with systems of radically different configuration apparently competing to serve the same objectives. Practicising engineers make choices which are sometimes perhaps too influenced by personal experience or long standing company preference, when technological advance has opened new options or removed the drawbacks of earlier and insufficiently exploited developments.

In mechatronic applications, most sensing and control functions are applied to relatively few control parameters:

Mechanical systems   Force; position or torque; speed.
Electrical systems   Voltage; current.
Hydraulic systems   Volumetric flow rate; pressure.

It is thus logical to begin a consideration of drives and actuators by classifying the ways in which motion can be produced against the objectives of the motion, as may be set out in the system specification or statement of requirements.

The output of the actuator will be related to the measurement made by the sensor according to the algorithms and calculation processes within the controller, but subject to the limitations imposed by the processing rate of the controller and the effects of saturation if, for example, the demanded rates of acceleration are beyond the capacity of the system to provide.

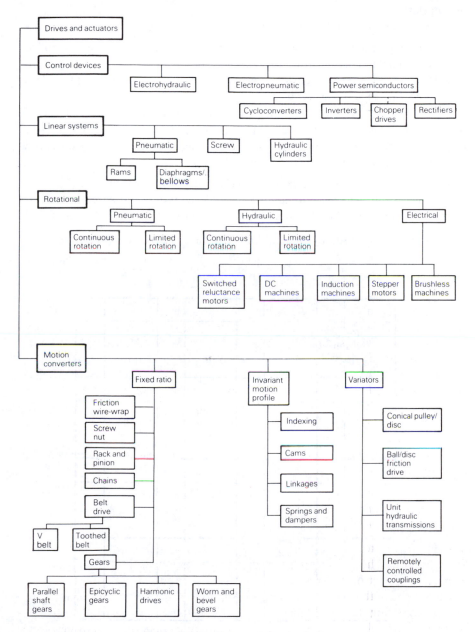

Figure 16.2   Drives and actuators.

The principal division lies between rotational and linear devices, with motion converters such as gearboxes or rack and pinion systems being added where appropriate.

Selection criteria are based first on:

- The magnitude and range of the primary parameters to be controlled. Secondary but significant consideration such as:
- Bulk, weight, cost, accuracy, resolution, rate of response. Figure 16.2 illustrates the scope of this section.

The approach can be facilitated by means of charts which, based on manufacturers' published data, show the areas of application of particular types of power, energy or motion conversion device against the primary operating parameters as follows:

Services energy converters, Linear   Based on force and distance.
Services energy converters, Rotational   Based on torque and speed.
Motion converters   Based on changing input/output speed or direction.

The charts may be used to give guidance in the following chapters.

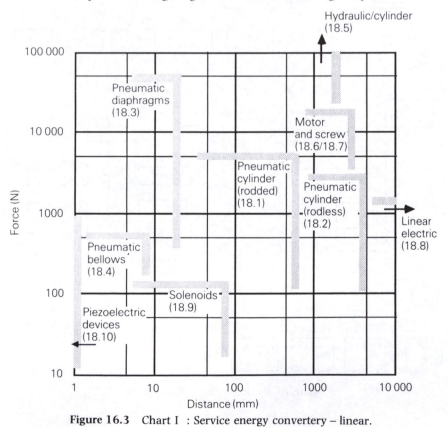

Figure 16.3   Chart I : Service energy convertery – linear.

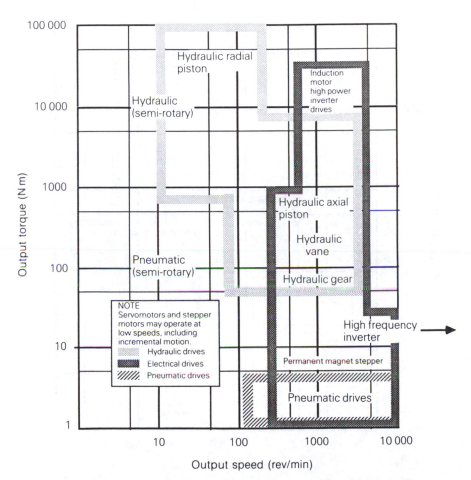

Figure 16.4 Chart II: Service energy converter – rotational.

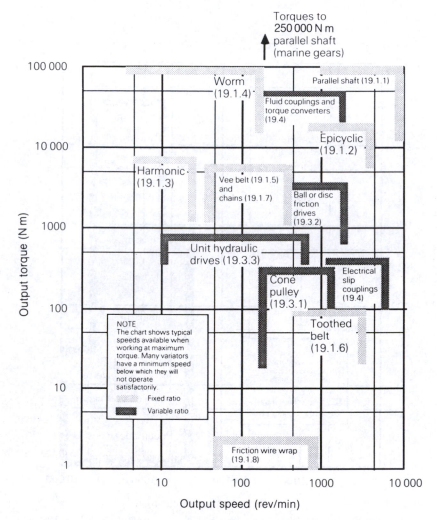

Figure 16.5   Chart III: Motion converters.

Chapter 17

Control devices

17.1 Electrohydraulic control devices

The development of reliable electrohydraulic elements interfaced to microcontrollers has been the most significant factor in the renaissance of hydraulic systems. The mechanical elements have been existence since the late 1920s, where they were first commonly incorporated as spool valves with mechanical position feedback in early versions of power steering on automobiles. Spool valve elements have been refined extensively and are used with manual input, or are fitted with solenoids for remote operation. The most common configuration is the 4/3 (4 port/3 position) directional control valve (Fig. 17.1) where in manual versions it may be possible to obtain a flow rate proportional to position. These are seen in everyday use on earth moving equipment. For electrical operation, solenoids are most commonly employed to drive the spool in light duty applications directly, and in higher pressure and flow applications through the intermediary of a pilot valve (Fig. 17.2). It may be sufficient to control the valve in a bang-bang mode, where the spool moves from the null position to the fully open position without any means of holding at an intermediate point. However, for more precise control of output parameters such as the velocity or position of a cylinder, it is necessary to have control of the spool position in proportion to the input signal.

Referring to Fig. 17.3, spool valve operating parameters are related by the following expression:

$$Q = C_d \pi D x \sqrt{[(P_s - P_1)/\rho]} \qquad (17.1)$$

where Q is the volumetric flow rate, C_d is the discharge coefficient for the orifice, ρ is the fluid density, P_s is the supply pressure and P_1 is the cylinder pressure. This is the familiar expression for the flow through a sharp edged orifice in which the viscosity term does not appear. In practice, changes in viscosity do have an effect upon the flow rate due to fluid friction losses in the port areas in the vicinity of the control orifice. Because of the very large changes in the viscosity of mineral oils with temperature it is necessary to exercise reasonably

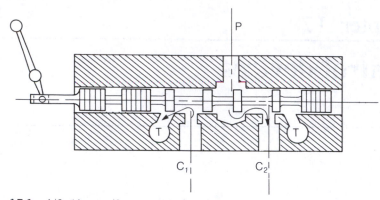

Figure 17.1   4/3 (4 post/3 position) directional control valve; where C refers to connection to cylinder, P to connection to supply passive and T refers to connection to tank return.

close control over the system operating temperature to limit variations to say ±10 °C once the normal duty cycle has been established and the system has warmed up. Measures discussed below will compensate for temperature variations, but temperature control will ease the demands made upon them.

Spool valves may be overlapped, underlapped or line-on-line as shown in Fig. 17.3, together with their corresponding characteristics. An underlapped valve can be used for the position control of a hydraulic cylinder. In this case the cylinder remains pressurized at both ends when the valve is at null position and a high degree of sensitivity is obtained, but the cylinder is not locked and will drift unless the loop is closed by position feedback. Too much overlap will increase dead band and may lead to instability. Most valves are supplied in a line-on-line spool configuration. It should be understood, however, that no

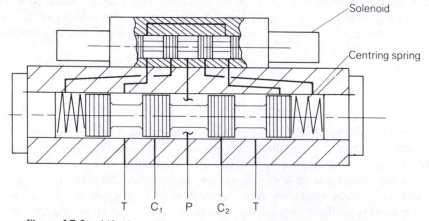

Figure 17.2   4/3 (4 post/3 position) directional control valve with pilot.

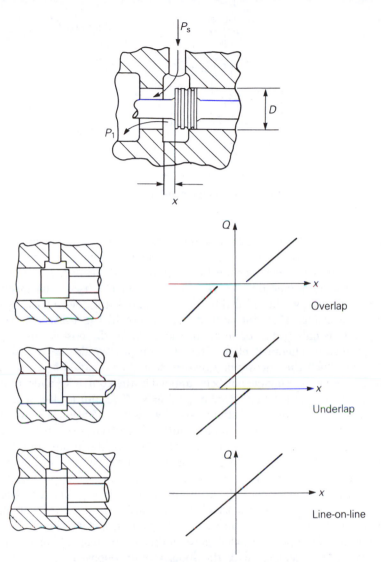

Figure 17.3   Underlapped and overlapped valves.

sealess low friction spool valve can provide positive locking of an actuator without the interposition of a leaktight shut-off valve.

The development of proportional electrohydraulic control valves to the degree of sophistication which they have reached today has been the result of the integration of a number of separate developments and inventions, some of which have led to new areas of technology in their own right.

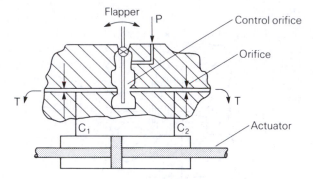

Figure 17.4   Simple flapper orifice valve.

During the 1930s pneumatic systems came into wider use in process plant for control, signal transmission and to an elementary extent for computation. There thus emerged a need for mechanical/pneumatic transducers and control valves. Pressure/momentum devices were extensively used – for example by Askania in Germany – but probably the most significant device was the flapper orifice valve of Fig. 17.4, introduced by Foxboro in the USA. This could be employed in its half bridge, or single nozzle, form for the positioning of process control valves in chemical plant. Here the output would be to a diaphram actuator in which the pneumatic pressure would be opposed by a spring.

The flapper valve principle has been applied in the control of hydraulic fluids, either directly or as the first stage of a spool valve. The spur for the development of high integrity servovalves came in the 1940s and 1950s with the increase in the speed and size of aircraft. The control forces required on flight control surfaces could no longer be exerted by the pilot via the traditional system of mechanical linkages and cables. Mechanical/hydraulic and eventually electrohydraulic systems were rapidly introduced, further encouraged by the requirements of fully automatic unmanned systems for missiles. The fairly primitive flapper nozzle devices used for directional control of the jet efflux guide vanes on the V2 missile were an early example of this development.

Preliminary attempts to provide closed loop control of spool valves employed direct acting DC solenoids, with the spool motion opposed by a spring. Thus a variation in solenoid current caused a proportional change in spool position and hence in flow rate. The loop be closed in such cases by measuring the output parameter.

Valve stiction, brought about or accentuated by corrosion or particulate contamination in the hydraulic fluid, was the Achilles heel of the single stage, direct acting spool valve. In all spool valves, the achievement of linearity of output and repeatability depends upon the maintenance of close working clearances. The smaller the valve, the more stringent do the dimensional requirements become. Tolerances on mating parts such as the spool/bushing assembiy are typically within $10\,\mu$m. Thus filtration of the working fluid is of vital importance and must eliminate all particles larger than $5\,\mu$m. Even finer filtration will be required in small systems. The limitations of the single stage valve become of special significance in safety critical systems. In aircraft controls, for example, the valve *must* move even if in so doing it suffers damage such as scoring and its performance degrades. This requirement can be regarded as the chip shearing capability of the valve, where a fragment of swarf – which should not be in the system – trapped in the control orifice is simply cut through by the sharp edges of the valve land and bushing. The forces available from solenoids were insufficient to do this or to provide low hysteresis control of a high flow spool, and therefore the two-stage valve was developed by Tinsley in the UK in 1946. Here, the second stage spool was driven by differential pressures produced by a small direct solenoid driven first stage.

At about this time, work at the Massachusetts Institute of Technology led to the incorporation of two important developments. The first was the use of a permanent magnet variable reluctance torque motor instead of a solenoid; this gave better linearity, with a reduction in the signal power needed to operate the valve. The second was the use of a spool position transducer at the first stage of the valve in order to detect spool stiction. In the mid 1950s a number of further developments by Carson, Wolpin, Atchley and W. C. Moog Jr, all in the USA, brought the electrohydraulic servovalve to the position it holds today. Moog Controls Inc. is the principal supplier worldwide of servovalves based on a flapper orifice first stage with mechanical feedback of the spool position, and has licensed the design for manufacture by other companies, principally in the UK and Germany. However, it is a tribute to the efficacy of the concept that engineers tend to refer to all electrohydraulic valves of this configuration by the generic term 'Moog valves'. In looking at servovalves in more detail it is thus logical to start with this type.

17.1.1 FLAPPER-CONTROLLED ELECTROHYDRAULIC SERVOVALVE

A sectional view of this type of valve is shown in Fig. 17.5, with an operational diagram in Fig. 17.6. The valve comprises:

1. A torque motor (essentially an electric motor with a limited and often very small angle of rotation) which receives the input signal. This typically has a DC voltage in the range $\pm 5\,°$V. The motor consists of an armature aligned between permanent magnet pole pieces and carried on a thin walled flexure tube. Direct current in the coils causes an increase in the attraction force at the diagonally opposite air gaps, thereby tilting the armature through a small angle, 0.5 degrees or less, in a see-saw fashion.
2. A pilot stage hydraulic flapper amplifier which produces an axial imbalance on the power stage valve spool. The flapper element consists of a stiff rod extending from the armature and working between two opposed orifices.

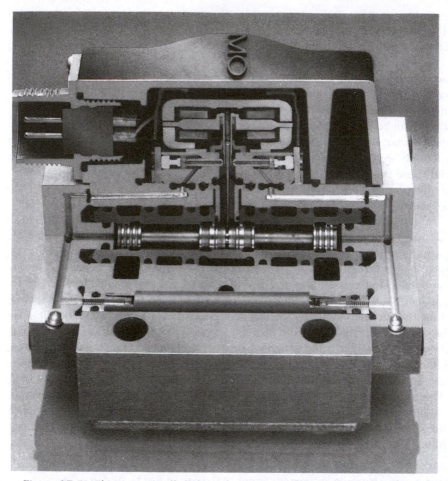

Figure 17.5   Flapper controlled electrohydraulic servovalve (Moog Controls).

3. A power stage spool which controls the main flow to and from the user machine (e.g. a double acting cylinder). Displacement of the spool is initiated by the differential pressure produced by the flapper. The position of the spool is detected by the profiled ball ended feedback wire which is cantilevered from the end of the flapper rod.

As shown in Fig 17.6, an equilibrium position is reached in which a balance exists at the armature between the torque due to the input current and the torque due to the bending of the feedback wire. The offset of the main spool, and therefore the controlled flow rate, is therefore directly proportional to the armature current. An alternative design which has similar characteristics is shown in Fig. 17.10.

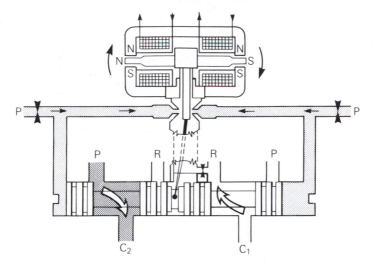

Figure 17.6   Electrohydraulic servovalve operational diagram: valve condition following change.

Valves of this type are found in every application where it is necessary to exercise control of output parameters such as the position, velocity, acceleration or force of hydraulic rams. Flight controls, machine tool slides, hydraulic pump stroke controls, testing machines and ship steering gear are a few examples. A particularly interesting application is in the use of electrohydraulic vibrators for geological survey work. Here a large vertical cylinder and platen are used to impart vibrations to the ground over a range of frequencies. By measuring the reflected response at different locations, the characteristics of the substrata can be revealed.

Typical output characteristics are shown in Fig. 17.7 where, within the high flow saturation limits, the divergence from linearity is less than 1%. Resolution and hysteresis are 2% or better. As implied by the last example quoted above, frequency response is an important criterion in servovalves. The amplitude ratio − 3 dB roll-off does not occur until an input signal frequency of around 150 Hz is reached; for special high response servovalves the corresponding figure can be over 500 Hz. Quiescent flow due to internal leakage is small and is accounted for primarily in the first stage flapper nozzles.

A good example of an integrated electrohydraulic position controller is shown in Fig. 17.8, where a hydraulic cylinder is controlled via a spool valve and the ram position is sensed by an internal inductive sensor. Single axis or multi-axis systems can readily be built up with independent or grouped supervisory control. Valves are also available with elements of control functions such as proportional integral derivative (PID) algorithms and required numbers of repetitions resident in hardware and externally reprogrammable software within the valve housing itself.

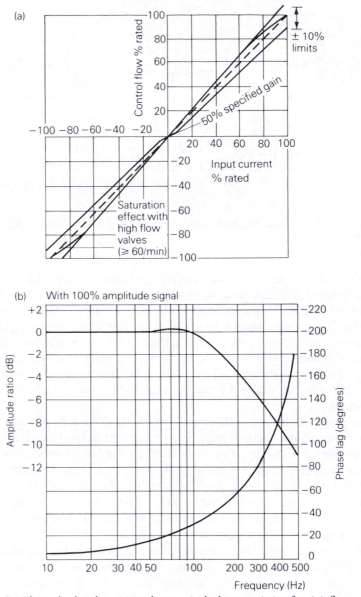

Figure 17.7 Electrohydraulic servovalve: typical characteristics for (a) flow gain (b) frequency response.

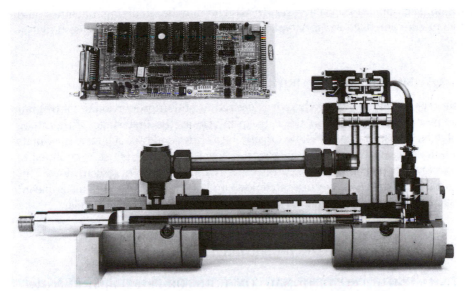

Figure 17.8   Electrohydraulic position controller (Sacol Powerline).

17.1.2   PROPORTIONAL SOLENOID CONTROLLED ELECTROHYDRAULIC VALVE

Advances in the design of solenoids, and in particular the incorporation of rare earth magnets, have made direct driven valves a more attractive proposition than before. Valve stiction and hysteresis are less of a problem, though the flapper controlled valve probably still has the advantage in dealing with particulate contamination. The direct solenoid valve is, however, significantly cheaper owing to its less complex construction. The cost is around half to two-thirds of the flapper controlled valve, an important consideration for an

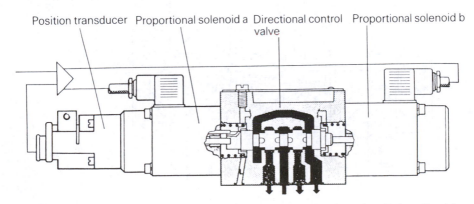

Figure 17.9   Proportional solenoid controlled electrohydraulic valve (Robert Bosch).

original equipment manufacturer (OEM) who may be supplying machines with five or six controlled axes. A typical arrangement of such a valve is shown in Fig. 17.9.

17.1.3   SIMPLE FLAPPER ORIFICE

The type of valve is used where it is desired to obtain proportional control but at low cost. It operates on the same principle as the first stage of the Moog valve but is used to control flow directly into either end of a cylinder. It exhibits similar characteristics to an underlapped spool valve and its use is confined to low power applications on account of its relatively large quiescent flow. This is up to 40% of the maximum rated flow, and of course represents a significant loss of power in the system. Figure 17.4 shows the working principle.

17.2   Electropneumatic proportional controls

17.2.1   DIRECT PROPORTIONAL CONTROLS FOR PRESSURE AND FLOW

Manually operated pressure regulating valves have been used on compressed air supply systems for many years to control the pressure available at air driven

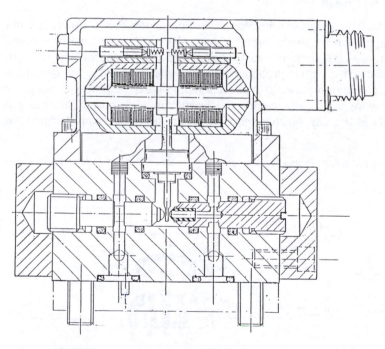

Figure 17.10   Flapper orifice valve (Servotel controls).

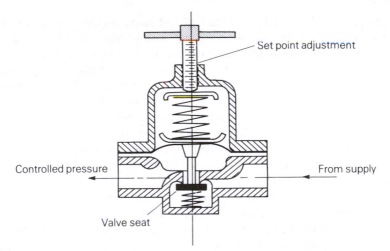

Figure 17.11   Pressure regulating valve.

tools such as controlled torque spanners. The controlled output pressure acts on a diaphragm in such a way as to provide a force balance, with the pressure acting on the supply side, restricting and eventually closing off the supply across an annular orifice (Fig. 17.11). The set point is adjusted by means of a screw and a spring. Valves are available in which the screw may be driven remotely, for example by a stepper motor. Faster acting spool and piston/poppet valves have been developed which are suitable for the control of force or the position of cylinders. In comparison with hydraulic systems, the control of air cylinders is made more difficult on account of the much greater compressibility of air. A hydraulic cylinder with inlet and outlet closed can be regarded for most practical purposes as locked, whereas an air cylinder, initially held at an intermediate position in balance between its internal pressure and the externally applied force, is highly compliant. This is why air cylinders have traditionally been used principally for two-position bang-bang applications.

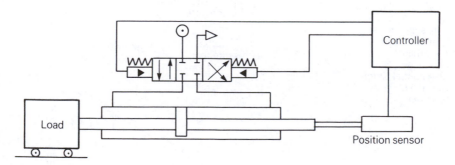

Figure 17.12   Pneumatic position controlled cylinder.

To hold position away from mechanical end stops, a control loop is required incorporating position sensing and a rapid acting valve. In Fig. 17.12, if an increase in the externally applied load occurs, then the valve acts to permit an additional mass of air to flow into the cylinder, thus increasing the pressure at constant volume, and restoring or maintaining the position. Small single stage spool valves with direct acting solenoids are now available which can respond within a few milliseconds. They must be placed as close as possible to the actuator in order to avoid problems due to the compliance of the air in the pipeline. However, it is unrealistic to expect a pneumatic actuator to hold position against alternating loads such as the out-of-balance forces generated at the frequencies commonly encountered in industrial rotating equipment.

An alternative configuration of valve is the high flow piston valve shown in Fig. 17.13. The value of the input pressure setting is compared with the output pressure and modulated in real time. Within the valve, the magnetic force from the solenoid is opposed by a feedback force which is proportional to the output pressure. For the position control of a double acting cylinder, two such valves would be required.

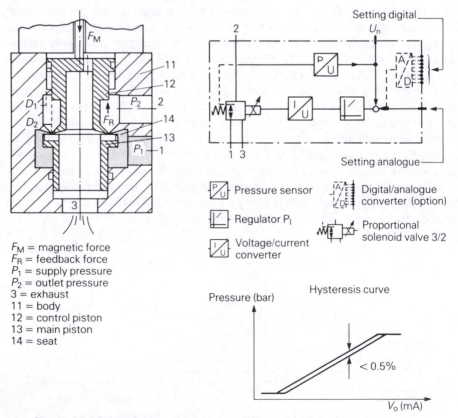

F_M = magnetic force
F_R = feedback force
P_1 = supply pressure
P_2 = outlet pressure
3 = exhaust
11 = body
12 = control piston
13 = main piston
14 = seat

Figure 17.13   High flow proportional piston valve (Joucomatic Controls).

Applications exist, however, in which the control of force rather than position is the main objective. Commercial vehicle braking systems are a good example, where the degree of retardation obtained is directly related to the air diaphragm pressure acting via the brake shoe wedge expander. In an anti-lock braking (ABS) system the pressure must be modulated rapidly to release and restore the braking effect. Within vehicles a modular approach can be adopted in which similar, if not identical, control elements can be incorporated for different purposes. The following example is hypothetical, but all the elements are individually in place on different vehicles. Referring to Fig. 17.14, four applications of force control are shown, in two of which it is also necessary to control position. The values given are based on a service air pressure of 5.5 bar and flow rates at standard conditions of 1.0 bar and 0 °C.

(a) CLUTCH CONTROL

Maximum actuator pressure	4.0 bar
Maximum flow rate	400 l min^{-1}
System time constant	20 ms

Other inputs in this case come from the engine management system, chiefly engine speed and gearbox input shaft speed. From these the relative slip and engagement time is derived. The air pressure opposes clutch operating spring force and therefore controls the force on the clutch plate directly. The position in the linkage travel at which this occurs is not of great significance, at least within broad limits.

(b) CAB RIDE RESPONSE

Maximum actuator pressure	3.0 bar
Maximum flow rate	200 l min^{-1}
System time constant	20 ms or less

The inputs here are cab/chassis relative displacement and cab vertical acceleration. This system also has value in controlling events of longer time constant, such as preventing the cab from dipping forward under heavy braking.

(c) VEHICLE ROLL

Maximum actuator pressure	5.0 bar
Maximum flow rate	400 l min^{-1}
System time constant	200 to 1000 ms

The inputs here may be lateral acceleration, derived from steering wheel angle and road speed and/or measurements of the relative displacement of the axle and body on opposite sides of the vehicle.

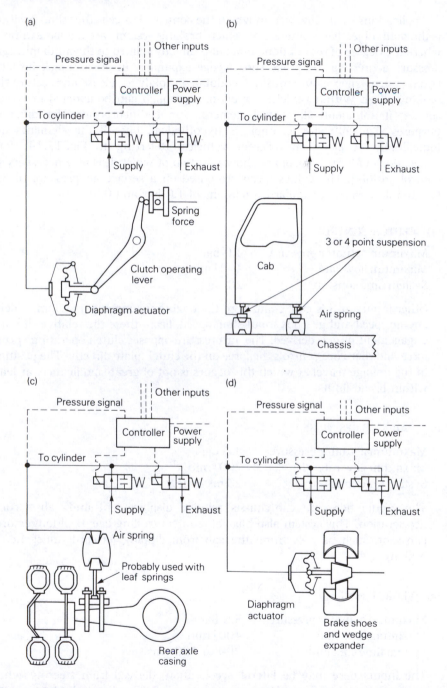

Figure 17.14   Force control pneumatic actuators, vehicle applications: (a) clutch control (b) cab ride response (c) vehicle roll (d) anti-lock braking.

(d) ANTI-LOCK BRAKING

Maximum actuator pressure	full service pressure
Maximum flow rate	$400 \, l \, min^{-1}$ or more
System time constant	20 ms or less

Inputs are derivatives of wheel velocity, including interwheel comparisons. The aim is to keep the braking effort available in an emergency as near as possible to the limit of slip.

A similar if not identical type of air valve could serve all these functions, and could be a 3/2 proportional valve of the type shown in Fig. 17.15. Alternatively, pairs of 2/2 spool or poppet valves could be employed, but because of their limited flow capacity these would be used as the pilot stages of larger diaphragm or disc valves.

A technique which may be usefully applied to the control of small air valves, as an alternative to direct proportional action, is pulse width modulation.

17.2.2 PULSE WIDTH MODULATION CONTROL OF SOLENOID VALVES

The essential features of a pulse width modulation (PWM) control for a valve are shown in Fig. 17.16, in which the valve is continually cycled open/shut, the proportion of time for which it is open being determined by the mark/space ratio of the input signal. The average flow rate through the valve, at constant pressure difference, is then given by

$$Q_{avg} = (T_1/T)Q_{max} \tag{17.2}$$

where T_1 is the on time and T is the period. At low pulse frequencies (up to 80 Hz) the valve can exercise a full open–close cycle, and so the average flow

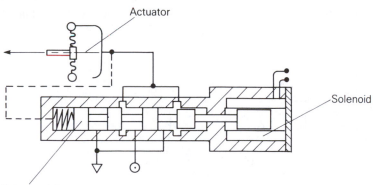

Figure 17.15 3/2 (3 port/2 position) proportional solenoid valve.

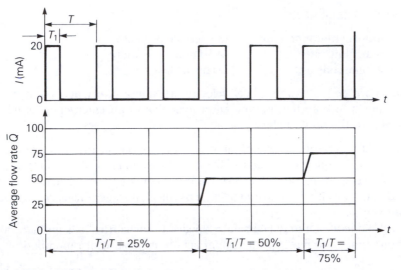

Figure 17.16   Pulse width modulation (PWM) control of a 3/2 (3 port/2 position) poppet valve (Honeywell Controls).

may be represented as in Fig. 17.16, which illustrates a case where the cyclic ratios vary from 25% through 50% to 75%.

Even if a true square wave is available from the input, the flow response of the valve is modified on account of the rise/fall time of the coil current and the inertia of the plunger, leading to a departure from linearity at the ends of the range where T_1/T approaches 0 and 1.

PWM control of this type is more suitable for pneumatic circuits than hydraulic because, with the relatively much higher density and bulk modulus of fluids used in the latter, undesirable water hammer effects would result. With air the high compressibility is advantageous in smoothing pulsations provided that the excitation frequency does not correspond with low harmonics in the piping. The first question which will occur to any user concerns the reliability of a component which may have a target life of 10^{10} shifts or more.

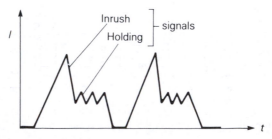

Figure 17.17   PWM at higher frequencies.

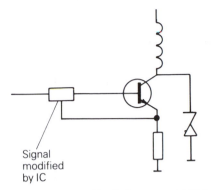

Figure 17.18   Generation of high frequency PWM waveform.

At higher input frequencies, the form of the input signal may be modified as in Fig. 17.17 by means of the circuit of Fig. 17.18 to give an average value of holding current. The velocity of the plunger may be reduced shortly before closure, thus reducing wear and noise, or it may be held in dither at an intermediate position using a high frequency, low amplitude signal about the mean. In the second case the effect on the fluid flow is similar to that of a proportional spool valve.

17.3   Control of electrical drives: power semiconductor devices

The principal power switching devices are diodes, thyristors, gate turn-off thyristors, triacs, power transistors, power MOSFETs and insulated gate bipolar transistors.

17.3.1   DIODES

The diode is the simplest of the power switching devices and has the characteristics shown in Fig. 17.19, allowing significant current in one direction only.

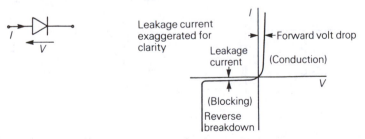

Figure 17.19   Diode characteristic.

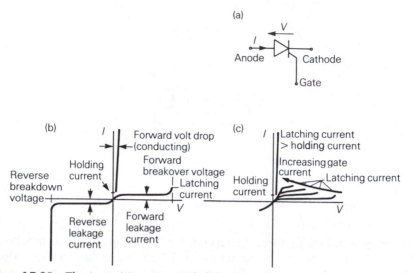

Figure 17.20 Thyristor: (a) circuit symbol (b) characteristic with zero gate current (c) switching characteristic.

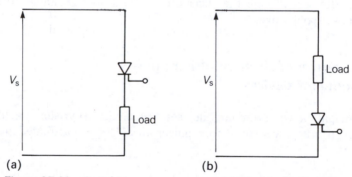

Figure 17.21 Switching arrangements (a) high side (b) low side.

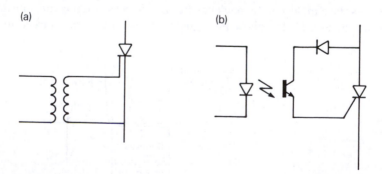

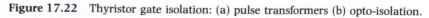

Figure 17.22 Thyristor gate isolation: (a) pulse transformers (b) opto-isolation.

Once a diode has been placed into its conducting mode by a forward voltage it will remain in that condition until the forward current falls to zero, at which point it will turn off.

17.3.2   THYRISTORS

The thyristor can turn on when forward biased by a current via the gate terminal, producing the characteristics of Fig. 17.20. Typically, thyristor firing circuits use pulse techniques which allow precise control of the point-on-wave at which the thyristor is fired and which dissipate less energy in the gate than a continuous current. Reliance is not usually placed on a single pulse to fire the thyristor, and the firing circuit is usually arranged to provide a train of pulses.

   A particular problem in many circuits employing thyristors is that the gate may be at some potential with respect to ground. This may be illustrated by considering the low side and high side switching arrangements of Fig. 17.21a and b. In the case of low side switching, the gate circuit can be referenced to ground but the load remains directly connected to the supply even when not conducting. With high side switching, normally the preferred mode of operation, the load is only directly connected to the supply when the thyristor is conducting but the gate is now at some potential relative to ground. An isolated gate signal can be provided in this latter case by the use of pulse transformers and opto-isolated electronic switches in the gate circuit, as illustrated by Fig. 17.22.

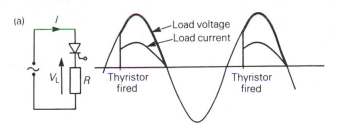

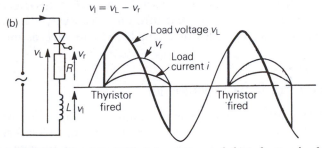

Figure 17.23   Thyristor with (a) resistive and (b) inductive loads.

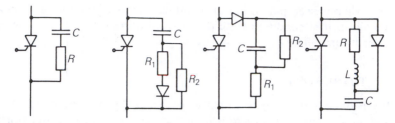

Figure 17.24   Snubber circuits.

The thyristor, once conducting, will remain in this state until the forward current falls below the holding current level, at which point it will begin to turn off. This has important implications when a thyristor is used to control an inductive load such as a motor, as can be seen from Fig. 17.23a and b, which show a thyristor being used to control a resistive and an inductive load respectively. With the resistive load, the load current follows the applied voltage once the thyristor has been fired, turning off at the voltage zero. With the inductive load, the effect of inductance is to sustain the load current – and hence the thyristor current – beyond the voltage zero.

The conditions for turn-off occur automatically in naturally commutated converters. However, there is a range of circuits which are operated from a DC source voltage in which additional, external circuitry must be employed in order to turn the thyristor off. These forced commutation circuits drive a reverse current through the thyristor to reduce the forward current below the holding level and then maintain the reverse voltage for the time interval necessary to complete the turn-off (see also section 17.4).

The operation of a thyristor is defined by a series of ratings which define the operational boundaries for the device. Rating parameters include the peak, average and RMS currents, the peak forward and reverse voltages and the gate circuit limits. In addition, there are a number of transient limits which may need to be taken into account, particularly when duty cycle operation is being considered.

Turn-on of a thyristor has been considered in terms of the forward breakover voltage and the effect of the gate current. A thyristor may also be turned on by an excessively high rate of rise of forward voltage. For this reason a thyristor has a dv/dt rating chosen to prevent turn-on in this way. The magnitude of imposed dv/dt can be controlled by a snubber circuit such as those shown in Fig. 17.24.

17.3.3 GATE TURN-OFF THYRISTORS

The gate turn-off (GTO) thyristor is a variant of the basic thyristor in which the internal structure has been modified to enable the turn-off of forward current to be achieved by the application of a reverse or negative gate current. This

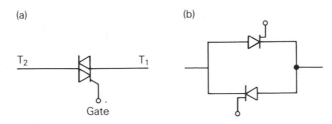

Figure 17.25   The triac: (a) circuit symbol (b) thyristor equivalent.

ability of the GTO thyristor to be both turned off and turned on under the control of an external signal has led to its extensive use in a range of power switching circuits.

On turn-on a GTO thyristor must be supplied with a higher forward gate current than an equivalently rated conventional thyristor to ensure that the forward current is established. Once conduction is established it may be necessary to maintain a small forward gate current to prevent any tendency to drop out from the conducting state.

The turn-off of the GTO thyristor requires the introduction of a reverse gate current of sufficient magnitude to divert the forward current.

17.3.4   TRIACS

A triac is electrically equivalent to a pair of thyristors connected in reverse parallel on the same chip (Fig. 17.25), and is capable of being turned on in either the forward or the reverse direction by the application of an appropriate signal via the gate circuit. Referring to Fig. 17.25, a triac would normally be arranged to turn on with the application of a forward current when terminal T_2 is positive with respect to terminal T_1, and a negative current when T_1 is positive with respect to T_2. Triacs are used extensively for the control of motors and heating loads.

17.3.5   POWER TRANSISTORS

In power applications the transistor is used as a controlled switch, with transitions between the ON or saturated condition with a high base current and the off condition with zero base current, as in Fig. 17.26. The transition between the on and off states can introduce high instantaneous losses and forms the major source of loss in high speed switching circuits such as those used in inverters. To reduce turn-on times, an excess base current is used to speed up the transition and force the transistor into saturation. Once the transistor is conducting, its base current is reduced to the level required to maintain it in saturation. On turn-off the base current must be reduced as rapidly as possible consistent with avoiding the occurrence of secondary breakdown in the

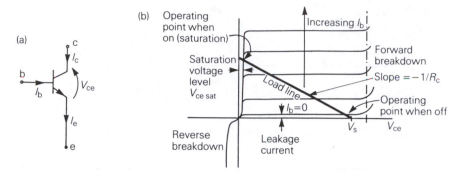

Figure 17.26   The power transistor: (a) npn transistor symbol (b) $I_c - V_{ce}$ characteristic (forward and reverse voltage scales are unequal).

transistor. In some instances a reverse base current may be used to enhance turn-off, and this reverse bias is maintained in the OFF condition.

Transistors are capable of switching more rapidly than a thyristor, with times of the order of a few microseconds being achievable. However, the need to supply the transistor with a continuous base current to maintain it in the on condition means that the requirements of the base drive circuit are more severe than those of the thyristor gate circuit. In some devices, the base drive circuit is incorporated on to the same chip as the power transistor, an arrangement which reduces the external base current requirements but increases the switching time of the power transistor.

17.3.6   POWER MOSFETS

The increasing demand for high speed power switching devices has led to the introduction of power MOSFETs. These provide very short switching times with low switching losses and require significantly lower levels of gate current to maintain them in the on condition than equivalently rated power transistors.

17.3.7   INSULATED GATE BIPOLAR TRANSISTORS

The insulated gate bipolar transistor (IGBT) combines MOSFET and bipolar technology on the same chip to produce devices which offer improved switching performance with a lower demand for control current.

17.3.8   SMART POWER DEVICES

Developments in semiconductor technology have led to the introduction of devices which combine logical and control elements alongside the power switching devices, either on the same chip as in Fig. 17.27 or as part of a hybrid circuit. These smart power devices have the ability to monitor and control their

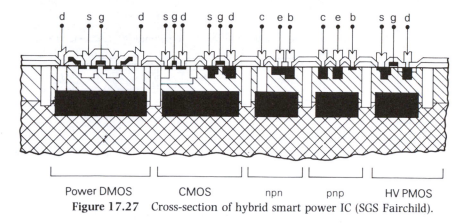

Figure 17.27   Cross-section of hybrid smart power IC (SGS Fairchild).

own operation and to provide status reporting on system behaviour, increasing the flexibility of the overall system. Figure 17.28 shows the layout of a typical smart power chip for motor control, incorporating acceleration control and protection.

17.3.9   HEAT TRANSFER AND COOLING

The heat generated in a power semiconductor device due to its internal losses must be removed from the device and dissipated to an ambient temperature

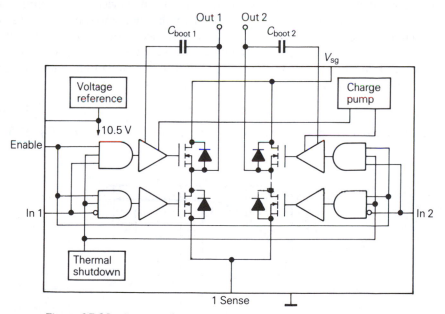

Figure 17.28   Layout of a typical smart power chip (SGS Fairchild).

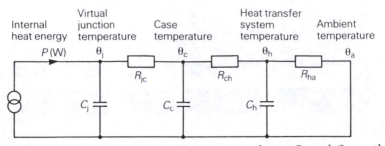

Figure 17.29 Semiconductor heat transfer, circuit analogue: Rs and Cs are thermal resistances and capacitances.

heat sink. The heat transfer path is considered to be as follows:

1. From an internal junction to the case of the device;
2. From the case to a heat transfer system such as a fin;
3. From the heat transfer system to the ambient temperature heat sink.

The thermal characteristics of the heat transfer process can be represented by the circuit analogue of Fig. 17.29. Under steady state conditions, the thermal capacitance has no effect and the heat flow can be represented by the series combination of the thermal resistances as

$$\theta_j = \theta_a + P(R_{jc} + R_{ch} + R_{ha}) \tag{17.3}$$

During transient conditions, the temperature rise in the device can be estimated using a parameter referred to as the transient thermal impedance Z_t, which incorporates the effects of the thermal capacitance. The transient thermal impedance is a time dependent function as shown by Fig. 17.30, and relates the device temperature rise to the energy input in a defined time interval. For a step change in energy input from 0 to P_t at time $t = 0$,

$$P_t Z_t(t) = \delta\theta(t) \tag{17.4}$$

For a power semiconductor providing a continuous pulse train as in Fig. 17.31,

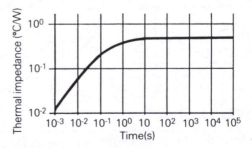

Figure 17.30 Transient thermal impedance curve for a 100 A thyristor.

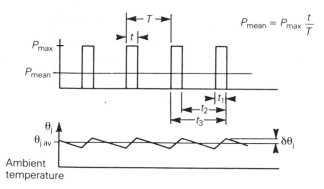

Figure 17.31   Device temperature variation with a pulsed load.

the mean junction temperature can be expressed in terms of the mean power loss and the transient thermal impedance (junction to ambient) at $t = \infty$ as

$$\theta_j = \theta_a + P_{mean} Z_{ja}(\infty) = \theta_a + P_{mean} R_{ja} \qquad (17.5)$$

The junction temperature varies about this mean by an amount $\delta\theta_j$, an approximation to which is given by considering the last two pulses when, referring to Fig. 17.31,

$$\delta\theta_j = [(P_{max} - P_{mean}) Z_{ja}(t_3)] - P_{max}[Z_{ja}(t_2) - Z_{ja}(t_1)] \qquad (17.6)$$

More complex waveforms can be analysed by representing them as a series of step functions and applying the principle of superposition.

17.3.10   PROTECTION

(a)   OVERCURRENT PROTECTION

As a power semiconductor has a restricted overcurrent capacity, special fast acting fuses are usually provided for overcurrent protection. The selection of the appropriate fuse must take account of:

1. The need to permit the continuous passage of steady state current;
2. Permitted overload conditions including transients and duty cycle loads;
3. Prospective fault conditions;
4. The i^2t rating of the device and the need to clear the fault before this is exceeded;
5. Peak current levels during faults due to current asymmetry;
6. Fuse voltage rating;
7. Ambient temperature conditions.

The overcurrent protection of transistors presents particular problems. A fault condition can cause an effective reduction in transistor load which will cause

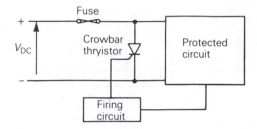

Figure 17.32   Crowbar protection.

the transistor to come out of saturation and into a region of high internal energy dissipation. Though this increase in internal energy may damage the transistor, the increase in collector current may not be sufficient to blow a series fuse. To overcome this problem a crowbar thyristor is used as in Fig. 17.32. This uses dedicated circuitry to monitor the collector-emitter voltage (V_{ce}) of the transistor, and fires the thyristor if any increase in voltage is detected, causing the fuse to blow.

(b)   OVERVOLTAGE PROTECTION

Given that the power semiconductor device is rated to withstand the normal voltages expected, it remains necessary to provide protection against transient overvoltages and excessive dv/dt values. This is typically achieved by placing a non-linear surge suppressor with a characteristic of the form of Fig. 17.33 in parallel with the power semiconductor, along with the snubber circuit shown in Fig. 17.24.

17.4   Converters, choppers, inverters and cycloconverters

Power electronic circuits employing thyristors can be divided into two principal categories depending on whether natural or forced commutation is required.

17.4.1   NATURALLY COMMUTATED THYRISTOR CONVERTERS

Naturally commutated converter circuits can be separated by their construction and operation into three categories: uncontrolled, half controlled and fully controlled.

An uncontrolled converter such as the three-phase bridge rectifier shown in Fig. 17.34 uses diodes only and operates as a rectifier, permitting power flow from the AC system to the DC system. The DC output voltage is fixed by the amplitude of the AC supply.

The fully controlled three-phase bridge converter of Fig. 17.35 uses thyristors as the switching element throughout, allowing control of the DC output voltage

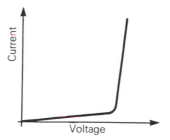

Figure 17.33   General non-linear surge suppressor characteristic.

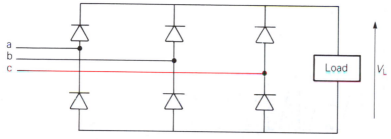

Figure 17.34   Three-phase diode bridge.

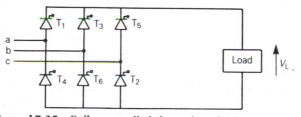

Figure 17.35   Fully controlled three-phase bridge converter.

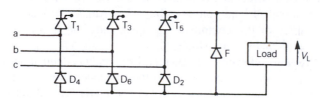

Figure 17.36   Half controlled three-phase bridge converter.

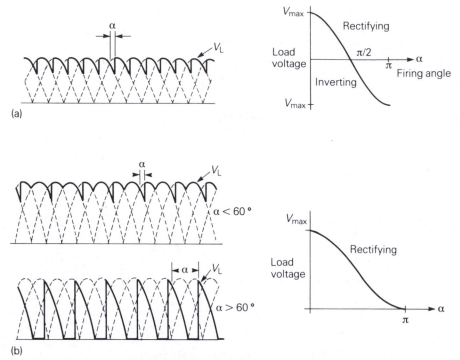

Figure 17.37 Converter load voltage waveforms for (a) fully controlled and (b) half controlled three-phase bridge converter.

in terms of the point-on-wave at which the thyristors are fired. Unlike the uncontrolled and half controlled converters, the fully controlled converter can transfer power not only from the AC system to the DC system – the rectifying mode – but also from the DC system to the AC system – the inverting mode. For this reason, the fully controlled converter is sometimes referred to as a bidirectional converter.

A half controlled three-phase bridge converter is shown in Fig. 17.36. This uses a combination of diodes and thyristors, allowing control of the DC output voltage by variation of the point-on-wave at which the thyristors are fired. Operation is only possible in the rectifying mode, with power flow from the AC system to the DC system. The commutating or freewheeling diode is often included in the half controlled converter to aid the commutation process by preventing a reversal of load voltage when the transfer of load current from the main converter to the commutating diode aids the main thyristors and diodes to recover their blocking states.

Converters are also referred to by their pulse number, defined as the number of discrete switching operations involving load transfer that take place during one cycle of the AC supply. The pulse number is directly related to the repetition period of the DC voltage ripple.

The point-on-wave at which an individual thyristor is fired is defined by the firing angle α, which is measured from the point at which an ideal diode replacing the thyristor would begin to conduct. Figures 17.37a and b show the relationship between firing angle and output voltage for the fully controlled and half controlled three-phase bridge converters of Figs. 17.35 and 17.36 respectively.

(a)   OVERLAP

In practice, the commutation of load current between thyristors does not occur instantaneously but requires a finite time during which two thyristors are conducting simultaneously. This time interval is expressed by the overlap angle γ and is determined by the magnitude of the DC load I_L and the inductance of the AC supply. The effect of overlap on the output waveform is illustrated by Fig. 17.38.

(b)   CONVERTER EQUATIONS

For a general p-pulse fully controlled converter, the defining equations in the rectifying and inverting modes are as follows:

Rectifying
Ignoring overlap:

$$V_o = \frac{p}{\pi} V_m \sin(\pi/p) \cos \alpha \qquad (17.7)$$

Including overlap:

$$V_{mean} = \frac{p}{2\pi} V_m \sin(\pi/p) [\cos \alpha + \cos(\alpha + \gamma)]$$

$$= \frac{p}{\pi} V_m \sin(\pi/p) \cos \alpha - \frac{p\omega L}{2\pi} I_L \qquad (17.8)$$

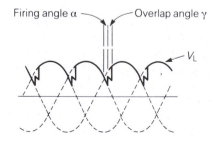

Figure 17.38   Effect of overlap on converter output voltage waveform.

Inverting
Ignoring overlap:

$$V_o = \frac{p}{\pi} V_m \sin(\pi/p) \cos\beta \qquad (17.9)$$

Including overlap:

$$V_{mean} = \frac{p}{2\pi} V_m \sin(\pi/p)[\cos\beta + \cos(\beta + \gamma)]$$

$$= \frac{p}{\pi} V_m \sin(\pi/p)\cos\beta + \frac{p\omega L}{2\pi} I_L \qquad (17.10)$$

where p is the pulse number of the converter,
   α is the firing angle,
   β is the delay angle,
   γ is the overlap angle,
   V_m is the maximum value of the supply voltage,
   V_0 is the converter voltage without overlap and
   V_{mean} is the converter voltage with overlap.

(c) POWER FACTOR

Although a converter transmits only real power (watts) it has a power factor
to the AC system, which must therefore supply both real power and reactive
power (VARs) to the converter. The general expression for power factor is

$$\text{power factor} = \frac{\text{mean power}}{V_{rms}I_{rms}} = \frac{(1/T)\int_0^T vi\,dt}{V_{rms}I_{rms}} \qquad (17.11)$$

The discontinuous current drawn by a converter from the AC supply can be
represented by a fundamental component at the supply frequency together with
a series of harmonics. Assuming that the AC system voltage remains sinusoidal,
the power transferred to the DC system will be derived from the supply
(fundamental) frequency components only. Therefore

$$\text{power} = V_{1.rms}I_{1.rms}\cos\phi_1 \qquad (17.12)$$

where ϕ_1 is the phase angle between $V_{1.rms}$ and $I_{1.rms}$. Substituting this
relationship in equation 17.11:

$$\text{power factor} = \frac{V_{1.rms}I_{1.rms}\cos\phi_1}{V_{1.rms}I_{rms}}$$

$$= \frac{I_{1.rms}}{I_{rms}}\cos\phi_1 = \mu\cos\phi_1 \qquad (17.13)$$

where $\mu = (I_{1.rms}/I_{rms})$ is referred to as the current distortion factor.

For a fully controlled converter in the absence of overlap, the phase angle ϕ_1 will be equivalent to the firing angle α.

(d) CONVERTER WITH DISCONTINUOUS CURRENT

Where the load inductance is insufficient to maintain a steady DC, this current will contain a ripple component which will be reflected in the AC, as illustrated by Fig. 17.39a and b, and may become discontinuous, as in Fig. 17.39c, under conditions of light load. A discontinuous AC will also result when capacitive smoothing of the output voltage is used, as in Fig. 17.40. In this case, as the

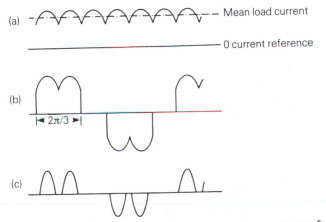

Figure 17.39   Effect of low inductance load on converter current waveforms: (a) DC current with ripple (b) phase current (c) discontinuous phase current.

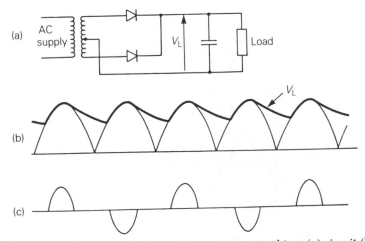

Figure 17.40   Operation of a rectifier with capacitive smoothing: (a) circuit (b) output of a two-phase-half-wave bridge with capacitive smoothing (c) supply current waveform.

load is increased the load current becomes more peaky in order to meet the
energy requirements of the load.

(e)   CONVERTER WITH VOLTAGE BIAS

When a converter is used to supply a load such as a DC motor, the back EMF
across the motor appears as a bias voltage on the DC side of the converter. The
performance of the converter under these conditions is determined by the
relationship between the firing angle of the thyristors, the point at which the
AC source voltage exceeds the bias voltage and the point at which the thyristors
stop conducting. The effect on the operation of a fully controlled single-phase
bridge converter under one specific set of conditions is shown in Fig. 17.41.

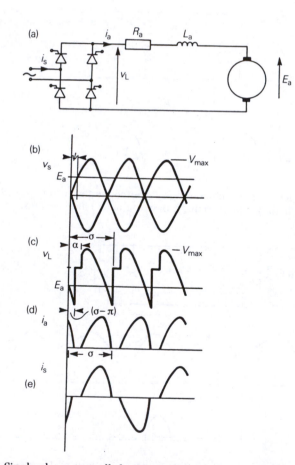

Figure 17.41   Single-phase controlled converter with a DC machine load. Firing angle
α is greater than the cut-off angle ψ, which is in turn greater than $\sigma - \pi$. (a) Circuit
(b) supply voltage (c) load voltage (d) load current (e) supply current.

17.4.2   DC CHOPPERS

A DC chopper provides a means of achieving a controllable DC level from a DC source voltage by switching the source on and off the load. In order to achieve this, a forced commutation circuit must be provided in order to turn the thyristor off after conduction has been initiated. Referring to Fig. 17.42, the mean and RMS load voltages are given by

$$V_{L,\text{mean}} = V_s t_1 / T \qquad (17.14)$$

$$V_{L,\text{rms}} = V_s \sqrt{(t_1 / T)} \qquad (17.15)$$

respectively. If the period T is much less than the load time constant, or if smoothing is used, the load current will vary linearly as shown in Fig. 17.42c, with an RMS ripple current of

$$I_{r,\text{rms}} = (I_1 - I_2)/(2\sqrt{3}) \qquad (17.16)$$

Should the period T be of the order of the load time constant a linear variation of current can no longer be assumed, and the output current varies exponentially during the conducting and non-conducting intervals as shown in Fig. 17.43.

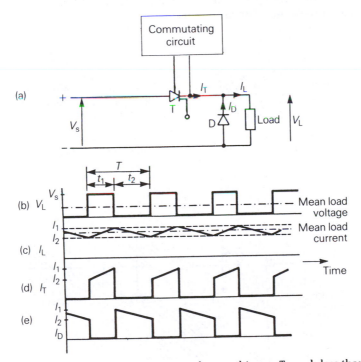

Figure 17.42   Operation of a basic chopper with smoothing or T much less than system time constant: (a) circuit (b) load voltage (c) load current (d) thyristor current (e) diode current.

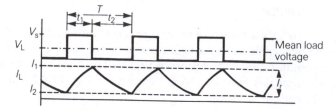

Figure 17.43   Operation of a basic chopper with unsmoothed output where T is of the order of the system time constant.

The circuit shown in Fig. 17.42 allows power flow only from the supply to the load; it is referred to as a class A or single quadrant chopper, as it allows operation only in the first quadrant of the V_L–I_L diagram of Fig. 17.44a. Figure 17.45 shows the arrangement of a class B step-up chopper in which energy is returned from the load to the source. Combining the circuits of Fig. 17.42 and 17.45 gives rise to the class C two-quadrant chopper of Fig. 17.46.

The further combination of two class C chopper circuits produces a class E full four-quadrant chopper circuit capable of producing a bidirectional current in the load; this is the basis of the inverter.

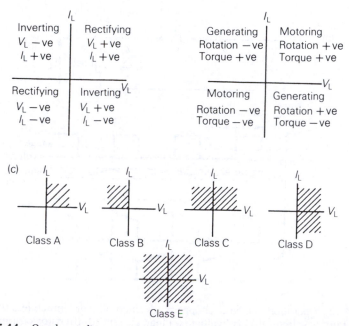

Figure 17.44   Quadrant diagrams: (a) converter (b) motor (c) chopper classification.

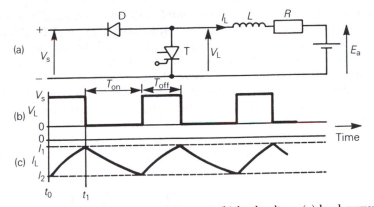

Figure 17.45   Class B chopper: (a) circuit (b) load voltage (c) load current.

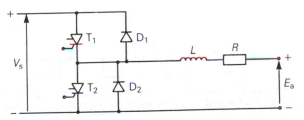

Figure 17.46   Class C chopper.

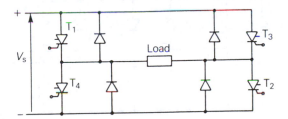

Figure 17.47   Basic single-phase voltage sourced inverter.

17.4.3   INVERTERS

Inverters are used to provide a variable frequency supply from a DC source and form the output stage of AC variable speed drives.

(a)   VOLTAGE SOURCED INVERTERS

Figure 17.47 shows the configuration of a basic voltage sourced single-phase bridge inverter, including reverse diodes across the main force commutated thyristors to accommodate the phase angle difference between current and voltage of an inductive load. If the pairs of thyristors T_1/T_2 and T_3/T_4 are fired at equal intervals then the load voltage waveform will be a square wave, as shown in Fig. 17.48a. Though variable in frequency, the mean voltage of this waveform is fixed; the quasi-square wave of Fig. 17.48b, obtained by delaying the firing of successive pairs of thyristors, is therefore used to provide some control of load voltage amplitude. A three-phase bridge inverter is shown in Fig. 17.49.

If GTO thyristors, power transistors or power MOSFETs are used as the switching elements of the inverter as in Fig. 17.50, the need for forced

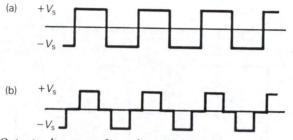

Figure 18.48   Output voltage waveforms for a single-phase voltage sourced inverter: (a) load voltage with equal interval firing (b) quasi-square wave output produced by delayed firing.

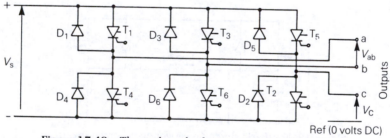

Figure 17.49   Three-phase bridge inverter using thyristors.

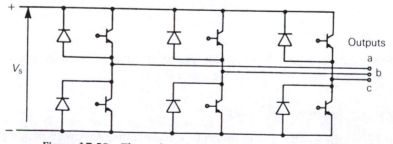

Figure 17.50   Three-phase bridge inverter using transistors.

commutation is removed, enabling the inverter to operate at the much higher switching frequencies needed for pulse width modulated inverters.

(b)   PULSE WIDTH MODULATED INVERTERS

In its simplest form, the pulse width modulated (PWM) inverter switches the supply voltage at regular intervals throughout each half cycle to produce an output voltage waveform of the form shown in Fig. 1.7.51, enabling control of both output frequency and output voltage. By varying the pulse width throughout each half cycle an improvement in performance is obtained through the associated reduction of the harmonic content of the output current waveform.

The type of pulse width modulation applied is usually determined by means of a custom integrated circuit which determines both the shape of the output pulses and the switching frequency used. Typically, the switching frequency is varied in line with the required output frequency.

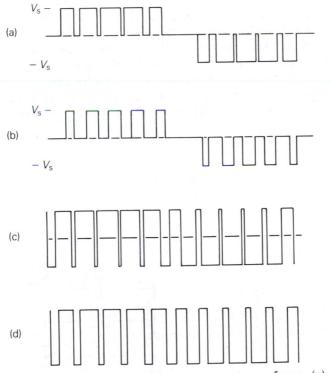

Figure 17.51   Pulse width modulated inverter, output waveforms: (a) type A, full amplitude (b) type A, reduced amplitude (c) type B, full amplitude (d) type B, reduced amplitude.

(c)   CURRENT SOURCED INVERTERS

In a current sourced inverter, the current from the DC source is fed to the inverter via a large inductance, as in Fig. 17.52, to maintain it at an approximately constant value independent of load conditions.

Current sourced inverters are typically used to supply higher power factor loads, with operation in the frequency range from 5 Hz to between 50 and 100 Hz. The upper limit is set by the time required for commutation.

(d)   INVERTER PERFORMANCE

In addition to load dependent losses and the losses in protection circuits such as the snubber circuits, forced commutation inverters suffer from additional losses in the commutation circuits. In a PWM inverter the major source of loss is likely to be the switching losses incurred during the transitions between the on and off states.

Typically, efficiencies are of the order of 96% for a quasi-square wave inverter and for a current sourced inverter, 94% for a PWM inverter using GTO thyristors or power transistors, and 91% for a PWM inverter using thyristors.

Thyristor quasi-square wave inverters tend to operate at frequencies in the range from a few hertz to around 100 Hz, with transistor versions extending to 500 Hz or more. The majority of PWM inverters used for motor control operate with switching frequencies of the order of 20 kHz for output frequencies up to 100 Hz in PWM form, with outputs above this frequency in quasi-square wave form. For higher frequency operation, inverters using power MOSFETs operating with switching frequencies in excess of 100 kHz have been developed.

17.4.4   CYCLOCONVERTERS

A cycloconverter synthesizes its output voltage waveform by switching between the phases of an AC supply, and provides an alternative to an inverter where

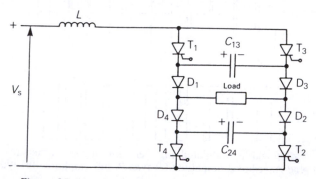

Figure 17.52   Single-phase current sourced inverter.

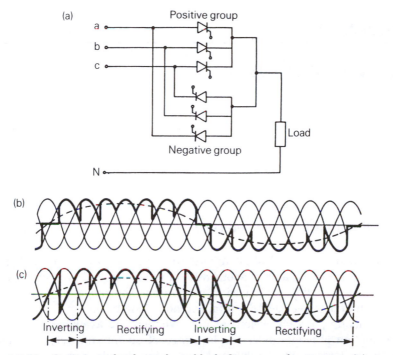

Figure 17.53 Operation of a three-phase blocked group cycloconverter: (a) circuit (b) supplying resistive load (c) supplying indirective load.

only frequencies below the AC supply frequency are required. By using a multiphase supply and varying the point-on-wave at which the individual power switching devices are turned on or off, an output can be obtained which emphasizes the fundamental component. Figure 17.53 shows a typical output from one phase of a three-phase blocked group cycloconverter – so called because of the need to prevent the positive and negative groups from conducting simultaneously and short-circuiting the supply. As an alternative to blocked group operation, the circulating current mode can be used in which a reactor is connected to limit circulating current.

Cycloconverters tend to require more complex control systems than voltage or current sourced inverters, and their applications tend to be limited to high power systems.

Chapter 18

Linear systems

18.1 Pneumatic rams: rod type

The construction of a typical pneumatic ram is shown in Fig. 18.1. Typically the annulus (rod end) area is about 80% of the piston end area and the force available on retraction is reduced in proportion. The range of motion cannot be more than about 45% of the total extended length.

Single acting push (extension) or pull cylinders are available in which the return motion is made under the influence of an internal return spring, or imparted by forces from the external load when the air supply is cut off and the cylinder exhausted to atmosphere. Single acting cylinders can be used as position controllers with simple pressure control in applications such as the flue damper of Fig. 18.2, where fine positioning is not required. In such cases the external forces acting on the system, such as pressure or aerodynamic loads, should as far as possible be balanced about the operating axis so that the predominant force seen at the actuator is that due to the return spring only.

Rod-type pneumatic rams are most commonly used in two-position applications working between end stops, and for hold on load it is necessary that the air supply remain connected to the appropriate end of the cylinder. To avoid shock loads at the end of the travel, such cylinders may incorporate a cushioning device as shown in Fig. 18.1.

The favourable attributes of rod pneumatic rams are as follows:

1. Inexpensive;
2. Commonly available;
3. Light alloy construction;
4. Safe, clean working medium;
5. Range of mounting configurations;
6. System protected by inherent force limitation;
7. Shock resistant;
8. Rapid acting;
9. Ideal for two-position sequence control.

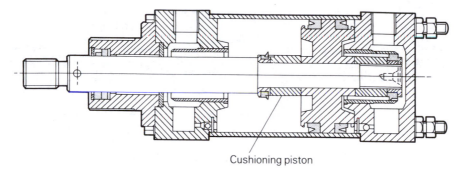

Cushioning piston

Figure 18.1   Example of a pneumatic ram with cushioning (Martonair).

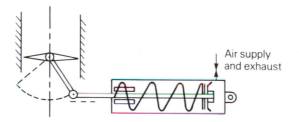

Air supply
and exhaust

Figure 18.2   Single acting air cylinder with return spring.

Less favourable attributes are as follows:

1. Costly working medium (if continuous operation);
2. Length of travel limited by stability of ram;
3. Lack of stiffness, i.e. cannot hold position against alternating loads unless holding on end stop;
4. Unsuited for accurate control of speed.

18.2   Pneumatic rams: rodless type

These devices have been introduced within the last few years. They are reminiscent of Brunel's atmospheric railway but include the benefits of modern materials technology.

Their construction is as shown in Fig. 18.3. The slot in the side of the cylinder is covered by an elastomeric seal backed by a flexible stainless steel strip which is displaced by the passage of the piston and carrier. The escape of air is negligible.

Rodless cylinders are competing with conventional rams in many applications and are extending the range of pneumatics into new areas formerly held by chain drives and lead screws.

The favourable attributes of rodless pneumatic rams are as follows:

1. The range of motion is 80% or more of the device length;

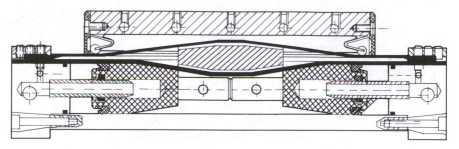

Figure 18.3   Rodless air cylinder (Martonair).

2. They can function as combined actuator and guideway for the load;
3. Very long travel is possible – 5 m or more;
4. Equal force is available in both directions.

Their main disadvantage is that the mounting possibilities are more restricted.

18.3   Pneumatic diaphragms

Diaphragm actuators are used where compressed air is the preferred working medium, or for reasons of safety in environments where a hazard may exist

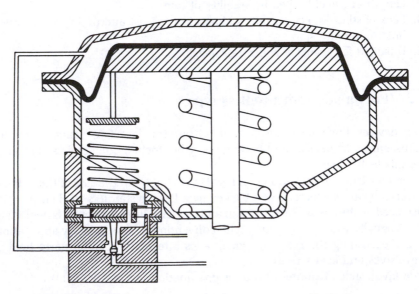

Figure 18.4   Pneumatic diaphragm actuator (Foxbro).

due to the presence of flammable gases. The commonest application is as a positioner for the actuation of fluid control valves on pipelines in chemical process plants. The substantial area of the diaphragm makes large forces available to overcome those on the fluid valve spindle. A typical construction is shown in Fig. 18.4, where the special configuration of the controller is apparent.

The main disadvantages of pneumatic diaphragms are their restricted range of movement and bulky topworks.

18.4   Pneumatic bellows

Bellows are used for light loads and small ranges of movement, typically for instrumentation applications. Most commonly they are employed as pressure sensing devices in which pressure is transformed into mechanical movement.

18.5   Hydraulic cylinders

These are the among the commonest devices used for linear positioning for higher loads than are possible with pneumatic rams. The principle distinctions are the much higher working pressure and the consequently much heavier construction. Furthermore, because the working fluid is for most purposes regarded as incompressible, precise positioning of loads can be accomplished to within $\pm 0.01\,\text{mm}$, with an appropriate control system.

Typical construction is shown in Fig. 18.5. The working pressures for hydraulic systems are typically between 100 and 300 bar, and may be higher in military equipment where economy of space is a major consideration. Thus construction is almost always in steel.

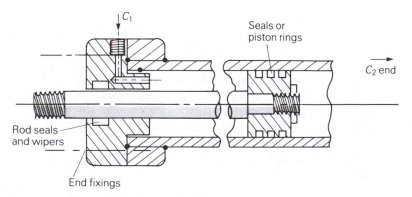

Figure 18.5   Typical hydraulic cylinder construction.

Hydraulic cylinders are rarely single acting, except for simple applications such as jacks where the external loads can always be depended upon to provide sufficient force to retract the ram and force the working fluid through the return line.

Where hydraulic cylinders are to be used for high loads and long strokes in the push direction, there is a danger of strut instability in the ram itself. The consequences of this appear more commonly in premature wear on seals and guides than in actual buckling. Cylinders designed for such applications will have large diameter rods and hence an annulus area of as little as 20% of the piston end area. The main jib cylinders used on mobile hydraulic cranes are a good example. For more stocky cylinders the annulus area may be around 50%, but this must still be taken into consideration when assessing the forces and rates of movement available from the cylinder in each direction.

The favourable attributes of hydraulic cylinders are as follows:

1. Very high loads are available in compact format;
2. Hold on load can be obtained with leaktight piston and valve seals; note, however, that many types of spool valve (section 17.1) are not leaktight on power off, and so the actuator may slump unless attention is directed to this during design;
3. Precise control of rate of movement and position is readily accomplished with the appropriate control system (section 17.1);
4. They can be obtained in a variety of mounting configurations from a wide range of supppliers;
5. They are easily protected against overload by means of a relief valve.

However, at the application boundary with pneumatic actuators, hydraulic cylinders are generally more expensive and heavier, and require a working fluid which is itself more costly and more objectionable when leakage occurs.

18.6   Motor and ball screw

Typically these linear actuators consist of an electric motor (e.g. a DC servomotor: see section 19.5.3) and a screw jack with a recirculating ball nut. They are a complex assembly, incorporating a rotary to linear motion converter, but can be considered as a single actuator and as a competitor with other linear actuators, especially hydraulic rams. A typical construction is shown in Fig. 18.6, which is of an actuator used for positioning flaps on aircraft wings. They can be employed in either push or pull mode, but strut instability limits the extensions which can be achieved in push.

In industrial applications, ball screws are frequently employed where the working load is predominantly a unidirectional pull. In this case it is possible, where the nut is arranged to move with the load, to have 95% or more of the overall length of the device occupied by usable travel.

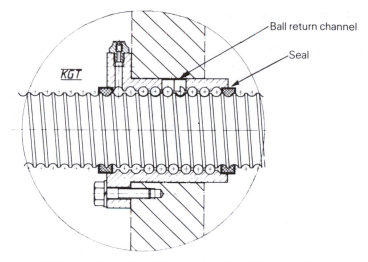

Figure 18.6   Recirculating ball screw and nut (Unimatic Engineers).

Alternatives to the electric motor drive are pneumatic or hydraulic motors (sections 19.1 and 19.3), but these would only be considered in exceptional cases, for example in areas with a hazard classification where there was already a commitment to fluid power as the working medium.

The favourable attributes of ball screws are as follows:

1. High forces are available;
2. They compare favourably with hydraulic rams on the force/bulk criterion;
3. The hold on load with power off is normally intrinsic to the design of ball screws; however, runback is possible because of the very low friction if the pitch is sufficiently coarse.

Less favourable attributes are as follows:

1. They tend to be expensive and less widely available than hydraulic rams;
2. They are slower than hydraulic rams, and much slower than pneumatic rams;
3. They can be subjected to inadvertent overloads if driven to a physical end stop or otherwise to a mechanically jammed condition.

The third of these disadvantages is characteristic of any mechanical system involving a large reduction ratio between a motor and a load. Unless the motor speed has been ramped down, the residual kinetic energy available in the spinning rotor, even after tripping out or switching off, can impart sufficient strain energy into the downstream system to jam it irretrievably if not actually to cause breakage.

18.7   Motor and lead screw

This system is similar in many respects to the motor and ball screw described in section 18.6. The principal difference is the higher friction generated at the nut/screw interface. This confers both advantages and disadvantages. For all practical purposes with single start screws, this means that hold on load with power off is guaranteed. The penalty, which can be accepted in most applications, is of lower efficiency, which may be under 25%.

In machine tool drives, the most obvious application, the lead screw will propel headstock or table motions and be driven via a gear train by a stepper motor or DC servomotor.

18.8   Direct linear electrical actuators

Direct linear electrical propulsion systems have been in use for some time, a well known example being the maglev passenger transport system at Birmingham International Airport. A linear induction motor is located under the floor structure of each of the 6 m long passenger vehicles as shown in the sectional view of Fig. 18.7. The motor is a short stator, single sided, axial flux linear induction motor driven by an inverter of the pulse width modulated type using transistor switching. It is rated at 240 kVA continuously, or 325 kVA for short periods. The reaction rail is a steel beam capped by an aluminium plate. This

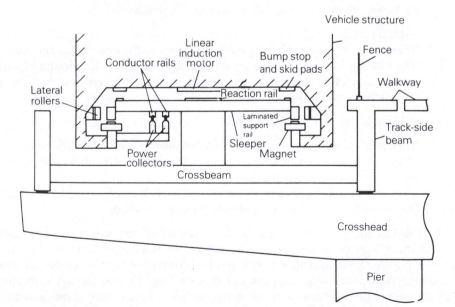

Figure 18.7   Maglev system (GEC Traction: Van Nostrand Reinhold).

system provides a tractive force of 2 kN for steady propulsion at $15 \, \text{m s}^{-1}$, and up to 4 kN for acceleration or retardation. The control of the propulsion system itself, including waveform generation and inverter control and protection, is assigned to a microprocessor resident in the vehicle.

The system as a whole is managed by a supervisory system which handles automatic train operation (ATO) vehicle propulsion functions and automatic train protection (ATP) safety functions. The vehicle cycles can be run to a schedule set up on the ATO controller by the control room operator, or can be put 'on demand' by passengers at the stations. Track based sensors are used to initiate braking by adjusting the motor slip, and an accuracy of positioning of $\pm 100 \, \text{mm}$ is achieved at stops.

This example demonstrates what is possible in terms of power and speed in linear electrical drives. However, industrial actuators also normally require positional control to 0.1 mm or better, together with high stiffness at rest. While DC linear motors with permanent magnet slider and wound stator configurations can meet these criteria, the mechanical problems associated with commutation are a disadvantage in normal industrial environments.

An alternative approach has been recently developed for linear drive units comprising a fixed thrust rod on which travels a carrier or thrust block. As shown in Fig. 18.8, the thrust rod extends over the length to be travelled and contains a series of embedded permanent magnets or continuously energized coils. The thrust block or armature comprises a number of annular field coils which are supplied with current from the driver circuit. The coils are switched in sequence to give controlled acceleration and velocity and may be energized to provide a holding force at specified positions. The software associated with proprietary systems may also provide a teach facility in which program points can be stored by moving the thrust block through the desired sequence. At present, speeds of up to $10 \, \text{m s}^{-1}$ may be provided with a travel of 3 m, under thrust loads of 500 N. In smaller units an acceleration of 100 g has been achieved,

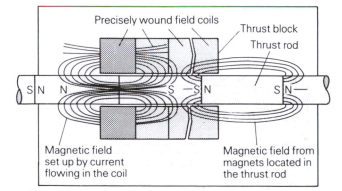

Figure 18.8   Linear drive unit: motion is provided by attraction or repulsion between magnetic poles (Linear Technology Limited).

while in larger units the thrust may be up to 10 kN. These devices have evidently only recently begun to be explored, and they are potentially serious competitors with other forms of high speed, long travel actuators.

18.9   Solenoids

Solenoids are normally employed for short stroke actuators, and were first used for the operation of signalling devices and relays. The construction of a typical solenoid is shown in Fig. 18.9. The force available from a simple solenoid is given by the expression

$$\text{force} = \frac{1}{2}\frac{N^2 I^2}{x^2} A\mu_0 \tag{18.1}$$

where N is the number of turns on the coil, I is the current through the coil, A is the cross-sectional area of the air gap, x is the length of the air gap and μ_0 is the permeability of air.

The force/current characteristic available from a simple DC solenoid is non-linear and depends upon the length of engagement of the plunger within the coil. This is of no particular importance in on/off two-position applications, but a linear characteristic is needed for the proportional solenoids used in electrohydraulic servovalves (Chapter 17), as in Fig. 18.10. This is obtained by designing the armature, pole piece and core tube with appropriate shapes to produce a more nearly constant force over the working stroke. Thus at any position within the working stroke, the solenoid force is proportional to the coil current. In this application the solenoid force is opposed by a spring at the valve spool, or within the solenoid body itself; thus the overall action of the valve is proportional in terms of the flow/current relationship. Because of the effect of flow induced axial forces in spool valves, it is normal in higher performance valves to include a potentiometer and feedback loop to detect and control the solenoid and spool position.

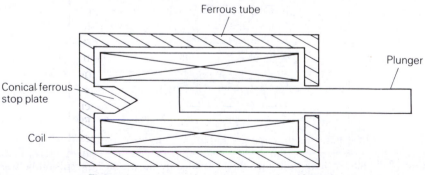

Figure 18.9   Typical construction of a solenoid.

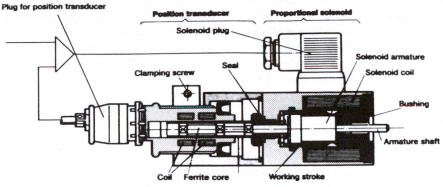

Figure 18.10   Solenoid for a servovalve (Robert Bosch).

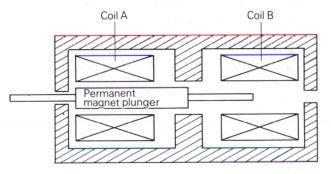

Figure 18.11   Pulsed latching solenoid.

Solenoids are essentially short stroke (to 25 mm) unidirectional devices in which the action is always to draw the plunger into the coil, irrespective of the polarity of the current. Hence they can only be used in conjunction with a returning force such as a spring or by using two opposed solenoids. A ferrous stop plate may be used at the end of the solenoid stroke to increase the flux density and reduce the current, and hence the heating effect, to hold on load. In the selection of solenoids for a given duty, care must be taken with regard to the heat dissipation requirements. In some applications, however, the magnetic flux which remains after switching off may cause undesirable sticking of the plunger to the pole piece. To prevent this, non-ferrous washers may be incorporated in some designs to inhibit the flux path at the end of stroke.

An interesting concept is the pulsed latching solenoid of Fig. 18.11. In the configuration shown, the moving element is a cylindrical permanent magnet carried on an actuator rod and able to move between two coils. In the position shown, no current is flowing in the coils and the magnet is latched in a stable state to the left pole piece. If the coils are then pulsed with current in such a direction as to create adjacent like poles between the magnet face and pole piece, then the magnet will be repelled and move across to the right pole piece. On decay of the current, it will remain there in a stable latched condition until the current is pulsed in the reverse direction. Actual or potential applications for such devices include central door locking systems for cars and the remote operation of on/off fluid control valves.

18.10   Other forms of electrical actuator

Actuators with strokes of a few millimetres have been manufactured based on using electrical heaters to expand fluid or wax filled capsules. However, these suffer from a relatively long time constant and provide forces of a few newtons. Much higher forces are available from electro/thermomechanical devices based on memory metals. The most promising developments, however, are in piezoelectric actuators where substantial forces of 1 kN or more can be developed at very high cyclic frequencies. However, the range of movement is small, usually much less than 1.0 mm, though new developments will undoubtedly increase this with advances in materials and fabrication techniques.

Chapter 19

Rotational drives

19.1   Pneumatic motors: continuous rotation

Pneumatic motors are available in a range of configurations, but are not widely used for continuous rotation in mechatronic systems because of their intrinsically low efficiency and the difficulty in controlling their speed. Two basic types are in common use: vane and piston. Vane motors are met in their most familiar form as the drivers in pneumatic tools such as nut runners, where their high operating speed of 2000 to 10 000 rev min^{-1} and compact format require the use of a multistage reduction gear. A typical vane motor is shown in Fig. 19.1.

Piston-type pneumatic motors are used quite widely in V or in-line configuration as starter motors for diesel engines, an environment where they do not need to be subject to any form of speed control beyond that which can be provided by a simple pressure control or flow restriction. Pneumatic motors in the radial piston configuration shown in Fig. 19.2 are used in applications where compressed air is the preferred working medium, such as in quarry rock drills where air can be used to clear the bit, or in driving agitators in chemical plant where there is a good supply of compressed air and where their intrinsic safety is an added advantage.

The favourable attributes of these pneumatic motors are as follows:

1. Intrinsically safe;
2. Inherently torque limited by the available supply pressure;
3. Robust against mechanical shock or jamming, especially when directly connected without the intermediary of a reduction gear.

Less favourable attributes are as follows:

1. Very inefficient in terms of energy utilization; as little as 20% of the input energy is available as useful shaft work;
2. Require exhaust silencer;
3. Speed is highly load dependent and control of speed can only be achieved by throttling:

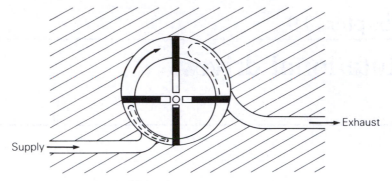

Figure 19.1   Pneumatic vane motor.

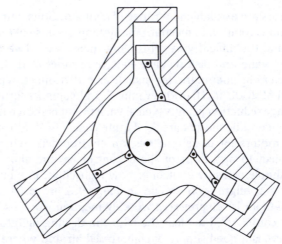

Figure 19.2   Pneumatic piston motor.

Figure 19.3   Vane actuator.

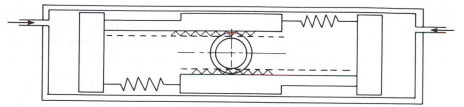

Figure 19.4   Balanced rack and pinion air actuator.

19.2   Pneumatic motors: limited rotation

This family of machines is important, especially in the generic configuration of a quarter turn actuator. These are used to provide rotations of 90° or slightly more, and are often used in process plant to operate fluid control valves such as ball valves and plug cocks where the movement range is between two fixed positions, typically on and off. Two types are available: vane (Fig. 19.3) and rack and pinion (Fig. 19.4). The choice is a matter of preference. A somewhat higher torque is available from the vane type within a given size, but at higher cost.

19.3   Hydraulic motors: continuous rotation

Hydraulic motors are very widely used where high torques are to be transmitted within a restricted space. The configurations available are exceedingly varied and reflect the efforts of manufacturers to get round each other's patents over the years. In terms of applications we can amplify Fig. 16.4 as follows:

Application	Motor type
Low torque, low power	Gear; vane
Low torque, medium power	Vane; axial piston
Medium torque and/or power	Axial piston; Radial piston (inward working)
High power	Axial piston; radial piston
Very high torque	(outward working)

Before considering the relative merits of hydraulic motors *vis-à-vis* competing systems, it is necessary to outline the main features of the types as summarized above.

First and most important, all hydraulic motors which operate hydrostatically (as opposed to hydrokinetic machines such as torque converters) are reversible: the same geometrical configuration can be employed equally satisfactorily as a pump or as a motor. All hydraulic motors have a specific displacement which is usually expressed in cm^3 per revolution or similar volume per angle terms. It must be noted that this is an *ideal* or geometric displacement, which is a function, as set out in the more detailed description of the axial piston machine (section 19.3.3), of such parameters as the number of cylinders and their bore and displacement. Hydraulic motors depend for their satisfactory operation upon the creation of close working clearances during manufacture, and the maintenance of these during the lifetime of the product. All hydraulic motors however are subject to internal leakage through the working clearances, with the effect that the actual displacement is always less than the ideal. Since the output speed is proportional to the volumetric flow rate, it follows that the effect of internal leakage is to reduce the output speed to below what would be expected from the geometric displacement. This is expressed as a hydraulic efficiency:

$$\eta_H = N_{actual}/N_{ideal} \tag{19.1}$$

According to the type of motor, the hydraulic efficiency may vary from 75% in the case of small gear motors to over 95% for axial piston machines. In general, the efficiency is less for smaller machines and reduces with an increasing pressure difference across the motor. Efficiency is also reduced as the output speed falls. The mechanical losses in hydraulic motors, due to factors such as bearing friction, vary according to type, but are usually less significant than the hydraulic losses.

There is a wide variety of configurations available in hydraulic motors, some of which compete for the same market slot and which demonstrate considerable mechanical ingenuity, presumably in order to avoid infringement of earlier patents.

The principal varieties are, however, those briefly listed above, which will now be described in more detail.

19.3.1   GEAR MOTORS

The gear motor shown in Fig. 19.5 is the simplest form of hydraulic motor. They are employed on low power drives up to about 20 kW, such as for cooling fans on diesel locomotives, located in the roof panels where space is at a premium. They are of fixed displacement and therefore control of speed can be exercised only by controlling the oil flow rate to the motor. Internal leakage of oil across the tips of the teeth, through the mesh and especially across the end faces of the gears make these the least efficient of hydraulic motors, but their employment is normally in applications where this can be tolerated.

19.3.2   VANE MOTORS

Vane motors have a higher speed and output power rating than gear motors and are also offered by some manufacturers in a variable displacement configuration (Fig. 19.6). This is achieved by varying the eccentricity of the rotor where, by moving the rotor through centre, it is also possible to reverse the direction of rotation.

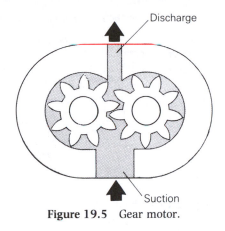

Figure 19.5   Gear motor.

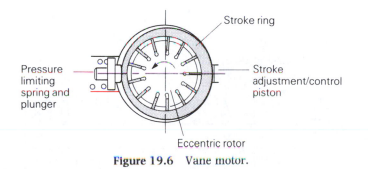

Figure 19.6   Vane motor.

19.3.3   AXIAL PISTON MOTORS

Axial piston motors are undoubtedly of the most interest for employment in a mechatronic system for interfacing to electronic controls. Two varieties are in common use: the bent axis machine of Fig. 19.7, which is usually of fixed displacement, and the variable displacement swash plate motor of Fig. 19.8.

A brief discussion of the construction and working principles of these machines will be useful for the later consideration of how hydraulic transmissions may be controlled. From the diagram of a variable displacement swash plate machine (Fig. 19.8) it may be observed that each piston, when exposed to the supply pressure via the inlet port, exerts an axial force upon the swash plate which thus produces a tangential reaction because of the inclination. The pistons are contained within a cylindrical block, accommodating 7, 9 and 11 or more bores. In most axial piston machines the swash plate does not rotate around its centre, but is able to tilt about a diameter. The whole cylinder block is free to rotate about its axis of symmetry, the admission and exhaust of the working fluid taking place via banana-shaped ports in a fine ground or lapped valve plate

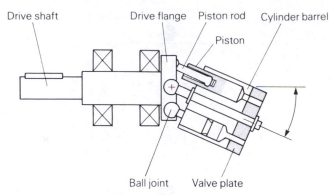

Figure 19.7   Bent axis pump/motor.

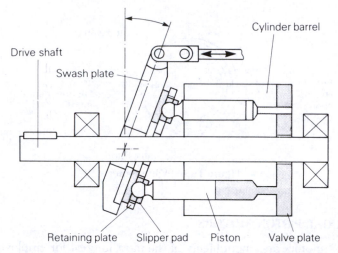

Figure 19.8   Variable displacement swash plate pump/motor.

which slides on the rear face of the block. The angle through which the inlet
port extends is approximately $\pi/2 - d/R$, where d is the cylinder diameter and
R is the pitch radius of the cylinder. Thus the number of pistons exposed to the
supply pressure fluctuates by one unit as each crosses the division between the
pressure and return ports. The net effect is to produce a torque reaction whereby
rotation is imparted to the cylinder block and thus to the output shaft.

Figure 19.9 shows the geometry. The swash plate is tilted through an angle
θ from the normal untilted position. ϕ is the angle made between the plane of
the *untilted* swash plate and a line on the tilted plate which is tangential to the
pitch circle of the cylinders, radius R.

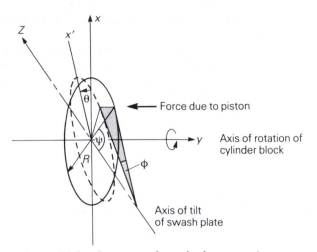

Figure 19.9   Geometry of swash plate pump/motor.

For an individual piston the reaction vector in the tangential direction is thus $p a \tan \phi$, where p is the system pressure and a is the piston area, and the corresponding torque contribution is $p a R \tan \phi$. However, ϕ is not constant but varies as the motor cylinder block rotates. ψ is the angle through which the cylinder block has rotated from the position where the piston crosses the swash plate tilt axis. It may be seen that

$$\tan \phi = \cos \psi \tan \theta \qquad (19.2)$$

and thus

$$T = p a R \cos \psi \tan \theta \qquad (19.3)$$

where T is the torque contribution of a single piston. For a piston at its full stroke,

$$\psi = \pi/2 \qquad (19.4)$$
$$\phi = 0 \qquad (19.5)$$

Thus

$$T = 0 \qquad (19.6)$$

For a piston at its full outstroke,

$$\psi = 3\pi/2 \qquad (19.7)$$

For a piston on the swash plate axis,

$$\psi = 0 \text{ or } \pi \qquad (19.8)$$
$$\phi = \theta \qquad (19.9)$$

This last is the position of maximum torque, where

$$T = p a R \tan \theta \qquad (19.10)$$

This can only occur at either $\psi = 0$ or $\psi = \pi$, as one of these position must be on the induction stroke.

The total torque output of an axial piston motor is a function of the number of cylinders and is subject to fluctuation. The fluctuation, which may be troublesome in exciting torsional vibrations, becomes of smaller amplitude with an increasing number of cylinders. For example, a motor with seven pistons has the following torque output:

$$\text{maximum torque} = 2.18\, p\, a\, R \tan \theta \tag{19.11}$$

$$\text{minimum torque} = 2.11\, p\, a\, R \tan \theta \tag{19.12}$$

in which case the fluctuation has a half cycle extending over 25.7° of rotation.

The volumetric flow rate demanded by an axial piston machine is a function of the number of piston strokes per unit of time, and can therefore be expressed as

$$Q = f N a R \tan \theta \tag{19.13}$$

where f is the rotational frequency and N is the number of cylinders.

It will be observed that the geometric displacement (otherwise termed the ideal or specific displacement) per revolution can be expressed as

$$D = N a R \tan \theta \tag{19.14}$$

Normally in manufacturers' literature, D would be given for maximum values of θ.

Thus we observe that both the output torque and the flow rate of an axial piston motor (or the equivalent input parameters for a pump) are proportional to the tangent of the swash plate angle. For zero angle, no flow is possible, but if the swash plate is driven overcentre, through and beyond the null position, a reversal of rotation may be obtained.

The position control of the swash plate is provided by a linear actuator built as an integral part of the machine. This uses a potentiometer or linear variable inductive transformer (LVDT) to determine the swash plate position.

19.3.4 Radial piston motors

Radial piston motors are used for comparatively low speeds, up to about $400\,\text{rev min}^{-1}$, but, depending on their size, are capable of very high torque outputs of $50\,000\,\text{N m}$ or more. A typical example is shown in Fig. 19.10 of the inward working piston type, which now finds many applications particularly on mobile plant such as cranes. The motion of the pistons in these machines is transferred to the output shaft using sliding pads or short connecting rods and an eccentric. Oil inlet and outlet are provided using ported sleeves on the non-output end of the shaft. The very largest hydraulic motors are of the cam ring type of Fig. 19.11 with the pistons working outwardly.

Radial piston machines are usually of fixed displacement. However, in double banked versions of the inward working type, or in cam ring motors having

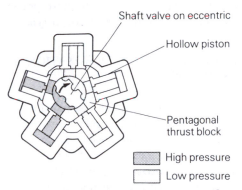

Figure 19.10   Hydraulic motor, radial piston inward working type.

even numbers of cylinders, it may be possible by means of suitable valving to make say half of the cylinders inoperative, thereby doubling the output speed for the same input flow rate. Radial piston motors have been demonstrated in which it is possible to vary the position of the eccentric steplessly and thereby achieve fully variable control of the motor output characteristics. This is no easy task to accomplish mechanically, where the radial position of a highly loaded rotating machine element must be precisely controlled within the motor casing.

All variable displacement hydraulic motors become very inefficient when approaching zero swash or rotor eccentricity. This is because the torque due to mechanical friction becomes a larger proportion of the ideal output torque, until eventually the motor stalls. A swash plate angle of 15° is regarded as a practical minimum.

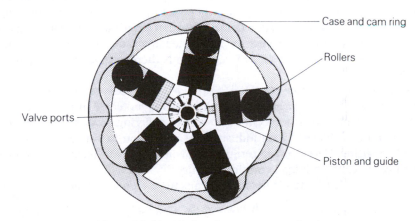

Figure 19.11   Hydraulic motor, cam ring type.

Within the field of hydraulic motors summarized above there are evidently substantial overlaps and exceptions. The highest power hydraulic motors available in the world today are of the axial piston type.

19.3.5   GENERAL CHARACTERISTICS OF ROTATIONAL HYDRAULIC TRANSMISSIONS

Undoubtedly the principal advantage conferred by hydraulic motors is their high torque/weight ratio. Chapple and Dorey (see Bibliography) have published data showing output characteristics of a range of commercially available hydraulic motors in terms of N m per kilogram of machine mass. In bent axis or axial piston motors, the type which corresponds most closely with conventional AC induction motors in terms of their normal operating speed, the specific torque is in the region of $15 \, \mathrm{N \, m \, kg^{-1}}$. This is ten times or more higher than is available from typical induction motors, on the basis of full load torque directly at the machine shaft. Hence a hydraulic motor can often be used to drive a mechanical system directly without the use of a reduction gearbox. This confers a particular advantage in applications where there is a possibility of the driven machine being suddenly arrested by jamming, examples being agitators and mixers. Where an electrical drive is used with a substantial gear reduction, it is possible to show in specific cases that up to 95% of the rotational kinetic energy of the whole system resides in the rotor of the electrical machine, when the $I\omega^2$ contribution of each part is considered.

Conventional overload trips may not respond sufficiently quickly. Even after the current has been switched off, the residual kinetic energy in the rotor may impart sufficient strain energy into the downstream part of the system to cause mechanical breakage. In such environments electrical drives are often mechanically protected by means of slipping clutches or fluid couplings. These however introduce additional complication and space requirements and, in the mechanical variety, are notoriously difficult to set up. Long term inaction may also prejudice the operation of such devices when at length they are called upon to perform. By contrast, the directly' connected hydraulic motor has a low rotational kinetic energy and the whole system is thereby more robust against shock loads.

Protection against normal overload is provided by system or cross-line relief valves as in Fig. 19.12. However, hydraulic systems protected in this way lack the ability to 'bump start' a mechanical system suffering from stiction in the way that can be achieved by an industrial AC induction motor with direct on line (DOL) starting. Here starting torques, allowable for short durations, can be up to three times the full load running torque.

The favourable attributes of hydraulic transmissions are as follows:

1. High torque and power with low weight and bulk;
2. Very resistant to mechanical shock;

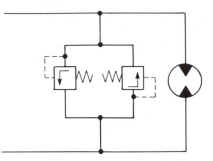

Figure 19.12   Cross-line relief valves.

3. Acceptable in areas with a hazard classification;
4. Low speed varieties have a low noise level;
5. The ability to remove, to a substantial degree, sources of both noise (in the case of low speed, high torque motors only) and heat generation from the working environment. This requires the pump unit and power pack to be remotely positioned and suitably insulated and serviced to contain noise and remove heat developed within the hydraulic system.

Less favourable attributes are as follows:

1. They require provision of a power pack or hydraulic main;
2. High speed hydraulic motors (over $1000\,\mathrm{rev\,min^{-1}}$) tend to produce a high noise level;
3. Apart from direct drive, high torque, low speed motors, system costs are higher than for electrical drives.

The majority of hydraulic transmissions employ one pump per hydraulic motor in a transmission circuit, as shown in Fig. 19.13. It is possible to employ variable capacity in both machines, in which case the generalized output characteristic is as shown in Fig. 19.14. However, it is usual to base operation on the assumption of a constant output power, though such conditions rarely occur in practice.

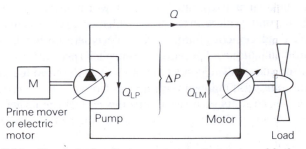

Figure 19.13   Simple hydraulic transmission: Q_L is internal leakage flow.

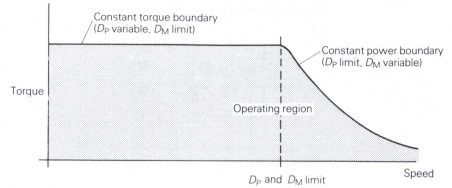

Figure 19.14   Torque/speed characteristic for a combination of variable displacement pump and motor with pressure limitation.

For any particular settings of displacement D of the two machines, the speed and torque ratios are given respectively by

$$N_M/N_P = D_P/D_M \tag{19.15}$$

$$T_M/T_P = D_M/D_P \tag{19.16}$$

where subscripts P and M refer to the pump and the motor. These are ideal characteristics. Whereas the torque ratio is little affected by the small mechanical losses, apart from the case where the motor is near its minimum displacement, the speed ratio is significantly altered by the internal leakage flows Q_L. Thus

$$N_M/N_P = \eta_{hM}\eta_{hP}D_M/D_P \tag{19.17}$$

where

$$\eta_{hM} = (Q - Q_{LM})/Q$$

$$\eta_{hP} = Q/(Q + Q_{LP})$$

The simple open loop form of speed control is employed on applications such as winch drum drives and is obtained by variation of the pump displacement only, the motor being of fixed displacement. The actual output speed is therefore load dependent, influenced by the internal leakage. In this respect the drive is not positively held in a fixed ratio, as in a worm drive, but has a loss of synchronization comparable with the creep which occurs in V-belt pulley drives.

The variable displacement pump, fixed displacement motor drive (or PV–MF drive) is without difficulty configured as a controlled speed drive by the inclusion of a speed sensor and control loop as in Fig. 19.15. Here the integral hydraulic actuator in the pump is equipped with an electrohydraulic servovalve. This is a mechatronic device in its own right, forming part of the larger system.

The characteristics of the driven machine or application are a particularly important consideration when selecting the control strategy for a hydraulic transmission. These may be divided into four broad categories:

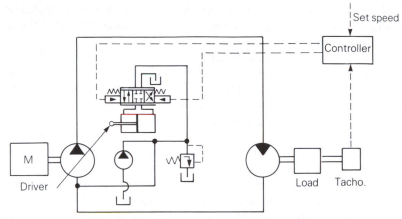

Figure 19.15   Hydraulic transmission with speed control.

1. Winching and reeling operations (assuming no significant change in drum diameter): T = constant.
2. Rapid acceleration applications (e.g. a batch centrifuge): $T = C/N$: constant power.
3. Constant product of speed and force (e.g. a high production lathe): constant power.
4. The drive of hydrodynamic machines (e.g. centrifugal pumps, fans and propeller-type mixers): $T = CN^2$.

There is increasing interest in the use of hydraulic transmission in ring or distributed systems since it is evidently logical to group services such as tankage, filtration and oil conditioning and to include the whole in a noise attentuating enclosure. Thus the pumps may be grouped to discharge to a common main held at constant pressure. Each pump will be equipped with a pressure override device whose set point may be routinely adjusted. A reduction in system flow demand resulting in an increase in system pressure will cause the pump stroke to back off. The converse follows in the event of an increase in flow demand.

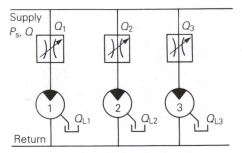

Figure 19.16   Hydraulic ring circuit.

Parallel connections are used in almost all practical industrial applications of hydraulic ring mains in order to ensure that individual control of the user machines can be assured.

A typical arrangement of a ring circuit is shown in Fig. 19.16, in which fixed displacement hydraulic motors are each controlled by a separate pressure compensated flow control valve. These are well proven mechanical devices, such as that of Fig. 19.17, with either a manual change of set point or remote positioning using, for example, a pneumatic bellows or stepper motor. Their effect is to maintain a constant flow rate against changes in the pressure required at the user machine as a result of load variations. This is achieved provided that the main system pressure, upstream of the valve, remains constant.

In selecting the motors, regard must be paid to the corner power, or the product of the maximum speed and the maximum torque which the motor may have to produce, even though these maxima may never coincide. The motor specific displacement will be chosen to produce an output torque which matches the maximum torque required by the driven machine during its anticipated operating cycle. In practice, however, the torque demand is often likely to be less than the design maximum.

Since $T_{\mathrm{motor}} = f\Delta P$, speed control is exercised by introducing a pressure drop across the flow control valve. For systems which have a variable duty cycle, as is often encountered in batch process chemical plant, the power utilization diagram may appear as shown in the hypothetical example of Fig. 19.18. A particularly adverse situation occurs when a number of machines run at high speed but at low torque levels, while others working in 'stickier' conditions need high torque and thus determine the system set pressure. The hatched area in Fig. 19.18 represents the aggregate of the power lost at the control valves, plus of course a smaller loss due to the internal leakage flow.

Measures which may be adopted to ameliorate the problem have included a degree of self-tuning control whereby the system pressure is controlled so as to provide only a small margin above that required at the most heavily loaded machine. However, this does require that the torque/speed characteristics of the driven machines be predictable and that speed and/or torque sensing be provided at each. Nevertheless, this does not address the situation of mismatch

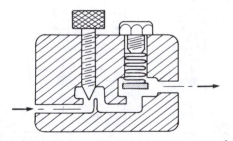

Figure 19.17   Pressure compensated flow control valve.

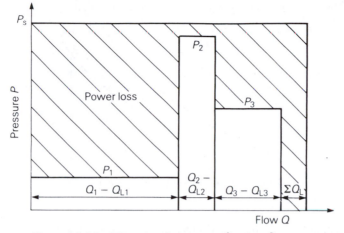

Figure 19.18   Ring circuit power utilization diagram.

described above. A more novel approach involves using variable displacement hydraulic motors, driven from the constant pressure hydraulic main, without the interposition of loss producing flow control valves. On/off valves would be retained for reasons of directional control or safety. The operational diagram is shown in Fig. 19.19. Implementation of the mechanical hardware is straightforward on axial piston motors, as pump/motor units are available with integrated swash plate positioners driven via an electrohydraulic servovalve. The operation of this system may be visualized with reference to Fig. 19.20. Here, the driven machine characteristic under a given set of conditions is represented by curve (a). Suppose the conditions change – for example, in a

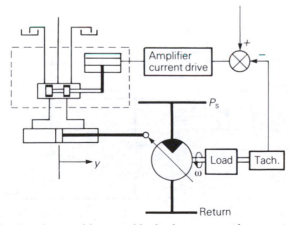

Figure 19.19   Speed control by variable displacement with motor in ring circuit.

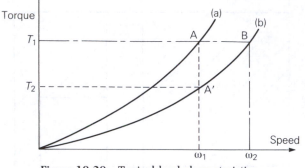

Figure 19.20   Typical load characteristics.

stirred chemical vessel the mix viscosity reduces – and the characteristic curve therefore changes to (b). In the absence of control action, the speed would increase from ω_1 to ω_2, i.e. the state of the system would move from A to B. In order to maintain the speed at the set point ω_1, the torque output of the hydraulic motor must be reduced by reducing the displacement, and thereby moving to the state A′.

Experimental investigations have demonstrated that it is possible to achieve stable control by this means, and also by the simple hydromechanical systems outlined in Fig. 19.21. However, although ring main systems with pressure compensated flow control valves are quite common, there is little evidence to date of industrial systems incorporating more sophisticated control strategies. A few demonstration installations are needed to inspire confidence.

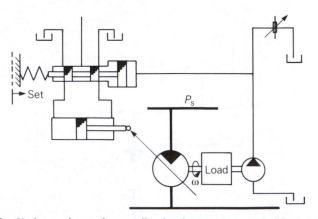

Figure 19.21   Hydromechanical controller for the speed control of hydraulic motors in ring circuits.

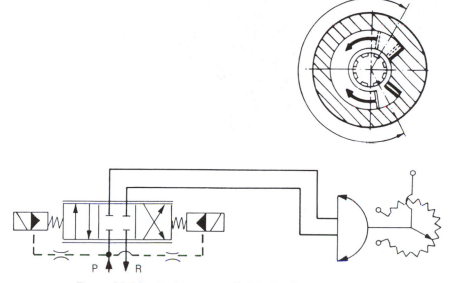

Figure 19.22   Position controlled hydraulic vane actuator.

19.4   Hydraulic motors: limited rotation

These outwardly resemble the pneumatic limited rotation actuators described above, but produce a much larger torque output in relation to their physical size. The vane type (Fig. 19.22) may be used with a shaft encoder and an electrohydraulic valve to enable precise control of position to be obtained.

19.5   Electric motors

19.5.1   DC MACHINES

A DC machine is constructed with a stationary field winding and a rotating armature winding as in Fig. 19.23. The field winding is supplied with DC to produce a static magnetic field pattern within the machine. This magnetic field then interacts with the current in the armature conductors to produce a torque which rotates the armature. In order to sustain this torque, the distribution of current in the armature conductors must be maintained constant relative to the field irrespective of the rotation of the armature. This is achieved by means of the commutator, which acts to reverse the current in the armature conductors as they pass from under one field pole to the next. Also, as the armature conductors are moving relative to the magnetic field they have induced in them a voltage, referred to as the back EMF of the armature and which appears as a DC voltage at the commutator.

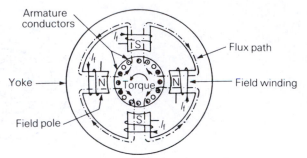

Figure 19.23   Four-pole DC machine.

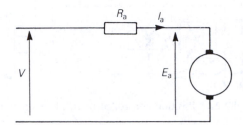

Figure 19.24   DC machine armature circuit (motoring).

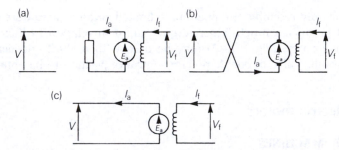

Figure 19.25   DC machine braking modes: (a) dynamic or resistive braking (b) reverse current braking or plugging (c) regenerative braking, $E_a > V$.

When motoring, the DC machine draws power from the DC supply and delivers mechanical power at its shaft. When used as a generator, mechanical power is input via the shaft which is converted to an electrical power output in the armature. The defining equations are, referring to Fig. 19.24:

Motoring: $E_a = V - I_a R_a$ (19.18)

Generating: $E_a = V + I_a R_a$ (19.19)

Back EMF: $E_a = K\phi\omega$ (19.20)

where ϕ is the flux per pole ω is the rotational speed in radians per second.

In addition,

$$\text{internal mechanical power} = T\omega = E_a I_a \tag{19.21}$$

$$\text{useful mechanical power} = \text{internal mechanical power}$$
$$- \text{mechanical losses} \tag{19.22}$$

$$\text{torque} = K\phi I_a \tag{19.23}$$

Electrical braking of a DC machine can be achieved by any of the techniques shown in Fig. 19.25.

(a)   DYNAMIC OR RESISTIVE BRAKING

If a resistor is connected across the armature of the DC machine in the presence of a field current, the energy stored in the inertia of the mechanical system is dissipated in driving a current through this resistor.

(b)   REVERSE CURRENT BRAKING OR PLUGGING

The reversal of the armature connections of the machine results in a corresponding change in the direction of armature current. This results in a torque reversal which rapidly decelerates the machine. This is a very severe condition, since the machine equation under these conditions can be written as

$$I_a R_a = V + E_a \tag{19.24}$$

A current limiting resistor is often introduced to limit the magnitude of the armature current.

The armature voltage must be removed once the machine is at standstill to avoid it accelerating and running in the reverse direction.

(c)   REGENERATIVE BRAKING

If the field of the machine is adjusted such that $E_a > V$, the machine will act as a generator returning energy to the DC supply. To maintain regeneration as the machine slows down, the field current must be increased accordingly within the limit set by the maximum permitted field current.

19.5.2   DC VARIABLE SPEED DRIVES

Referring to the defining equations 19.18 to 19.20 for a DC machine then, neglecting armature resistance

$$V = E_a = K\phi\omega \tag{19.25}$$

If V is kept constant it can be seen from equation 19.25 that the motor speed

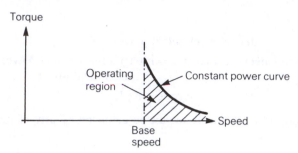

Figure 19.26   Operating region of a DC motor with field control: constant power curve set by rated armature current.

ω can be controlled by varying the field such that

$$1/\omega \propto \phi \qquad (19.26)$$

This is referred to as field control. It gives rise to the operating region on the torque/speed characteristic shown in Fig. 19.26, the boundary of which is defined by the maximum continuous current that the armature winding can carry, that is the rated current of the machine, and is a constant power curve. Base speed is defined by the combination of maximum armature voltage and maximum field current, and speed can be varied over a range of the order of 2 to 1 for machines above a few kW and 4 or 5 to 1 for smaller machines, with the upper limit set by commutation performance.

Referring again to equation 19.25, if the field ϕ is kept constant then the motor speed can be controlled by varying the applied armature voltage when

$$v \propto \omega \qquad (19.27)$$

Applying the same restriction on armature current as before, this results in the torque/speed envelope of Fig. 19.27, which has a constant torque boundary over its range. Speed variations of the order of 100 to 1 can be achieved by this means.

By combining the curves of Fig. 19.26 and 19.27 the overall torque/speed envelope of Fig. 19.28 is obtained, which may be compared with that of Fig. 19.14 for a hydraulic drive.

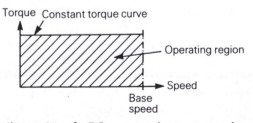

Figure 19.27   Operating region of a DC motor with armature voltage control: constant torque curve set by rated armature current.

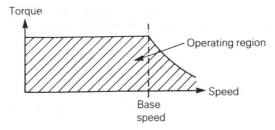

Figure 19.28 Operating region for a DC motor with both armature and field control.

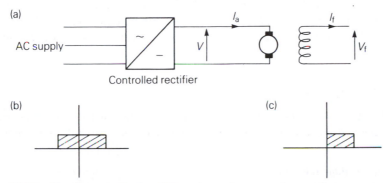

Figure 19.29 Speed control of a DC motor using armature voltage control from an AC supply: (a) controlled rectifier supplying a variable armature voltage (b) two-quadrant operation with a fully controlled converter (c) single-quadrant operation with a half controlled converter.

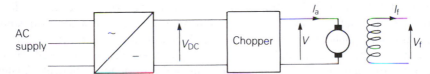

Figure 19.30 Speed control of a DC motor using a chopper drive from a DC supply.

The most common implementation of DC variable speed drive is that of Fig. 19.29a, in which a controlled rectifier is used to supply a variable DC voltage to the armature of the machine. If a fully controlled converter is used then regeneration is possible, giving two-quadrant operation as indicated by Fig. 19.29b. The use of a half controlled converter would prohibit any regeneration, and operation would then be in the first quadrant only as shown by Fig. 19.29c. The field of the machine is supplied by an additional rectifier, which may be controlled or uncontrolled as required.

Alternatively the configuration of Fig. 19.30 could be used, where an uncontrolled converter produces a DC voltage which is then supplied to a DC chopper circuit and thence to the motor. A DC chopper would also be used where a DC supply was already available, as for example on a battery powered electric vehicle.

The speed of the motor is determined by the mean armature voltage applied. The mean value of armature current defines the average torque, with any fluctuations tending to be smoothed by the inertia of the system.

For larger motors (> 2.5 kW) the armature inductance is normally sufficient to maintain an effectively constant DC under all but light load conditions. Machines intended to be used with converters are, however, normally designed with increased values of armature inductance.

Some derating of the machine is required because of the presence of the ripple currents in the motor supply which increase the machine losses. This derating is less with larger machines because of the smoothing effect of the armature inductance.

(a) REVERSING DRIVES

The direction of rotation of a DC machine can be reversed by reversing the either the armature voltage, and hence the armature current, or the field.

Armature voltage reversal

The simplest cheapest means of reversing the sense of the applied armature voltage is by the use of a contactor as shown in Fig. 19.31a. This requires that the armature current is first reduced to zero before the contactor is switched over, and introduces a time delay of the order of 0.2 seconds into the time required for reversal. For this reason, contactor reversal is used for drive applications such as hoists, presses, marine propulsion systems and some machine tool applications where speed of reversal is not important.

Where more frequent and more rapid torque reversals are required, as for example in paper making or steel strip mills, a dual bridge configuration such as those of Fig. 19.31b and c can be used. In the arrangement of Fig. 19.31b, only one bridge may conduct at any instant to avoid short-circuiting the supply. This requires that the armature current is brought to zero before load is transferred to the second bridge, introducing a delay of a few milliseconds into the reversal. By including current limiting reactors as in Fig. 19.31c, both bridges can be allowed to conduct simultaneously. This allows the transfer of armature current between bridges to take place without the introduction of any delays.

Field reversal

Field reversal can be achieved either by a contactor as in Fig. 19.32a, or by a dual bridge configuration as in Fig. 19.32b. Reversal of the field current is

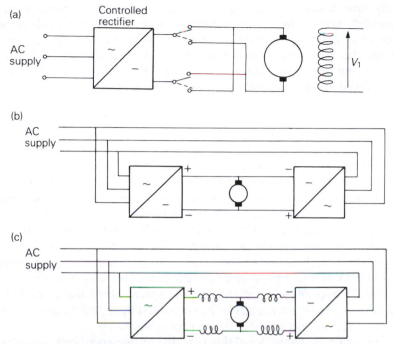

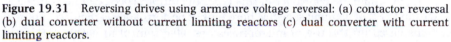

Figure 19.31 Reversing drives using armature voltage reversal: (a) contactor reversal (b) dual converter without current limiting reactors (c) dual converter with current limiting reactors.

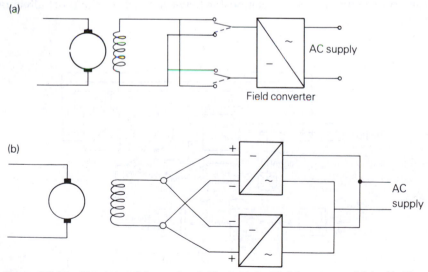

Figure 19.32 Reversing drives using field reversal with (a) contactor (b) dual bridges.

relatively slow because of the need to remove the stored energy from the magnetic field prior to field current reversal, with the introduced delay being of the order of 1 second. In order to speed up the reversal, field forcing is used to increase the rate of rise of field current by applying an initial field voltage in excess of that needed to maintain the required field current. The field voltage is then reduced as the field current increases to the appropriate value.

(b)   CONTROL

The basic control configuration for a single quadrant variable speed DC drive is shown in Fig. 19.33. The demand speed is fed to the comparator via a ramp generator in order to smooth out the effect of any sudden changes in setting. At the comparator the demand speed is compared with the actual speed as derived from a tachogenerator on the machine shaft, and the error signal is taken as input to a current limiting amplifier which acts to restrict the error signal and hence the maximum current in the machine. The output of the current limiting amplifier is then compared with the system current from either the AC or the DC side of the converter and the output is used to control the firing angle of the converter to match the demand speed. This basic controller will operate with a speed error which can be eliminated by the introduction of a full proportional integral derivative (PID) control scheme.

Using an analogue controller of this type, the output speed can be controlled to within 0.1% of demand speed. Where higher performance is required, control systems based on the use of microprocessors offer control to within 0.01% of demand speed. Digital systems also permit the precise control of the speed of a number of machines relative to a common speed reference, and can include the ability to control the phase relationship between the drive shafts of individual motors.

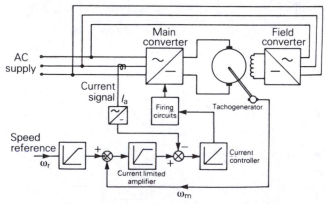

Figure 19.33   DC machine controller.

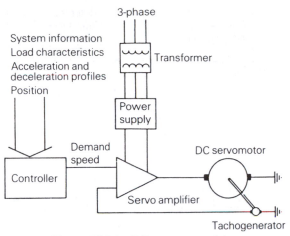

Figure 19.34 DC servomotor drive.

19.5.3 DC SERVOMOTORS

DC servomotors are usually constructed with a permanent magnet field and a wound armature, giving them characteristics similar to those of a shunt motor. Servomotors are designed to provide controlled accelerations and decelerations as well as motion control and are used in a wide variety of positioning applications. Figure 19.34 shows a typical drive configuration. The basic DC servomotor is being replaced in many applications by brushless machines.

19.5.4 INDUCTION MACHINES

Induction machines form the bulk of the electrical drives used in industry by virtue of their combination of simplicity and robustness. In an induction machine a rotating field is produced in the air gap by the AC in the stator windings. This field then interacts with the rotor windings to induce a voltage and hence a current into the rotor conductors. Torque is then produced by the interaction of the field and the rotor currents.

In order for a voltage to be produced in the rotor windings, there must be some relative motion between the rotor conductors and the magnetic field produced by the stator. The speed of rotation of the field is defined by the frequency f of the AC supply and the number of poles p on the machine, and is referred to as the synchronous speed:

$$\text{synchronous speed} = 120 f/p \qquad (19.28)$$

The difference between the synchronous speed and the actual speed of the rotor is expressed by the slip:

$$\text{slip } s = \frac{\text{synchronous speed-actual speed}}{\text{synchronous speed}} \qquad (19.29)$$

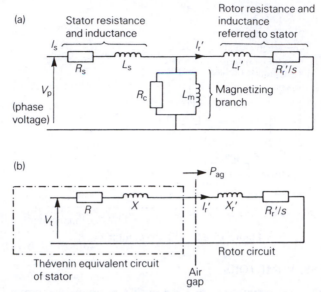

Figure 19.35   Induction machine equivalent circuits: (a) equivalent circuit of one phase of a polyphase induction machine (b) Thévenin equivalent circuit.

Figure 19.35a shows the equivalent circuit for one phase of a polyphase induction machine, usually simplified to the Thévenin form of Fig. 19.35b.

The total internal mechanical power developed can be expressed in terms of the torque T and the mechanical angular velocity of the shaft ω_m as

$$P_m = T\omega_m \tag{19.30}$$

The power transferred across the air gap into the rotor is then

$$P_{ag} = I_r'^2 R_r'/s \tag{19.31}$$

when

$$\text{power loss in rotor resistance } P_r = I_r'^2 R_r' \tag{19.32}$$

The internal mechanical power is then

$$P_m = P_{ag} - P_r = P_{ag}(1 - s) \tag{19.33}$$

and

$$\omega_m = (1 - s)2\omega/p \tag{19.34}$$

Thus

$$T = \frac{p}{2\omega} \frac{V_t^2}{(R + R_r'/s)^2 + (X + X_r')^2} \frac{R_r'}{s} \tag{19.35}$$

or, for an m-phase machine,

$$T = \frac{mp}{2\omega} \frac{V_t^2}{(R + R_r'/s)^2 + (X + X_r')^2} \frac{R_r'}{s}$$

(19.36)

where ω is the supply frequencies term.

The torque/speed (or torque/slip) curve for the induction machine can now be plotted and has the general form of Fig. 19.36.

Electrical braking can be achieved with an induction motor in a variety of ways, as follows.

(a)   REVERSE CURRENT BRAKING OR PLUGGING

If the induction machine is connected so that the direction of rotation of the stator magnetic field is reversed relative to the direction of rotation of the rotor, the slip will be greater than 1 and operation will be in the braking region of Fig. 19.36. This is a very severe condition and results in high currents in the machine windings.

If the supply to the stator is not disconnected once the speed has been reduced to zero, then the machine will accelerate in the reverse direction to its original rotation.

(b)   REGENERATIVE BRAKING

If the frequency of the supply is reduced so that the slip becomes negative, the induction machine will operate in the generating region of Fig. 19.36, returning energy to the supply.

(c)   DC DYNAMIC BRAKING

By connecting a DC supply to the stator terminals as shown in Fig. 19.37a, a stationary magnetic field is produced in the air gap which interacts with the rotating rotor to produce a braking torque with the characteristic of Fig. 19.37b.

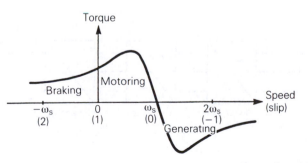

Figure 19.36   Torque/speed (torque/slip) characteristic of an induction machine.

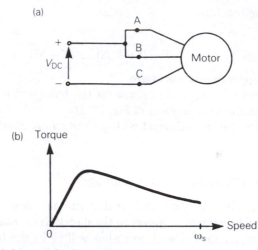

Figure 19.37   DC dynamic braking of an induction motor: (a) connection of DC supply (b) torque/speed characteristic.

19.5.5   AC VARIABLE SPEED DRIVES

The majority of AC variable speed drives are based around the use of an inverter to provide a variable frequency supply to an induction motor, as illustrated by Fig. 19.38. Inverters with ratings from fractions of a kilowatt to several megawatts are currently available, offering a wide range of features.

The maximum torque of an m-phase induction machine is given by

$$T_{max} = \frac{mp}{4\omega} \frac{V_t^2}{R + [R^2 + (X + X_r')^2]^{1/2}}$$

Ignoring the resistive term, this equation can be written in terms of the supply frequency as

$$T_{max} = \text{constant} \times V^2/f^2$$

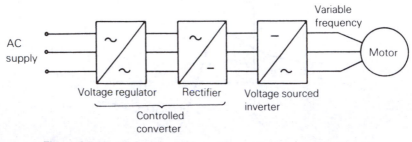

Figure 19.38   Induction motor with a voltage sourced inverter.

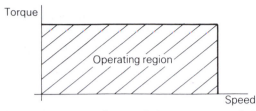

Figure 19.39   Induction motor speed control characteristic with V/f held constant.

Thus in order to maintain T_{max} constant the ratio V/f should be maintained constant. This would produce a torque/speed regime of the form shown in Fig. 19.39. In practice, the effect of the resistance is to decrease the torque available at lower speed, producing a characteristic of the form of Fig. 19.40a. To compensate for this a boost setting is included in the inverter to increase the applied voltage at these lower speeds; in this case V/f is no longer constant but varies in the manner of Fig. 19.40b.

Voltage and frequency control is usually employed up to the normal supply frequency, at which point full voltage is applied to the induction machine. For speeds, and hence supply frequencies, above this point a square wave signal with no voltage control is usually applied to the induction machine. The full torque/speed characteristic of the inverter drive is therefore as shown in Fig. 19.41, and should be compared with that of Fig. 19.28 for the DC machine and Fig. 19.14 for the hydraulic machine.

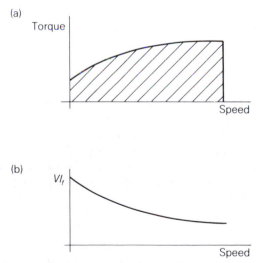

Figure 19.40   Resistance effects on speed control of an induction motor: (a) practical torque/speed envelope showing reduction in torque limit at low speed (b) effect of voltage boost on the V/f ratio at low speeds.

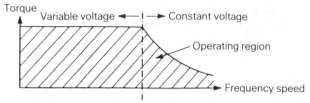

Figure 19.41 Complete operating envelope of an induction motor with variable frequency control.

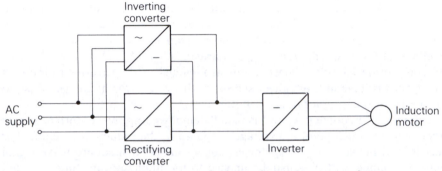

Figure 19.42 Regenerative braking of an induction machine.

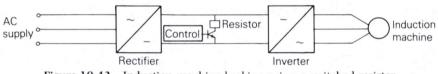

Figure 19.43 Induction machine braking using a switched resistor.

(a) BRAKING

Regenerative braking is obtained by controlling the inverter so that the induction machine operates with negative slip. This results in an increase in the DC voltage and allows energy to be returned to the supply by means of the arrangement of Fig. 19.42. Where this configuration is not possible, then a braking resistor can be included at the input to the inverter to absorb the regenerated power. Control is achieved by using a switching transistor in series with the braking resistor, as in Fig. 19.43.

(b) CONTROL

Figure 19.44 shows the structure of a typical controller for an inverter drive. The demand speed ω_r is compared with the signal from the tachogenerator ω_a to produce an error signal ω_e. This error signal is supplied to a regulator which

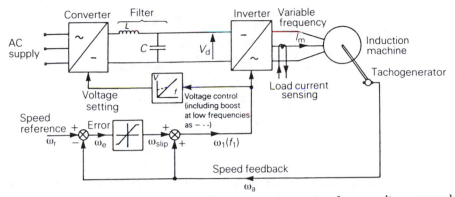

Figure 19.44   Control system for an induction motor operation from a voltage sourced inverter.

sets a slip speed ω_{slip} proportional to the actual speed. By limiting ω_{slip} to a value corresponding to the maximum torque, the motor is prevented from stalling. The required frequency is then determined by adding ω_{slip} and ω_a, and the resulting signal ω_1 is used to control the inverter.

(c)   SLIP RING INDUCTION MACHINES

Where the induction machine has a wound rotor, the ends of the rotor windings can be made available via slip rings, enabling access to the rotor circuit. Slip energy recovery schemes such as the static Kramer drive of Fig. 19.45 provide a means of speed control by extracting energy from the rotor and returning it to the supply.

19.5.6   STEPPER MOTORS

Stepper motors can be used to provide either a continuous controlled rotation or a series of discrete angular motions, and are very suited to computer control

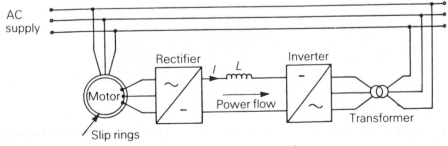

Figure 19.45   Static Kramer drive.

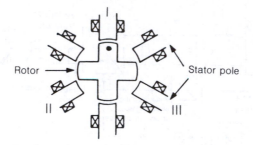

Figure 19.46   Single stack, variable reluctance stepper motor configuration.

in a variety of applications. A number of different types of stepper motor are available providing different characteristics.

(a)   VARIABLE RELUCTANCE STEPPER MOTOR

The construction of a simple, single stack, variable reluctance stepper motor is illustrated by Fig. 19.46. From this it will be seen that the stator, which carries the three pairs of energizing windings (phases), has six poles while the rotor has only four poles. With phase I energized, the rotor will be pulled into the position

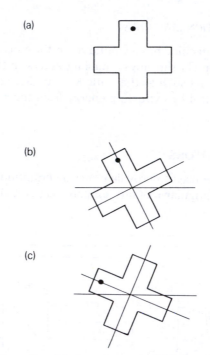

Figure 19.47   Stepping motion: (a) phase I energized (b) phase II energized (c) phase III energized.

shown in Fig. 19.47a. If phase I is now turned off and phase II is energized, then the rotor will move to the position shown in Fig. 19.47b, and then to the position of Fig. 19.47c with phase III energized. Repeating the sequence will cause the rotation to continue, with the rotor requiring a total of 12 steps to return to its original position. By energizing pairs of phases in sequence, intermediate positions can be obtained.

An alternative form of variable reluctance stepper motor is the multistack motor of Fig. 19.48. Operation is similar to that of the single stack motor but with each of the stacks being energized in turn.

It should be noted that the variable reluctance stepper motor only produces a torque when a phase is energized, and they are otherwise free to rotate.

(b)   PERMANENT MAGNET STEPPER MOTORS

The permanent magnet stepper motor, as its name suggests, has a permanent magnet in its rotor. Energizing each of the stator phases in turn will cause the rotor to rotate.

The permanent magnet stepper motor has the advantage that the rotor will align itself with one or other of the stator poles even when the motor is de-energized, and will provide a small holding or detent torque to maintain itself in that position. However, the stepping angle that can be obtained from the motor is limited by its construction.

(c)   HYBRID STEPPER MOTOR

The hybrid stepper motor combines features of the variable reluctance and permanent magnet stepper motors, incorporating a permanent magnet core with a toothed rotor.

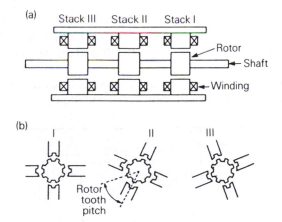

Figure 19.48   Four-pole, three-stack variable reluctance stepper motor: (a) construction (b) rotor stacks, stack I energized.

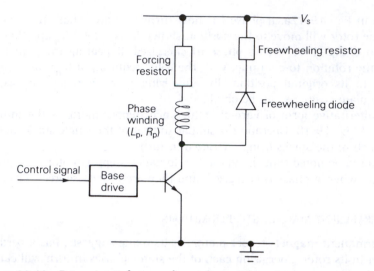

Figure 19.49   Drive circuit for one phase of a variable reluctance stepper motor.

Hybrid stepper motors generally operate with smaller step angles than the variable reluctance and permanent magnet steppers and have a higher torque/volume ratio. They also provide a detent torque when the stator windings are de-energized.

(d)   DRIVE CIRCUITS

A typical drive circuit for one phase of a stepper motor is shown in Fig. 19.49. The control pulses are generated by the control circuit, which may be a dedicated integrated circuit or a microprocessor, and used to switch the power transistor into the on state. In order to increase the rate of rise of the phase current, the supply voltage is increased and a forcing resistance is included in series with the phase winding. The energy stored in the magnetic field of the winding on turn-off is dissipated in the forcing and freewheeling resistances via the freewheeling diode.

Where a bidirectional drive is required, the circuit of Fig. 19.50 can be used with the transistors switched in pairs (T_1/T_2 and T_3/T_4).

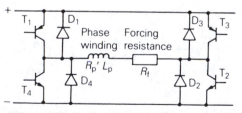

Figure 19.50   Bidirectional drive circuit.

(e)   CHARACTERISTICS

The generalized torque/stepping-rate characteristic for a stepper motor is shown in Fig. 19.51a.

The pull-in curve defines those combinations of torque and stepping rate against which the motor can start or stop without losing steps. It should be noted that there are some regions of low torque and low speed in which the motor is not capable of being started; these are a function of the motor construction.

The pull-out curve or slewing characteristic defines the maximum torque that the motor can provide once running without losing steps. The shape of the

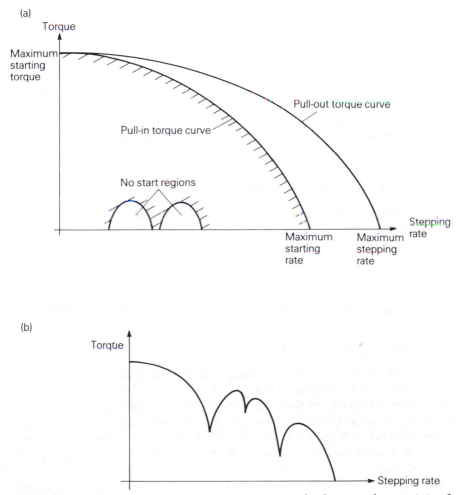

Figure 19.51   Stepper motor characteristics: (a) generalized torque characteristic of a stepper motor (b) pull-out curve with dips.

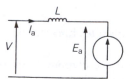

Figure 19.52   Per phase equivalent circuit of a synchronous machine.

pull-out curve is related to the drive circuit used. In addition, stepper motors can exhibit resonances at stepping rates below the maximum, resulting in dips in the pull-out curve as illustrated in Fig. 19.51b.

(f)   INERTIAL LOADS

Inertial loads present a particular problem for stepper motors on acceleration and deceleration. During acceleration the use of too high a stepping rate, or rate of change of stepping rate, will result in the motor missing steps and lagging behind the position suggested by the number of steps applied. During deceleration, too rapid a reduction in the stepping rate will result in the load overshooting the required final position.

19.5.7   SYNCHRONOUS MACHINES

A synchronous machine only produces torque when he rotor – carrying the DC field winding – is rotating at the same speed as the rotating magnetic field produced by the three-phase stator winding, i.e. at synchronous speed. Referring to Fig. 19.52, the power produced by a cylindrical rotor machine is given by

$$\text{power} = \sqrt{3} V_{\text{line}} I_a \cos \phi = \sqrt{3} V_{\text{line}} E_a \sin \delta / \chi \qquad (19.37)$$

where δ is the load angle and is the synchronous reactance of the machine.

19.5.8   BRUSHLESS MACHINES

In a brushless machine the field winding is replaced by high power, rare earth permanent magnets mounted on the rotor, removing the need for a brushgear. Construction is typically as shown in Fig. 19.53. The resulting machine offers advantages in terms of reduced maintenance, increased torque/volume ratio, increased torque at high speeds and simplified protection in comparison with more conventional machines. As a result they are increasingly being used as servomotors in a variety of applications.

A brushless machine can be operated either as a synchronous machine, receiving a multiphase variable frequency supply, or as a DC machine, in which case the commutation of the stator currents would be performed electrically.

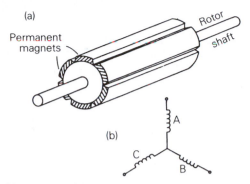

Figure 19.53   Brushless motor: (a) rotor construction (b) stator winding corrections.

The point at which commutation occurs is determined by reference to the position of the rotor, either by monitoring shaft position by means of an encoder or by sensing the magnetic field. Of the two modes, operation as a synchronous machine requires the more complex control system because of the need to generate a variable frequency supply. Overall, performance is more flexible than as a DC machine.

Brushless machines can exhibit a ripple in their output torque. This contains two main components: the reluctance ripple caused by the inherent magnetic asymmetry of the machine; and the drive current ripple at a frequency of $pn/2$, where p is the number of poles and n is the speed in rev min^{-1}. There is also a component at the speed of the machine, referred to as the once-round ripple; this arises from any non-alignment of the rotor within the stator.

19.5.9   SWITCHED RELUCTANCE MOTORS

The configuration of a simple switched reluctance motor is shown in Fig. 19.54. Operation is very similar to that of the variable reluctance stepper motor in that rotation is achieved by exciting pairs of stator poles in sequence.

If the stator is supplied with a constant voltage, and switching of the stator currents takes place at a fixed rotor position relative to the stator poles, the switched reluctance machine produces a torque/speed characteristic similar to that of a series DC machine. By controlling the stator currents and varying the position of the rotor relative to the stator at which switching occurs, characteristics corresponding to a DC shunt motor and a synchronous machine can be obtained.

Operation of the switched reluctance motor therefore requires that the controller makes reference to rotor position in order to establish the switching pattern to be used. As with the brushless machine, this can be achieved by the direct measurement of rotor position or by reference to the internal field pattern of the machine.

Table 19.1 Comparison of electrical variable speed drives: ****best, *worst

	DC machine		Slip ring induction motor	Squirrel cage induction motor			Synchronous motor
	Converter	Chopper	Slip energy	Voltage fed inverter	Current fed inverter	Cyclo-converter	Voltage fed inverter
Rating	10 MW	100 kW	20 MW	2 MW	2.5 MW	15 MW	2 MW
Speed range (typical)	>20:1	>20:1	5:1	>20:1	>10:1	10:1	15:1
Constant torque operation	**	**	**	**	**	**	**
Constant power operation	**	**	*	*	*	*	**
High speed operation	***		****	***	***		***
Low speed operation	****	***	*	**	**	***	**
Repetitive operation	****	***	*	**	*	***	***
Speed control	****	***	*	****	****	****	***
Costs	****	****	***	**	**	*	**

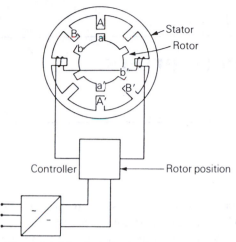

Figure 19.54   Switched reluctance motor.

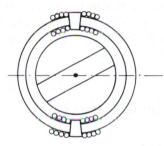

Figure 19.55   Toroidal torque motor.

19.5.10   TOROIDAL TORQUE MOTOR

A toroidal torque motor is a brushless DC machine which produces a ripple-free torque over a limited angular motion. Referring to Fig. 19.55 for a two-pole toroidal torque motor, the stator is formed from a solid or laminated cylinder with the windings located between the interpole barriers and within the air gap. The permanent magnet rotor then reacts with the field produced by the stator winding to provide the rotation.

19.5.11   ELECTRICAL VARIABLE SPEED DRIVE CHARACTERISTICS

Table 19.1 provides a comparison between various types of electrical variable speed drives.

Chapter 20

Motion converters

The range of available motion conversion (or power transmission) devices is extraordinarily wide and only a synoptic view will be given here. They are often incorporated into mechatronic systems with the purpose of matching driver speed or torque capacity to the requirements of the driven machine, and often provide passive or fixed ratio conversions of torque. Where a facility for variation of speed and torque is provided, the adjustment may be manual or remotely actuated through an electrohydraulic actuator.

More detail can be found in the profusion of texts on machine elements and design and in manufacturers' literature.

20.1 Fixed ratio motion converters

20.1.1 PARALLEL SHAFT GEARS

In most industrial applications the prime mover or drive motor will run at a higher speed than the driven machine. Speed increasing or step-up gears are comparatively rare and are confined to specialized machines such as turbo air compressors. Where a speed above the normal 2900 rev min^{-1} available from standard 50 Hz two-pole induction motors is required, consideration will be given to using high frequency motors. Industrial parallel shaft gears will therefore normally be used as speed reducers, the tooth forms being almost invariably involute and in single helical configuration. Typical reduction ratios per stage are not usually greater than 3:1. Two-stage speed reducers are common, but more than two is exceptional except in the case of very high power ratings, for example the triple reduction double helical gears used on marine steam turbines which range up to 50 000 kW. The practical limit for industrial applications is around 15:1 in double reduction form. Beyond this, the configuration becomes less economical in space utilization than other forms of speed reducer, notably epicyclic gears. The efficiency of involute gear transmissions is very high and can approach 98% per stage.

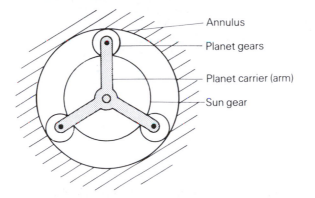

Figure 20.1   Epicyclic gear.

20.1.2   EPICYCLIC GEARS

These are essentially a subset of parallel shaft gears but lend themselves readily to compact arrangements in which the input and output shafts are coaxial. The basic layout of an epicyclic stage is shown in Fig. 20.1. For fixed ratio industrial applications the input is normally taken on the planet carrier and the output from the sun wheel while the annulus remains fixed in the casing. Solution methods are given in the standard texts, but practical reduction ratios are in the region of 1.4:1 per stage. For low power drives it is possible to exploit the short axial length occupied by a single stage using a stacked configuration with four or more stages in series, thus giving ratios of 3.8:1 or more.

Variants are available using differential drives and split power transmissions as in Fig. 20.2, which provide a variable ratio facility.

20.1.3   HARMONIC DRIVES

In some applications the advantages of compact epicyclic gears are offset by the effect of backlash, which becomes more apparent with each stage of

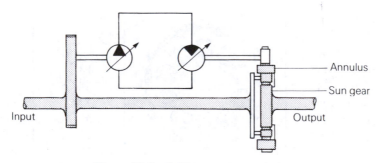

Figure 20.2   Split power transmission.

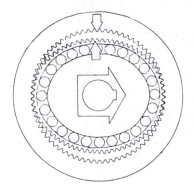

Figure 20.3 Harmonic drive with flexible element.

reduction. This presents no problem if the external load on the system is unidirectional but is plainly unacceptable where load reversal is to be expected as in the articulations of robot arms. Here, harmonic drives are commonly used. Their principal advantage is that a very substantial reduction ratio, up to 100:1, is available within a single stage. The form of construction is shown in Fig. 20.3, where a central input gear is eccentrically driven within the outer ring gear. The ring gear advances by one tooth per revolution of the input. An alternative is the 'Cyclo' drive in which a pin and disc arrangement replaces the conventional gear tooth profile (Fig. 20.4).

20.1.4 WORM AND BEVEL GEARS

Worm gears are the commonest form of speed reducer for ratios of 10:1 to 60:1. They involve a change in direction of the input/output axis of 90°, which may be used to advantage in many machine layouts. Single start worms used in ratios of 15:1 and above will reliably hold on load when the driver is de-energized, but the countervailing disadvantage is the lower efficiency obtained. In-service mechanical efficiencies of 80 to 85% are typical.

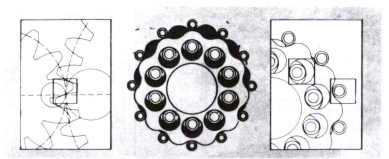

Figure 20.4 Pin and disc harmonic drive (Cyclo Transmission: 1 Centa Transmissions).

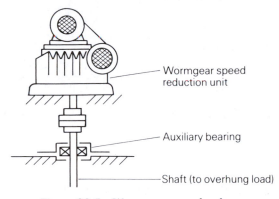

Figure 20.5   Worm gear speed reducer.

The output shafts of proprietary industrial worm gearboxes are often carried in substantial paired taper roller bearings, which make them well able to withstand heavy axial loads, side loads and bending moments. Thus a short output shaft may be supported via a rigid coupling entirely from the gearbox output. For applications requiring greater overhang, a single shaft end bearing may be used provided that due attention is paid to alignment (Fig. 20.5).

Bevel gears are often used for transmission of rotary motion through 90° for reduction ratios of up to 5:1, as illustrated by Fig. 20.6. The drive motors

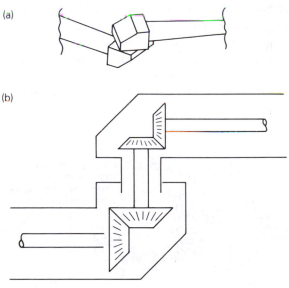

Figure 20.6   Manipulator using bevel gear drives: (a) manipulator configuration (b) bevel gear drive at joint.

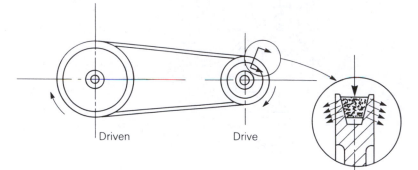

Figure 20.7   Action of a V belt drive.

are situated at the base, in order to reduce the dynamic inertia and provide a counterbalance.

20.1.5   V BELT DRIVES

V or wedge belt drives are a very convenient way of achieving ratio changes of around 3:1 either up or down. Multiple and poly V belt drives can be used up to high power levels; for example, in oil field slush pump drives, drive ratings in excess of 1000 kW are common. The effect of the V configuration is to produce a wedging action between the sides of the pulley groove and the belt. The greater the torque transmitted, the greater becomes the lateral pressure (Fig. 20.7) and hence the tractive force available. Belt drives of this type, however, do not give an absolutely constant ratio between the input and output shafts. A degree of creep occurs due to the elasticity of the belt material and the way in which it enters and leaves the groove. Periodic retensioning or automatic tensioning devices are needed to compensate for belt wear. The output speed is typically 1 to 2% less than would be expected from a positive drive.

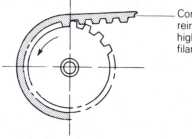

Composite belt, reinforced with high-strength polymer filament or steel wire

Figure 20.8   Toothed belt drive.

20.1.6 TOOTHED BELT DRIVES

These are often referred to as timing belt drives and are widely accepted even in the most rigorous of environments, as in the drive of automotive overhead camshafts. The general configuration is shown in Fig. 20.8. The back of the belt is reinforced with high tensile polymer or steel wire and provides a stiff and virtually backlash-free drive with an invariant ratio between input and output. They require no lubrication, are almost 100% efficient and are very compact. Their principal limitations are the power levels available – usually they are employed for drives of less than 20 kW – and the temperature limit of around 120°C. Miniature timing belt drives with belt widths of 3 mm or less and light alloy pulleys are at the heart of many mechatronic systems where low inertia and fast response are prerequisites. The most obvious example is in the traverse mechanism of printers and plotters.

20.1.7 CHAINS AND SPROCKETS

Though their position has been challenged successfully by toothed belt drives in many applications, roller chain drives still have a part to play in many higher power duties where no slip can be tolerated. With an appropriate level of pre-tension, backlash can be eliminated. Chains and toothed belts have a role in mechatronic systems for transmission of motion where, for space or weight reasons, the actuator cannot be placed at the point of action. An example is the modular robotic system of Fig. 20.9, where the linear carriage is driven via a chain by a motor located at one end of the gantry. This also exploits a further advantage of chain: the ease with which carriers can be attached to it.

20.1.8 FRICTION WIRE WRAP DRIVES

This configuration has been adopted in some small mechanisms where space is restricted and where the output is required to rotate through a limited number of revolutions before reversing. Traction is obtained by utilizing the bollard effect of the wire wrapping around the shafts. It is employed in plotter carriage drives as an alternative to timing belts and may also be used as a speed reducing device between shafts, as in Fig. 20.10. At the cross-over point the wire contacts the shafts on both sides.

20.1.9 RACK AND PINION

This is a most useful form of drive for converting from rotational to linear motion (Fig. 20.11), and may be used with a shaft encoder on the driver to provide fast and precise positioning of loads. In such applications the pinion may need to be split to eliminate backlash.

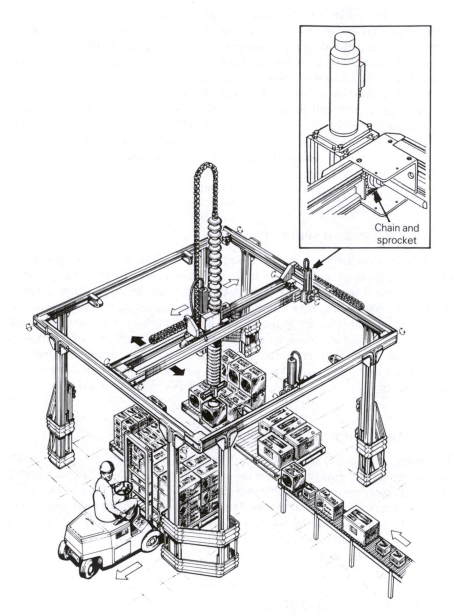

Chain and
sprocket

Figure 20.9   Chain and sprocket drive used in a modular unit for a robotic system
(Crocus).

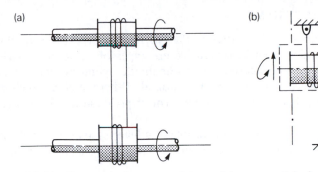

Figure 20.10   Friction wire wrap drives: (a) as parallel shaft drive (b) as linear carriagedrive.

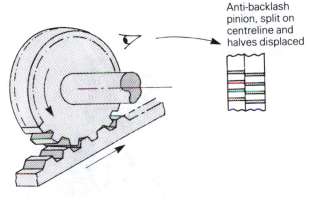

Figure 20.11   Rack and pinion drive.

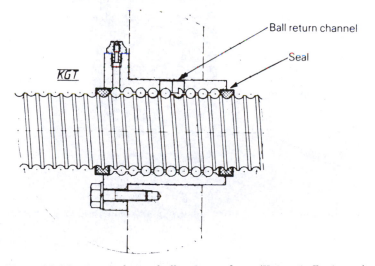

Figure 20.12   Recirculating ball screw and nut (Unimatic Engineers).

20.1.10   SCREW NUT SYSTEMS

A screw nut combination such as that of Fig. 20.12 converts rotary motion into linear motion, and is used in several recent types of electrically driven linear actuator. These are an alternative to pneumatic cylinders where close control of actuator speed and position is required. Where higher precision is needed, recirculating ball screws or some form of preloading is introduced to eliminate backlash.

The ASEA robot of Fig. 20.13 is an example of a lever system in which the upper arm and forearm are driven by screw nut systems, powered by DC motors

Figure 20.13   ASEA robot (ASEA).

via harmonic drive gearboxes. This is an interesting arrangement, incorporating several of the motion conversion systems described in this chapter, using rods instead of belts. Here the drive motors for the wrist motions are located near the base of the robot, in order to minimize the inertia of the moving arm. The rotation of the wrist is achieved by means of a system of transmission rods located between rotating discs on each joint axis of the robot.

20.2   Motion converter with invariant motion profile

20.2.1   CAMS

Cams have been traditional way of obtaining motion profiles in high speed machinery in which large forces need to be generated. Typical applications include textile and printing equipment, shoe making machinery and dedicated machines such as bolt headers. It is likely that a place will remain for cams in such environments, but with enhanced machine diagnostics and compensation for wear using techniques such as the automatic changing of datum positions and end stops. The variety of disc and face cams is infinite and includes intermittent motion and escapements such as the Geneva mechanism of Fig. 20.14. Where acceleration profiles are less stringent, consideration should now be given to replacing invariant hardware mechanisms with the software cam approach employing suitable electrohydraulic actuators or digital motor control. Reprogramming the system for different product configurations or output rates then becomes a simple matter by comparison with the substantial hardware stripdowns which are otherwise necessary.

20.2.2   INDEXING MECHANISMS

These can take many forms, from chain conveyors for indexing pallets round a materials handling system, to intricate film advance mechanisms in motor

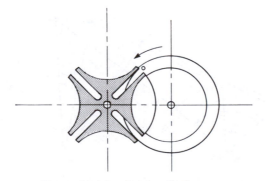

Figure 20.14   Geneva mechanism.

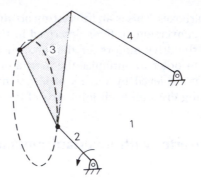

Figure 20.15   Four-bar linkage and coupler curve.

drive cameras. Their design will incorporate some of the basic types of mechanism considered above, together with such adaptations and innovations as may be yet be invented through the ingenuity of the design engineer.

20.2.3   LINKAGES

The four-bar linkage may be used to produce special motion profiles by using points on the coupler plane (Fig. 20.15) or as a function generator (Fig. 20.16). Agricultural machinery such as straw balers demonstrate many applications. The comments in regard to cams apply even more to linkages. Depending upon the market environment, the degree of acceptance of microelectronics and the class of maintenance labour available, linkages are an obvious target for substitution by intelligent systems wherever there is a need for on-line adjustment or process optimization.

20.2.4   SPRINGS AND DAMPERS

Although not immediately thought of as mechanisms, springs and dampers do transmit force and motion. All structures have a definable stiffness and an

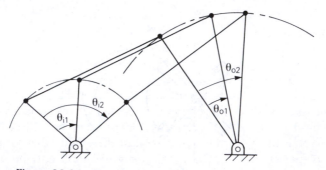

Figure 20.16   Four-bar linkage as a function generator.

ability to dissipate some energy when deformed, thus providing an intrinsic system of springs and dampers into which the structure must be apportioned before analysis of its behaviour can proceed. Analytical procedures such as finite element analysis will therefore be based on system models which comprise mass-spring-damper modules, whether these are representations of discrete hardware elements or intrinsic parts of the structure.

20.3   Variators (continuously variable transmissions)

The objective of obtaining a form of transmission system whose ratio is continuously adjustable, on load, has resulted in the development of a large number of variable ratio mechanical transmissions. A revival of interest has occurred in recent years due primarily to the need to find more efficient and less complex alternatives to the standard epicyclic transmission and torque converters used in automatic transmissions for vehicles. All mechanical variators can be readily configured for remote operation, for example by motorizing the adjustment handwheel, but achievable response rates are comparatively slow, usually a few seconds.

For industrial drives, therefore, the position of variators is challenged by the present developments in electric motor controls. However, where routine but not necessarily continuous adjustments are required to processes, for example a screw conveyor or paste pump, variators may still have a part to play as a robust and 'soldier-proof' alternative.

The following is a selection of the principal generic and proprietary types of continuously variable transmissions (CVTs).

20.3.1   CONICAL PULLEY/DISC SYSTEMS

The simplest of these works on the principle shown in Fig. 20.17, where the cone faced wheels on adjacent shafts are moved in opposition by a linkage system. The effective radius of action of the wedge-type belt on the two pulleys is therefore changed, altering the transmission ratio. Such simple systems when subject to varying torque do not hold speed constant at the set point. However, the wedging action may be used in more sophisticated arrangements to produce a controlled lateral separation of the conic elements by working against a spring or air diaphragm. With pneumatic (or hydraulic) control of the separation it is possible to arrange stepless load matching of the prime mover to the required output conditions.

Developments of such systems have given rise to the well known Van Doorne automotive transmission and its derivatives. Efforts have been directed towards two principal ends in the further refinement of these and other forms of CVT: increasing the power transmission capacity and life (primarily an automotive requirement); and producing a positive drive capable, when once set, of

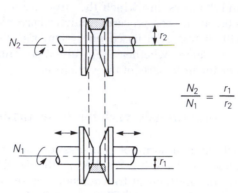

Figure 20.17   Cone pulley variable belt drive.

maintaining a timed relationship between the input and output shafts (chiefly an industrial requirement). In the first case, configurations have resulted in which the contact pressure has been greatly increased – involving the replacement of the polymer based belts by steel belts or solid rings, of which the Heynau drive (Fig. 20.18) is an early example. The second requirement has been met by devices such as the positive infinitely variable (PIV) Antrieb drive (Fig. 20.19), which uses a belt form in which stacks of thin steel slats can slide laterally to engage in staggered grooves in the opposing faces of the drive cones.

20.3.2   BALL/DISC FRICTION DRIVE SYSTEM

Probably the best known of these is the Kopp variator (Fig. 20.20), where the ratio change is effected by altering the inclination of the axes of rotation of the

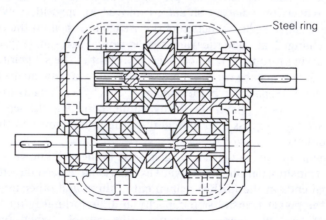

Figure 20.18   Heynau or H drive.

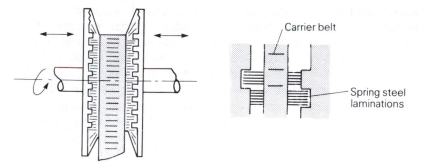

Figure 20.19   Positive infinitely variable (PIV) drive.

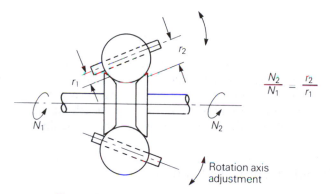

$$\frac{N_2}{N_1} = \frac{r_2}{r_1}$$

Figure 20.20   Principle of Kopp variator.

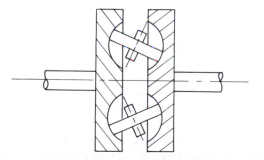

Figure 20.21   Toroidal (Perbury) continuously variable transmission.

balls. Higher power drives aimed at the automotive market, including truck transmissions of up to 350 kW, use rollers working between toroidally shaped discs (Fig. 20.21). This concept originates in the Perbury drive but has been enhanced by modern developments in lubrication and materials technology. These are, however, essentially high power friction drives in which the large interelement traction forces can only be developed by applying heavy contact pressures. Consequently, owing to the high Hertzian stresses which occur, fatigue is a danger. In the Kopp variator the axial loads (and hence contact pressures) are automatically adjusted by means of a simple but effective mechanical cam arrangement in proportion to the output torque developed. With automotive systems the axial load is applied by a pneumatic or hydraulic actuator under microprocessor control, and taking inputs from torque sensors or deriving these from parameters such as engine speed and throttle settings in relation to the known envelope of engine performance. By this means the highest stresses need only be experienced in the CVT for the small proportion of the operational time regime in which the maximum torque output is demanded. The prize of a high power, long life CVT is however still some way from being achieved.

20.3.3 UNIT HYDRAULIC TRANSMISSIONS

These are typified by the Carter gear, which contains within an integral unit a variable displacement radial piston pump and hydraulic motor. The characteristics are therefore those of the PV–MF drive discussed in section 19.3.5. Overloads and stalling of the output can be tolerated for short periods, and remote actuation of the adjustment wheel can be arranged.

20.4   Remotely controlled couplings

A wide variety exists of simple electromagnetically applied friction clutches, dog clutches and detents used in an on/off mode to connect a driver to the driven machine. They are often used in high speed machinery where the drive must be picked up very rapidly and the kinetic energy in the driver can be used to advantage. In small to medium power applications, from a few watts to perhaps 10–20 kW, magnetic powder couplings and, latterly, couplings using electroviscous fluids may have advantages. The special fluids developed for these purposes are ferromagnetic colloidal suspensions whose rheology can be modified by the application of a field. Thus they can provide a controlled slip facility.

Two classes of coupling are available with broadly similar characteristics but different operating principles: the eddy current coupling and the hydrodynamic fluid coupling. They are used in an uncontrolled form as simple torque limiters and to ease the starting conditions for AC induction motors when driving high inertia loads. The controllable eddy current coupling has an alloy rotor on one half coupling, working within a cylindrical wound armature on the other half in which the field strength can be controlled. The fluid coupling is controlled by varying the amount of oil in the circuit. The well known scoop controlled coupling is receiving some renewed attention for lower power drives, with the

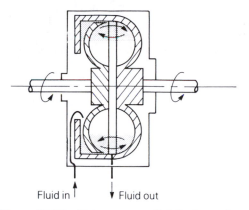

Fluid in ↑   ↓ Fluid out

Figure 20.22   Volume controlled fluid coupling.

replacement of the rather complicated scoop system by a auxiliary chamber which can remove or replace fluid from the working circuit as required. Figure 20.22 shows a typical arrangement.

By definition, slip couplings can only provide output speed ratios of less than unity. When used circumspectly they can provide considerable power savings in the drive of fluid impelling machinery such as fans and pumps, but it must be checked whether their thermal rating will allow continuous operation with the proportion of slip envisaged. The torque passes through the coupling unchanged, but reduction of speed represents a loss of power between the input and output and a thermal load generated within the coupling which is directly proportional to the degree of slip.

Systems and Design

Chapter 21

Mechanical systems and design

The purpose of this part of the book is to introduce the physical 'glue' which holds the components (sensors, microprocessors, actuators) of a mechatronic system together – namely the mechanisms, structures and packaging. These supportive elements are just as important to the functioning of the system as the primary elements mentioned above. They control the accuracy, reliability and safety of a product or process, and ultimately determine its viability.

The careful selection and design of these elements is a vital part of the mechatronic approach, aimed at optimizing the system performance. For example, when updating an existing product, the electronics and computing elements should not just be bolted on to existing mechanical hardware; rather, a complete re-evaluation of the design should be undertaken encompassing the new features to be introduced. This re-evaluation should examine the envelopes of movements, the accelerations and hence the dynamic forces involved, and the stresses generated. This will then lead to the appropriate selection or design of any mechanisms, structures and materials, with due regard to their costs and the manufacturing processes employed.

So a mechatronics approach to engineering design is a team game, relying on the varied experience of the players and also on the support of (computer aided) modelling to predict the behaviour of the system. As with any product design, the use of industrial designers and marketing executives as team members can enhance the skills of the straight engineering disciplines, in order to optimize the packaging and user interface. That is not to say, however, that a company can only succeed in mechatronics if it has an intellectual elite and a Cray computer. Rather, the people involved should be broadly based, able to talk across disciplines and willing to explore new areas.

In this part of the book the aims are to:

1. Present the basic mechanical concepts, devices, structures and systems;
2. Discuss the differences between the traditional and mechatronic approaches to engineering design;
3. Describe the possibilities for modelling the system prior to manufacture;

4. Examine the implications of the manufacturing process on the design of the elements, in terms of form and materials;
5. Consider the user interface and safety requirements of these automated products and processes.

The last two issues are of a substantial nature and form a major part of the process involved in bringing product or process into safe and economic operation. For a fuller treatment, the reader is referred to texts on industrial design and product development.

21.1   Tradition versus mechatronics

The transition from traditional to mechatronic products and systems has occurred gradually in various ways, but has been primarily associated with the introduction of increasingly sophisticated levels of control, ranging from no control, through sensor feedback to the operator and open loop control, to closed loop and adaptive control. However, one fundamental difference between a control approach and the mechatronics approach is the accompanying redesign of the mechanical system to take full advantage of the extra functionality provided by the microprocessor. Redesign is however not the whole story; many recent products have only come into existence because of mechatronic integration from the outset of conceptual design.

The various development opportunities afforded by the mechatronics approach are described in the next section, but in general terms the main differences between traditional and mechatronics engineering can be summarized as follows:

Traditional	*Mechatronic*
Bulky system	Compact
Complex mechanisms	Simplified mechanisms
Non-adjustable movement cycle	Programmable movements
Constant speed drives	Variable speed drives
Mechanical synchronization	Electronic synchronization
Rigid heavy structures	Lighter structures
Accuracy determined by tolerance of mechanism	Accuracy achieved by feedback
Manual controls	Automatic and programmable controls

21.2   The mechatronic approach

21.2.1   REPLACEMENT OF MECHANISMS

In certain limited cases the mechanism can be completely discarded, the functionality being replaced entirely by the microprocessor and actuators. Two examples are as follows.

(a) ELECTRONIC WATCHES

Here the precision (Swiss) mechanism has been replaced by an integrated circuit, together with a user interface in the form of a light emitting diode (LED), liquid crystal or analogue display and control buttons. The user may be conscious only that the integrated circuit has provided equivalent or enhanced functionality over the traditional watch, but at a much cheaper price, with features such as date, alarm, world times and stopwatch. Mechanical systems have however been substantially replaced: the potential energy of a spring by the electrical energy of a miniature lithium cell; the hairspring balance wheel and escapement by a crystal held in resonance; the gearing and pointers by a counter circuit and liquid crystal display. Here the mechatronic replacement has been radical.

(b) FLY-BY-WIRE

Here the linkages between the pilot's joystick and the control surfaces have been replaced by a microprocessor, a signal wire and an actuator. The large lever joystick is replaced by a small electronic unit mounted on the pilot's seat armrest; the force amplification is provided by the actuator in the wing. Additional operational safety is provided by the microprocessor, which can determine parameters such as the maximum angle of climb and rate of turn so as to optimize performance while protecting the airframe. In military aircraft, increased agility is obtained by designing the aircraft to be inherently unstable and using a microprocessor based controller to continuously modulate the control surfaces to maintain stable flight. The interposition of system intelligence between the pilot and the control surfaces has substantially modified the pilot's relationship to the airframe.

21.2.2 SIMPLIFICATION OF MECHANISMS

Many mechanisms can be simplified by adopting a mechatronics approach. The functionality of tasks such as profiling and speed and position control can be provided by the microprocessor and actuators. The following are two examples.

(a) TYPEWRITERS

These traditionally contained several mechanisms for keystroke, carriage shift and paper advance. The typewriter has evolved, with the addition of electronic control, through the golfball and the daisywheel integrated units, to the computer based word processors with separate dot matrix and laser jet printers with a capacity for both text and graphics.

(b) LARGE TELESCOPES

In order to track a particular object in the sky, as the earth rotated, the equatorial mounting was developed to synchronize with the earth's axis of rotation. On large telescopes, such as the 200 inch Mount Palomar unit, this resulted in large structures driven by constant speed motors. Using a mechatronics approach, a compact balanced support structure can be designed with just two degrees of freedom which will track any object in the heavens under computer control.

21.2.3   ENHANCEMENT OF MECHANISMS

By combining sound mechanical design with closed loop control, enhanced speed, accuracy and flexibility of movement can be achieved. The proportional control inherent in the closed loop produces less stress on components as compared with actuators operating between end stops. A consequence of this is that components may be made lighter and hence of lower inertia.

For example, some early robots worked between switchable end stops; others had complex mechanisms for achieving straight line paths for their grippers. Design then progressed to having DC servo drive motors and harmonic drive gearboxes on each axis of the machine, so providing proportional programmable speed for each axis. However, the inertia effects associated with moving these motors in space was significant. More recently, direct drive pancake motors located in the base of the robot have been used, which drive each axis by concentric tubes and steel belts.

21.2.4   SYNTHESIS OF MECHANISMS

The use of embedded microprocessor systems enables the synthesis of different mechanisms, as well as functions.

For example, if an automatic washing machine were driven by a twin speed (wash/spin) unidirectional motor, a mechanism would be required to generate the bidirectional wash action. By using a variable speed DC motor and controller, this action can be achieved using a direct drive.

21.3   Control

21.3.1   PROGRAM CONTROL

A feature of a mechatronic design is the ease with which the system functions can be redefined using software. In some cases this is achieved by the user selecting from a number of predefined and stored programs; in others a new program can be loaded from an external source. Two examples are as follows.

(a)   AUTOMATED MANUFACTURING CELL

Here, the product specification may be stored centrally and downloaded to the manufacturing units, CNC machines, robots and associated handling equipment, using local area and factory wide networks on request.

(b)   AUTOMATIC SEWING MACHINES

These incorporate the ability to preselect a particular stitch from a range of stored stitch types as well as special pattern elements such as buttonholing.

21.3.2   ADAPTIVE CONTROL

The availability of the local processing power provided by the introduction of embedded microcontrollers enables variation of the programmed path/speed/ setting to suit local conditions, in order to optimize the behaviour of the system. Examples are as follows.

(a)   ACTIVE SUSPENSIONS

Here, the effective spring and damper values can be automatically varied to suit the required style of driving, the load condition and the road surface, even to controlling the response of the system within the duration of single wheel bump events. This requires that the system be capable of effecting control within a few milliseconds.

(b)   CNC MACHINES

Advanced CNC machines incorporate the adaptive control of cutting speed, to suit variations in material properties. Thus the structures of machines can be reduced in weight compared with conventional machines, since cutting forces are more precisely controlled while optimizing the rate of metal removal.

(c)   AUTOMATIC SHEEP SHEARING

A sheep shearing robot has been developed in Australia – where else! This generates a path based on three-dimensional CAD model of a 'standard' sheep. The path is modified in real time by proximity sensors on the shears, in order to keep the shears a fixed distance from the sheep's skin. However, the sheep needs to be clamped in an elaborate fixture to keep it as still as possible. Apparently sheep remain peaceful and contented when inverted!

21.3.3   DISTRIBUTED SYSTEMS

Mechatronic devices need not be arranged as a serial mechanism like the industrial robot. Actuators and mechanisms can act in parallel, or be distributed over a wide area, synchronized electronically by the microprocessor. The following are two examples.

(a) ACTIVE SUSPENSIONS

These use four or more actuators working in parallel. The real-time characteristics of each one may be made dependent on the state of all the others for control of attitude, ride and braking.

(b) CONVEYOR SYSTEMS

Here, drive motors, sensors, lift tables and transfer stations all have to be coordinated, and tracking of individual loads through the system is often required.

21.4   The design process

The design process can be split into several stages as follows, each separated by a review and a continue, rework or abort decision point.

21.4.1   NEED

Development should be primarily market led and not technology led. Adequate market research and knowledge must be accumulated to achieve a clear picture of what the customer wants. That is not to say that a company cannot create a market need by producing an innovative product, but failures in this area can be crucial, as in the case of the Sinclair C5.

21.4.2   FEASIBILITY

The feasibility study must examine the commercial as well as technical aspects of the proposed product or process. In the case of a product, costings should be determined not only for the product itself – the component cost – but also for the production facilities and tooling required to produce the product, and the development costs prior to the launch of the product on to the market. Warranty costs and the future maintenance costs to the purchaser are also factors which will influence the profitability and sales of the product, and hence its feasibility.

21.4.3   SPECIFICATION

The initial, high level specification must be written so that everyone involved in the development is clearly informed as to the requirements and functions to be met. At this stage the needs (essential requirements) and wishes ('might be nice to have') should be separated and decisions made about their inclusion. Solutions to the required functions are not to be attempted or included, as these will be generated in the next stage of the design process.

The high level specification will encompass among others the following elements of information:

space requirements and constraints
weight
payload
types and range of motion required
velocities and speeds of motion, both linear and rotational
accelerations, both linear and rotational
accuracy and resolution
control functions
design life
input and output requirements of each element of the system
communications, both internal and external
power requirements
interfaces
operating environment
standards and codes of practice.

Once the high level specification has been produced and the functional areas defined, the design of the individual elements of the system can proceed from a secure base. Where the high level specification does not exist or is inadequately prepared, this base is not available and the design process is impaired or impeded. This is particularly so in the case of large systems involving a number of task groups, each of which is dealing with a different element or component of the overall design, when a failure to produce an adequately detailed initial specification will prevent the effective execution of the individual tasks.

21.4.4   CONCEPTUAL DESIGN

At the conceptual design stage, options should be generated for the solution to each of the functions required; it should be possible to think of at least six options for realizing each function. Preferred solutions can then be generated by a process of evaluating and combining these conceptual solutions.

21.4.5   ANALYSIS AND MODELLING

Analysis and modelling of the system and its components may be required to aid the selection of possible solutions, and also during the embodiment phase of the design process to determine parameters such as operating characteristics and sizes of components. By the use of the techniques of computer aided engineering (CAE) to test various models and options, the traditional prototype/ redesign/prototype loop can be much reduced, leading to shorter and hence cheaper times for product development.

21.4.6   EMBODIMENT AND OPTIMIZATION

Once an acceptable solution has been established, it is tempting to go straight to the detail design phase. However, in order to design the optimum product or system, parameters such as size, material, quantity and rating of the components of the chosen solution should be varied to determine their effect on the required characteristics of the system. This then enables fixing of the values of the parameters which give the maximum and minimum system performance characteristics.

For example, the solution chosen for a new design of DIY hand drill may require a motor, a gearbox and a controller. Variable parameters would be the diameter and length of the motor, the gearbox ratio and the output characteristics of the controller expressed in terms of current and voltage. The required performance parameter to be optimized may be the power, the weight or the size or some combination of these. Various prototypes could be made, with differing parameters, and tested to determine the values of the characteristics and trends determined; this is the basis of the Taguchi quality engineering method.

21.4.7   DETAIL DESIGN

Only when the design has been optimized should consideration be given to detail design and the production of manufacturing drawings showing the manufacturing tolerances for each component. The results of the Taguchi tests will indicate which tolerances have the most effect on variability of characteristics.

21.5   Types of design

Design for function is not the only element of the design process. The following should also be considered:

1. Design for strength or stiffness, taking into account factors such as static loading, fatigue performance and corrosion effects;
2. Design for reliability;
3. Design for maintainability; this takes into account any policy on features such as repair on replacement;
4. Design for manufacture, particularly where automatic assembly is concerned, when the machine capabilities must be taken into account;

also the rationalization of components, fixtures and fittings between similar products and processes.

21.6   Integrated product design

An essential element within mechatronics is the concept of designing the production facility in parallel with the product. Obviously this is the ideal; however, most products have to be designed with existing production facilities in mind. Nevertheless, the concept of an integrated product design team consisting of marketeers, design engineers, production engineers and field service engineers is one that every company could adopt. It is the establishment of such a team that then enables all the types of design outlined to be considered.

This introduction of an integrated design team often requires a major organizational rethink, since traditionally designers report to the technical director and production engineers to the manufacturing director. Ideally this design team should live in one office, under the leadership of a project manager.

21.6.1   PROJECT MANAGEMENT

In many organizations, the development of a product is not well coordinated. In general, the marketing organization presents their definition of a need to the designers, who generate concept drawings for the development engineers, who give test results back to designers, who give detail drawings to production, who then order components and eventually see that the product is made. In large organizations with multiproduct development, the result is that product lead times become large and it becomes necessary to appoint a product champion cum project manager to plan and control the development of a product. This leads to a matrix organization, with functional managers responsible for the performance and operation of their department and project managers responsible for the cost and timescale of their product.

By handing over the responsibility for product development to an integrated group involving all the necessary skills from marketing to production, many of the time consuming communications loops that would otherwise exist are eliminated, reducing lead times and tightening up control over the projects.

21.6.2   PLANNING AND IMPLEMENTATION OF FACILITIES

The planning and implementation of a manufacturing system should be a team activity involving management, marketing, design, development and production. It can be broken down into the following elements:

company strategy
production of functional specifications

Table 21.1 Models for the introduction of new technology

Phases	Model 0: technology led, muddling through approach	Model 1: task and technology centred approach	Model 2: organizational and end user centred approach
Initial review	Vague awareness of interesting technologies	Operating conditions; people are a costly resource, to be reduced if possible *Key actors*: top and senior management	Operating conditions; people are a costly resource, to be more fully utilized *Key actors*: initially from any part of the organization, then top management
Prior justi-fication	Fascination with the technology Expectation that it can be injected into present organization Short term improvements and return on invest-ment sought	Tightly prescribed planning objectives Central coordination and control Expert driven Most modern syndrome *Key actors*: manager-ial project team, including technical and financial experts	General policy formulation Decentralization, staff involvement Concern for end users System development potential rather than machine capability *Key actors*: a diverse and representative group; trade union involvement
Design of system	Technological development controlled by inherent laws Heavy reliance on system designers and promises of suppliers Machines to enable staff economies	Machines over people Task fragmentation 'Clean design' 'Final design' *Key actors*: engineers and technical consultants	People use machines Job enrichment/ teams Operator and maintenance needs Incremental and educative approach *Key actors*: design engineers, technical consul-tants, behavioral advisers
Imple-mentation	Compensation approach to unions	Machine capability; only minor modi-fications expected	User support; pilot projects used where possible

Table 21.1   *(Continued)*

Phases	Model 0: technology led, muddling through approach	Model 1: task and technology centred approach	Model 2: organizational and end user centred approach
	Unanticipated system debugging and organizational problems, e.g. staff motivation, demarcation Unanticipated need to undertake staff training	Once-off skill training Responsibility to line management	Continuing staff and organizational development Continuing reviews of operation and needs
		Key actors: as above, with line managers and end users; trade union negotiations on conditions	*Key actors*: as above, with line managers and end users; trade union negotiations on training/ grading

Source: Blackler, F. and Brown, C. (1987) Management, organisations and the new technologies. In Blacker, F. and Osboone, D. (eds) *Information Technology and People: Designing for the Future*, British Psychological Society.

hardware and software requirements
shortlisting and validating quotations
ordering
training and staff development
continuing evaluation.

The precise way each company approaches these stages will depend on the type of industry, but there are many common features in all systems. Each stage will now be considered in more detail.

(a)   COMPANY STRATEGY

For any new approach to be successfully integrated into an organization, it must have the full support of top management, with clearly defined objectives. In this context, it must be remembered that people are responsible for making innovations actually work, and if they are motivated within an organization they will gladly explore the capabilities of new systems and technological possibilities to the benefit of the whole company. Even when companies have clear objectives for a system, they tend to adopt a technology led approach. It is better to view the technology as an integral part of a social system. The adoption of new thinking and techniques should be managed by a team of experts and users, and the system should be allowed to develop as the people 'grow'. Various models have been suggested and researched by Blackler and Brown and are presented in Table 21.1. Though initially set out with regard to innovations in manufacturing systems and their management, they could just as easily be applied in generating new awareness and attitudes to advances in hardware and software technology.

(b)   FUNCTIONAL SPECIFICATION

This is a very important stage in the design process. It results in documentation setting out what is required in terms of the tasks the system has to carry out, error situations and user interfaces. The document will also be used as a tender document for potential suppliers to bid against. It must therefore be detailed enough to include all required standards – safety, electrical, paints, etc. – and interfaces with other equipment. Some companies employ consultants at this stage, retained either until the contract is placed or, increasingly, until the handover of the system. The simulation of complete systems using commercial packages is also finding increasing application in the design of factory processes.

(c)   HARDWARE AND SOFTWARE REQUIREMENTS

In any design environment there will almost certainly be links to other hardware and software systems, either existing or planned, within and possibly outside the company. A top down approach to planning the whole facility should be adopted, even though parts of it may not be implemented for some time. Hence the material and information flows for each part are defined from the start, and hardware compatibility is protected.

(d)   SHORTLISTING AND VALIDATING QUOTATIONS

The success of a project depends on many factors and not just price. The technical proposal, the delivery time, the quality of equipment and the personalities of the client and supplier all play important parts. Many companies decide on a supplier by using a quantitative assessment of each tenderer, including the above points along with others such as financial stability and the technical competence of staff. A weighting is applied to each factor, depending on its importance to the particular project. Some companies leave the price out of this evaluation until the end, dividing the total 'score' by the price to come up with a final figure.

(e)   ORDERING

The supplier should be chosen following a rigorous assessment of the form described. However, account must also be taken of the scope of supply, contractual cutpoints, performance guarantees, payment terms and maintenance agreements. It is through lack of attention to the last detail, and its associated costs, that much misery and disillusion results. It should also be clearly established before commitments are made as to who will hold software rights and other intellectual property, to which the client may have contributed, if the supplier defaults in some way or goes bankrupt.

(f) TRAINING AND STAFF DEVELOPMENT

Whenever technological change is contemplated in an organization, staff development has a pivotal role (see Table 21.1). This will sometimes need to go beyond training in specific skills and towards changing attitudes and providing experiences which are intellectually stretching. In this context a company might introduce staff to the concept of structured software design techniques, such as the controlled requirements expression (CORE) developed by British Aerospace, or the Taguchi approach to system parameter design. Staff can gain confidence by employing the methods firstly on simpler problems which are actually in the product plan. There is something to be said for using sledgehammers to crack nuts if the users thereby learn to use the sledgehammers with appropriate skill and discrimination.

(g) CONTINUING EVALUATION

Responsibility for a product or process does not end once the product is on the market or the process is in operation. Instead, it should be the subject of a continuous review process involving all elements from the designers to the users, with the resulting feedback used to define the next generation of systems.

Chapter 22

Mechanisms

Mechanisms have a number of roles or functions within a mechatronic system. The following examples do not form an exhaustive list, but are sufficient to give an appreciation of the concepts involved.

(a) CHANGE OF SPEED

This has been a necessary function of mechanisms and transmission systems in the past as a result of the use of constant speed motors, as on machine tools, or prime movers with limited torque envelopes such as automobile engines. Lever mechanisms may be employed to increase the velocity available from linear actuators such as pneumatic cylinders. The use of variable speed motors and actuators, operating under electronic control, has now reduced the requirement for this function.

(b) ACTION AT DISTANCE

The ability to provide action at a distance has arisen where the primary actuator is too bulky or heavy to be directly located at the point of action. The means of transmitting the motion has usually been through a Bourdon cable, rod, chain or hydraulic pipe. The development of compact, powerful electrical, mechanical and hydraulic actuators which can be placed close to the required point of action has reduced the need for this facility.

Demonstrating this change in capability is the master–slave manipulator of Fig. 22.1, used to enable an operator to manipulate dangerous chemicals and radioactive materials from outside an enclosure. Until recently these have been purely mechanical devices, with limited dexterity and capacity to exert force. Developments have included replacing the mechanical link between master and slave by an electronic one, and the use of joystick controlled robots either mounted on the wall of the enclosure or on a moving gantry as in Fig. 22.2. Such devices, though they may resemble robots and have some intelligence, should more properly be called telechirs (in effect a distant hand) since they need to be under the continuous control of a human operator in order to carry out their primary function.

(a)

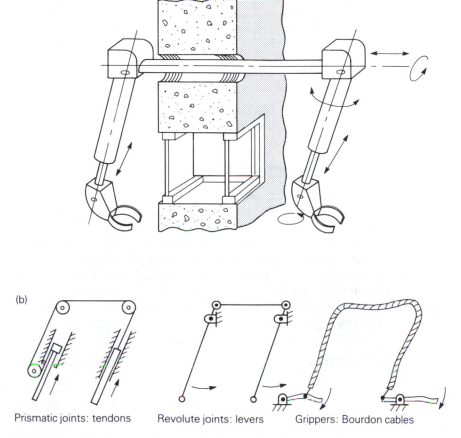

(b)

Prismatic joints: tendons   Revolute joints: levers   Grippers: Bourdon cables

Figure 22.1 (a) Mechanically linked telechir (b) some examples of motion transfer systems.

(c)   FORCE AMPLIFICATION AND FEEDBACK

Force amplification has often been needed between a user input and the point where the output force is to be applied, as in the case of a pilot's joystick, an automobile brake servo or a master–slave manipulator where the user cannot exert sufficient force directly. With the adoption of integrated electronic controls, such as fly-by-wire, direct force amplification is often no longer required.

Lever mechanisms may still be employed with benefit to provide a feedback which the user receives in the form of 'feel' as to the amount of effort being applied, as for example in the master–slave manipulator. In many such cases there remains the need to provide the user with 'feedback' to maintain his

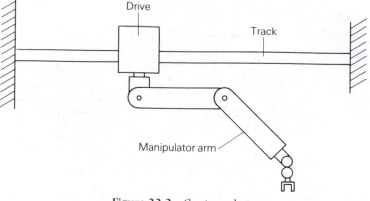

Figure 22.2   Gantry robot.

relationship with the operation of the system following the introduction of electronic controls.

An everyday example is the now increasingly common electronic governor used on automotive diesel engines. Here the mechanical linkage from the throttle pedal to the fuel rack on the injection pump has been entirely replaced by an electronically controlled rack positioner. The foot pedal hinge incorporates a potentiometer and a synthetic sense of force feedback is provided by a spring.

(a)   PARAMETER CONTROL

This includes the various mechanical governor devices for regulation of speed or force. This is illustrated by centrifugal flyweight devices used in the mechanical diesel engine governors mentioned above. Often these have been replaced by closed loop electronic controls.

(e)   SEQUENCING

Various mechanisms, such as the Geneva mechanism and walking beam conveyors, have been devised for this purpose. These may still be perfectly adequate for many materials handling applications, or where very high inertia forces are to be encountered.

(f)   SYNCHRONIZING MOTIONS

Mechanisms such as those used on a copy lathe are designed to describe fixed loci in space or to follow a template. This function can now be achieved by means of fully programmed motion control, or by optical tracking of a drawing, using only two or three actuators.

(g)   PROFILING MOTIONS AGAINST TIME

The Whitworth quick-return mechanism is typical of a family of mechanisms which were used principally in the machine tool industry to save time on unproductive motions. In the shaping machine it provides a rapid return stroke, but produces a lower speed and a larger available force on the cutting stroke. Where such facilities are required today, it is certain that an electronic control of the actuator would be the first choice, chiefly because of the ease of changing system parameters such as speed or stroke.

(h)   CHANGE OF REFERENCE

In certain applications a second mechanism is required to move the action mechanism to a new location, particularly where items are presented to a machine in random orientations. This transformation can now be performed in software, enabling a system such as a handling robot to automatically adjust to the orientation of the reference object as determined by location sensors such as vision systems. The need for the second mechanism can therefore be eliminated.

(i)   TRANSFER OPERATIONS

In most manufacturing systems there is a need to transfer a workpiece from location to location to enable different operations to be carried out. Transfer mechanisms for large scale movements have until quite recently been based on the use of conveyors and transfer lines. The introduction of mobile, automatic guided vehicles (AGVs) has significantly increased flexibility of operation in this area.

Localized transfer mechanisms are required for functions such as the loading and unloading of machine tools. An example of this is seen in a flexible manufacturing cell where robots are used to move workpieces between the various elements of the cell.

22.1   Load conditions

In order for the mechanism or transmission system to fulfil its intended function, it will be required to experience a variety of loads on both input and output, as well as inertia loads due to its own motion. The capacity of the driver in terms of torque or force must be sufficient to cover the full operational envelope.

The load paths must also be clearly established, paying special attention to areas where the load has to be diffused from one low stressed area to another via an area of stress concentration such as a bolted connection.

22.1.1   ACTUATOR REQUIREMENTS

The choice of actuator depends not only on the required dynamics of the payload, but also on the inertia and strength of the intervening mechanism. The mechanism design is likewise affected by the type, characteristics and size of the actuator; for example, the actuator may generate impulses on start-up.

22.1.2   ATTAINING PARTIAL STATIC BALANCE

If the actuator does not have to support the weight or directly drive the mass of the mechanism it can be made smaller. This can be achieved by counter-balance weights, as in Fig. 22.3, by the use of pneumatic cylinders, or by ensuring the support structure carries the load, as in the SCARA type robot of Fig. 22.4.

22.1.3   ARTICULATION REQUIREMENTS

In the first stages of design the kinematic requirements of the mechanism are the primary consideration; the question must be addressed of how the mechanism can cover the intended envelope of geometrical paths. The traditional graphical methods for the kinematic synthesis of linkages and the mathematical relationships such as Freudenstein's equation for four-bar chains have now been embodied in commercial software packages. These have evolved into integrated design procedures, starting with kinematics definition and leading through structural solid modelling to the display of nodal forces and eventually to the calculation and display of deflected shapes and stress distributions in the structural elements. A short account of two such software packages is given in section 22.4.

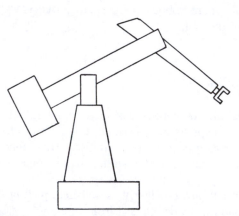

Figure 22.3   Partial static balance using a counterweight.

Figure 22.4   SCARA robot (UMI).

22.1.4   SPEED VERSUS ACCURACY

This is influenced by the control loop. However, if the inertia of the mechanism is high, driver overshoot or the flexibility of the mechanism will require a certain settling time before it is stationary in the correct position.

22.1.5   MINIMIZATION OF KINETIC ENERGY

This is a consequence of minimizing the inertia, and is also good practice for safety reasons in avoiding overruns likely to injure operators. The danger is most apparent where the end effector of a robot travels at high velocity. However, there are also real but less obvious dangers where a large reduction ratio is employed between a high speed driver and a mechanical output. Even after the operation of an emergency trip or 'panic button' the residual kinetic energy in the driver may still be able to generate very large forces at the output. In such circumstances severe crush injuries may result or the mechanism itself may be damaged.

22.1.6   POWER TRANSMISSION OVER A DISTANCE

Where power is to be transmitted over a distance, some flexibility is needed at either end to allow for movement of the structure and misalignment, otherwise large stresses could be set up in the mechanism. Particular care must be taken to avoid kinematic overconstraint where connected parts of the mechanism are 'fighting' each other. The transmission of rotary motion will set up twisting moments in some mechanisms. Mechatronics has not removed the need for sound mechanical design, and much valuable insight can be gained in this area from a consideration of the standard texts in fields such as machine tool and instrument mechanical design.

22.1.7   EFFECTS OF ASSEMBLY PLAY AND FRICTION

Any play or backlash in the mechanism can generate unwanted impulse loading as well as a loss of positional accuracy where the motion reverses. Friction can have a similar effect, and can enormously amplify the problems of kinematic overconstraint. A well known but still perpetrated example is of guide bushes which are too short in relation to their length and which can become irretrievably jammed. Friction, however, can also be beneficial in damping down vibrations.

22.1.8   INERTIA

It is obvious from the above that the inertia of a mechanism should be kept to a minimum, and this is especially so in the case of high speed machinery such as canning lines.

22.2   Design

Good design can allow products such as dot matrix printers and pen plotters to achieve rapid operation with remarkable accuracy by reducing the dynamic inertia to a minimum. The intelligent use of form rather than mass to achieve rigidity is a necessity in this instance. Thus the designer will look for opportunities to employ structural forms which create a high section modulus. This means taking mass out of the middle of structures and putting it only where it can be most effective in or close to the surface or skin. In such an approach it might

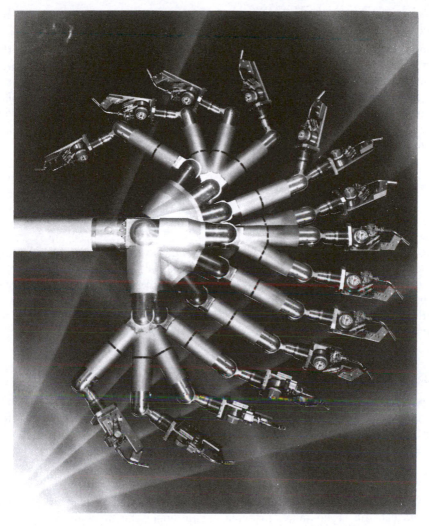

Figure 22.5   Prototype hydraulic manipulator (Taylor Hitec).

be said that the main function of what remains in the middle of a structure is simply to keep the surfaces apart.

22.2.1   MATERIALS

Materials with a high Young's modulus and a low density, such as composites and exotic metals, can achieve substantial reductions in inertia. An example is the Taylor Hitec in-core nuclear manipulator of Fig. 22.5, where titanium is used extensively to reduce the dynamic inertia of the arm.

22.2.2 SIZING OF ACTUATORS

Where the movement of a mechanism results in a change of its geometrical shape, as opposed to a simple translation, the effective inertia 'seen' at the actuator will change. An extreme example of this is the anthropomorphic jointed arm robot. The inertia experienced by each axis drive motor depends on the angular position of the succeeding axes. To compensate for this the controller may incorporate a compensation input to the control loop of each axis from every other axis. Simulation of mechanisms can determine the dynamic loads involved.

22.3 Flexibility

22.3.1 RESILIENCE

Each member of a mechanism will have some flexibility in tension, bending and torsion. The design of its cross-section must be appropriate to the type of loadng experienced; for example, a circular section is stiffer in torsion, whereas a box section of the same cross-sectional area and wall thickness is stiffer in bending. Loading can be determined by mathematical modelling, if not by classical calculation methods.

22.3.2 BACKLASH

Flexibility in a mechanism can also be caused by play in joints and backlash in gears. Special anti-backlash gears can be obtained, but provision at the design stage for future adjustment of joints is a wise decision.

22.3.3 VIBRATION

Flexibility in a mechanism subject to dynamic loading will lead to vibration unless care has been taken to avoid excitation of the mechanism's natural frequencies. It is usually the excitation frequencies which are an external constraint on the system design; once a configuration and structure have been proposed, modal analysis (usually in software) is undertaken to discover the natural frequencies. Even fourth or higher order harmonics may cause problems in fine mechanisms, for example in the polishing of optical components. Design measures may therefore involve moving the natural frequencies and harmonics away from the likely envelope of excitation frequencies. This is done by removing mass, having in mind that this may increase flexibility, or adding mass, recognizing that this will increase loadings. Alternatively, the natural damping inherent in bolted joints can be enhanced by judictious repositioning. As a final resort, external damping can be added in the form of telescopic hydraulic struts, viscous rotational dampers of the Lanchester type, or friction pads with added masses (the so-called inertial damper).

22.4   Modelling and simulation

22.4.1   GRASP

GRASP is a generalized robot simulation package initially developed at Nottingham University and now maintained and marketed by BYG Systems. Using GRASP, a kinematic model of an individual robot can be created incorporating all the necessary velocities and accelerations, while the solid modelling feature enables the incorporation of the structural elements – the 'flesh' of the robot.

By combining the robot model with the generated model of the operating environment – including machine tools, conveyors, AGVs and sensors – the operation of the complete cell can be simulated and performance assessed. As a result of the simulation an animated representation of the cell operation is provided which can be viewed on a variety of timescales. Incorporated at this stage of the simulation are clash detection routines which detect any interference or collisions between elements of the system. Once a satisfactory operating regime has been established then the appropriate code for the target robot can be generated.

An additional software module within GRASP is the Digital Simulator, which allows the user to access much more management-type information. A manufacturing system can be fully simulated within GRASP when the Digital Simulator can be used to identify bottlenecks and machine tool utilization as well as providing statistical information such as the amount of work in progress. This information can be reported by means of graphs, histograms and bar charts in a totally user defined manner.

The introduction of simulation and off-line programming for robotic cells has had a significant impact on the design of such cells. By means of the simulation the designer is able to assess the performance of any robot against the required operations while having the flexibility to rapidly reorganize the elements of the cell, something which would be impossible using real devices. In addition, the modelling feature enables novel robot geometries to be investigated, such as for robot cranes and excavators in the civil engineering and construction industries.

22.4.2   ADAMS

Another example of a commercial software suite is ADAMS originally developed by the University of Michigan, Ann Arbor, USA and marketed by TEDAS GmbH of Marburg, West Germany. The mnemonic stands for automatic dynamic analysis of mechanical systems, which is a succinct exposition of its capability. Features include:

1. Wire frame and solid modelling of the system configuration;
2. Real or scaled motion time display;
3. Calculation of nodal velocities, accelerations and forces;
4. User defined display of graphical relationships between operational parameters;
5. Software interfaces to further modules in modal analysis and finite element analysis.

Chapter 23

Structures

As with mechanisms, it is important to consider the purpose of a mechatronic structure before looking at the design aspects. The functions of the structure may be summarized under the following headings.

(a)   PROVISION OF LOAD PATH AND SUPPORT OF PAYLOAD

The load has to be transmitted, either directly or via the user, to some reference such as the ground. This is referred to as grounding.

(b)   ACCURACY AND REPEATABILITY

The human hand is very adaptable for a variety of tasks, of which assembly operations provide a good illustration. However, a machine must operate within tight tolerances with limited flexibility. The use of open loop control places very heavy demands on the mechanism and structure to ensure repeatability.

(c)   ISOLATION OF SYSTEM FROM THE ENVIRONMENT

In terms of stress, vibration and temperature, changes in the outside environment should not affect the performance of the device. The expected environment must, of course, be specified at the higher levels of the design process before detailed design commences.

(d)   SUPPORT OF SENSORS, CONTROLLERS AND ACTUATORS

The obvious requirement is that of physical support. It is also important that the design separations and clearances are maintained during the operation of the device in the specified environment.

(e)   SUPPORT AND GUIDANCE OF MECHANISMS

Individual mechanisms will require grounding with respect to the actuator or load, and will normally use the structure to maintain accurate travel by ensuring that the loads are transferred, through a path whose characteristics, such as stiffness, are know. 'Skyhooks' are not found in successful engineering!

(f) SUPPORT OF WIRING, CONTROLS AND DISPLAYS

Cables require support and protection from areas such as the moving and heat generating parts of the device. The man–machine interface may well require controls and displays to be mounted on the device.

(g) SUPPORT OF POWER PACK

Some systems, such as rechargeable hand power tools and automatic guided vehicles, carry their own on-board power pack. This power pack is often heavy relative to the rest of the device and thus the structure may need strengthening in that area.

(h) ENCLOSURE OR PACKAGING

The enclosure or packaging of a system may be purely cosmetic or intended to provide additional isolation, for example against the ingress of dust. The enclosure has to be fixed to or contain the structure in such a way as to allow easy removal for maintenance of the device. In some products, for instance hand held power tools, the structure and the enclosure are the same, and maintenance considerations are secondary to the manufacturer's requirements for achieving speed and consistency in the initial assembly.

(i) STORAGE OF INTERCHANGEABLE PARTS AND TOOLS

Sophisticated storage mechanisms such as carousels with automatic tool changers are used with CNC machines to enable a large number of tools to be held on the machine, permitting continuous unmanned operation. Problems can arise if such mechanisms can transfer vibrations into the main machine structure while machining is in progress, leading to unsatisfactory finish on fine machined surfaces.

23.1   Load conditions

In order to design the structure adequately for the requirements, the loading on it must be carefully analysed.

The stresses due to the load cases described below may be difficult to determine. Analysis is aided by the computer modelling of structures and mechanisms.

23.1.1 STATIC LOADING

In addition to the mechanism and actuator weight, there is also the self-weight of the structure and the enclosure to be considered. The fixing points of these components should be given careful consideration in the design as they are often the source of failure.

23.1.2 DYNAMIC AND CYCLIC LOADING

When the system is operating, extra inertia loading is applied to the structure. This can lead to loss of accuracy and even to fatigue failures in the structure if this not adequately designed. Similar effects result from the cyclic loading occurring as a result of vibration.

23.1.3 IMPULSE AND SHOCK LOADING

This may occur as a result of system operation, but can also be caused by dropping the device accidentally! Consideration must also be given to stresses expected in the process of transport, delivery or maintenance, which may also involve excessive static or cyclic loads. The effect of ships' vibration when transporting equipment containing ball bearings is a well known example. If the static loads are already high, as in a heavy rotor, brinnelling or fretting of the bearings may occur.

23.2 Flexibility

All structures have some flexibility, especially if joints are incorporated in the design.

23.2.1 FLEXIBLE STRUCTURES

With closed loop control, as long as the feedback sensor is very close to the point of action, any local flexibility in the structure will be taken care of by the microprocessor. This concept is being used in the design of large telescope dishes or mirrors. Here, instead of a heavy rigid structure, a lightweight structure is used incorporating a series of actuators which automatically maintain the desired profile of the dish or mirror under computer control.

Where the sensors and actuators are located remotely from the point of action then errors can result from the flexibility of the structure. An example of this is seen where a robot arm is used to apply a significant force at the extent of its reach, when a loss of positional accuracy can occur as a result of the deflections in the structure. For this reason, industrial robots are designed to be very stiff and the result is a heavy structure, the movement of which absorbs a significant amount of the robot's energy requirements. The development of adaptive control should enable the introduction of lighter structures.

23.2.2 VIBRATION EFFECTS

Any flexible structure will be prone to vibration, particularly if excited at its natural frequencies. Computer modelling of complex structures will reveal these natural frequencies and allow redesign of the structure to compensate if necessary.

23.2.3 MATERIALS

The suitable choice of material and manufactured form can greatly affect the flexibility and life of the structure. Aluminium extrusions are very useful and composites are being used increasingly.

23.3 Environmental isolation

Isolation of the device from the environment and the user from the device are important considerations, especially when this is the prime function of the device, as in the case of an active suspension.

The system isolation should ensure that none of the thermal or mechanical forces produced in or by the device should be transmitted to other connected structures or to the user. For example, a camera should not move around in the hands when operated, nor should it jam whether it is being used in the desert or at the North Pole.

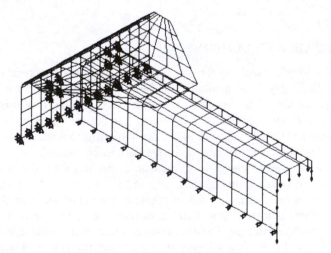

Figure 23.1 Finite elements.

Systems should also be isolated from external sources of disturbance. For example, portable compact disc players must operate when swinging from someone's shoulder. Closed loop control helps here, but in many cases traditional techniques of vibration isolation such as flexible mounts can be used.

There are, however, some cases where the user should not be isolated from external disturbances, as the feedback – or feelback – derived is an indication of the device's performance. A good example is seen in flight simulators, where the dynamic movement of the mock cabin is needed to maintain realism.

Intelligent systems such as are employed for inertial navigation may also have to distinguish between the normal and proper motion of the aircraft or ship in response to the control inputs, and the motion resulting from strong external disturbances such as turbulence, which may cause the controller to saturate.

23.4   Modelling

The computer modelling technique most often used for the analysis of structures is that of finite element analysis. In this technique the geometry of a complex structure is defined by a series of elements, the shape of which is determined by the form of the structure as in Fig. 23.1. The physical characteristics of each of these elements are defined by the properties of the materials used at each part of the structure. For each condition of loading – for example static, dynamic, thermal – the behaviour of each of the individual elements can be defined by means of a set of simultaneous differential equations. These equations are solved by the computer using a series of matrix operations through which the behaviours of all the individual elements are combined to give deformations and stresses within the structure.

The results obtained can be displayed as coloured images identifying areas of high and low stress. As in any analysis, the results are only as good as the model and must be checked by physical testing to confirm their validity.

23.5   Systems

In a mechatronic design, the structure will be designed taking account of the function and use of electronics. This integration is illustrated by a number of materials handling systems in which the actuators and sensors are incorporated into the design of the structure.

One such system, shown in Fig. 23.2, uses a high rigidity aluminium extrusion as the structural member. This structural member also contains the pre-tensioned, backlash-free chain drive to the moving carriages and locates the sensors as required. The primary positioning, however, is provided by sensors on the DC servomotors, thus making it possible for the end user to build his

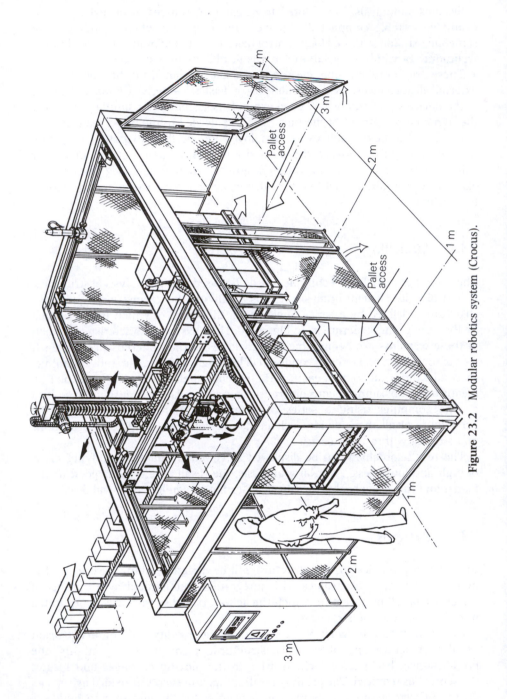

Figure 23.2 Modular robotics system (Crocus).

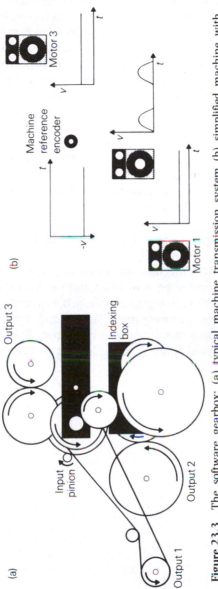

(a)

Output 3

Input pinion

Output 1

Output 2

Indexing box

(b)

Machine reference encoder

Motor 3

Motor 1

Figure 23.3 The software gearbox: (a) typical machine transmission system (b) simplified machine with programmable motions synchronized with software gearbox (Quin Systems).

own system from a set of self-contained axis drives. Interestingly, this product was brought to the market by an entrepreneur who observed that many conventional robotic systems were inappropriately specified for their tasks. In particular, the degree of accuracy achievable, at high cost, by arm-type robots was often unnecessary; the paramount need was for a high speed and load carrying capacity. These systems have found good markets in applications such as the loading of heat treatment furnaces, the laser cutting of sailcloth and the 'flying' of cameras in aircraft flight simulators using scale models of terrain.

An instance of how mechatronic thinking can influence mechanism design in this area is provided by the software gearbox approach developed for packaging machinery and similar uses. Traditionally, mechanical cams and linkages have been employed to provide the necessary controlled motion paths, with adjustment or reconfiguring only achievable during shutdown and often with difficulty. The software gearbox technique allows individual servomotors to be synchronized to mimic gears, cams or linkages in such a way that they can be easily changed in software to suit different machine requirements. This has substantial advantages over traditional mechanisms and aids the application of flexible manufacturing systems. A typical before-and-after example is shown in Fig. 23.3.

The extent to which such mechatronic systems can be applied depends upon the accelerations and forces to be controlled and the available slew rate of the driver. Applications where such parameters reach extremely high values, as in cigarette making machines, canware manufacture and chequebook printers will see traditional mechanical cams continuing in use. Even in such systems, however, mechatronics will have a valuable part to play in diagnostics and condition monitoring.

Chapter 24

Man–machine interface

The man–machine interface (MMI) is the facility by which the bidirectional transfer of information required for system operation is provided. The MMI is therefore primarily concerned with the ergonomic aspects of the design and covers areas such as the display of information, operator systems, environmental factors and safety.

24.1 Industrial design and ergonomics

Industrial design embraces those elements of the overall design function – such as aesthetics, style and ergonomics – which are concerned with the interaction of the machine or system with the human operator.

24.1.1 AESTHETICS AND STYLE

Aesthetics and style are concerned with that combination of shape, form, colour and materials which the system presents to the observer. They are therefore primarily associated, through the observer's visual and tactile responses, with the external appearance and feel of the system as represented by its enclosure or packaging. The role of industrial design lies in the achievement of the balance between the aesthetic and stylistic aspects of a design and engineering requirements such as safety or the need to reduce aerodynamic drag in cars.

Within this framework it is not sufficient simply to consider the visual aspects of the system and its presentation. Factors such as performance, reliability, choice of materials, safety and cost effectiveness must be taken into account.

24.1.2 ERGONOMICS

Ergonomics or human factors engineering is concerned with the application of the knowledge about the physical skills and limitations of a human being to provide a safe and effective working environment. The achievement of these objectives requires the designer to give consideration to areas such as:

1. The mechanisms by which humans can receive and transmit information;
2. The forces that humans can develop;
3. The environmental conditions in which humans can function;
4. Increasing user comfort;
5. The sharing of tasks between the human and the system.

In the context of mechatronics and the increasing use of embedded microprocessor systems the last point is of particular importance, as the design must take into account the abilities of the human operator and the system. Some of these are summarized in Table 24.1.

In ergonomics the human being is the focus of attention, and it is the responses and requirements of that human being in a particular environment which will determine the nature and form of the man–machine interface adopted. Table 24.2 shows some of the physical parameters influencing human performance.

Table 24.1   A comparison between man and machine

Humans	Machines
Respond to a limited range of stimuli	Respond to stimuli outside the range of humans, such as infrared or ultrasound
Response to stimuli can be variable	Consistent response to stimuli
Can detect unexpected or unusual variations in stimulus	Limited response to unexpected or unusual variations in stimulus
Can react to patterns within a wide range of stimuli	Limited response to patterns within a a wide range of stimuli
Capable of inductive and deductive reasoning	Limited reasoning capacity
Apply principles and concepts to the solution of a wide range of problems	Limited range of solutions generally associated with defined problem areas
Large learning capacity	Limited learning capacity
Use experience to aid in the solution of new problems	Limited in the application of experience
Highly adaptive physical responses	Limited physical responses defined by structure
Multifunctional	Dedicated functions
Limited strength	Powerful
Tire easily and lose concentration	Do not tire or lose concentration
Capable of being distracted	Dedicated, not capable of being distracted
Information retrieval limited in scope and accuracy	Highly efficient, accurate and reliable information retrieval
Poor long term storage of information	Good long term storage of information

Table 24.2 Some physical dimension for various percentile groups

	Mean (percentiles)			Women (percentiles)		
	5th	50th	95th	5th	50th	95th
Height (mm)	1640	1750	1860	1520	1625	1730
Eye level (mm)	1525	1640	1750	1400	1510	1615
Shoulder breadth (mm)	420	460	505	375	420	465
Seat height (mm)	370	415	460	340	380	425

24.2   Information transfer: from machine to man

In order for an operator to monitor and control the functioning of a machine or system, information regarding the current status of that machine or system must be made available. The levels of information transfer involved in meeting this requirement range from the simple – for example in a washing machine, where all that is needed for normal operation is an indication of the current status and the time required to complete the selected sequence – to the complex, as in aircraft where data needs to be made available covering the operation of a wide range of systems.

Indeed, in many complex environments the operator is faced with problems of information overload, particularly in emergencies, where the relevant information necessary to the continued safe operation of the system must be rapidly isolated from a large volume of data. This has led to the adoption of information management techniques in which only that information which is relevant to the current operation is available to the operator, who may however have access to other, more detailed information on demand. An example of this is seen in vehicle systems, where diagnostic information produced by the vehicle's own monitoring system is not normally accessible to the driver but can only be obtained using special equipment at its maintenance base.

The quality of the information is also important, and correct information displayed and presented incorrectly can be as damaging as incorrect information. Consider the situation that arose at the Three Mile Island nuclear plant, where the instrumentation monitored the signals that had been sent to a pressure relief valve and not the actual condition (open or closed) of the valve itself. The result was that although the instrumentation showed – correctly – that the operating signal had been sent to the valve, the fact that the valve had failed to move was not registered as its motion was not monitored directly.

24.2.1   HUMAN RESPONSES TO STIMULI

Human beings respond to a number of physical stimuli, of which the visible and audible regimes are the most important. Table 24.3 sets out some of these responses.

Table 24.3   Response of human beings to stimulus

Stimulus	Range	Amplitude range	Notes
Visible	380 to 720 nm	10^{-9} to 10^9 cd m^{-2}	Response varies with colour of light
Audible	20 Hz to < 20 kHz	0 to 140 dB	Optimum range 300 Hz to 6 kHz and 40 to 80 dB; physical damage possible at higher intensities
Tactile			Maximum sensitivity at fingertips
Vibration	0 Hz to 10 Hz		Physical damage at high intensities

(a)   VISUAL DISPLAYS

Visual displays, through the use of shapes, texture and colour as well as numerical data, offer the opportunity for the presentation of a wide variety of types and forms of information.

Where precise numerical information is to be transmitted, a digital format offers the best means of presentation. However, where a less precise value is required but where trends or a rapid assimilation of status is needed, an analogue display probably offers the best overall performance. It is worth noting in this respect that modern cockpit displays, such as that of the A320 Airbus described in section 2.4, generate an analogue output.

Where symbols or characters are used, their height, positioning and colour all influence the response to the display. Colour also serves to highlight individual features of the display and to provide, perhaps through a colour change, additional stimulus to the operator.

Visual displays are required to operate in a variety of ambient lighting and viewing conditions and the type of display chosen must be compatible with these ambient conditions. This means that consideration must be given to the viewing angle of the display, contrast levels and response times.

(b)   AUDIBLE SIGNALS

While audible signals can carry less information than a visual signal, they are significantly better at gaining the attention of an operator and are less affected by fatigue. For this reason audible signals, particularly including speech, are used to attract attention, for example as warnings and alarms. The growth of synthesized speech for this purpose is particularly notable.

(c)   TACTILE SIGNALS

Levers may be equipped with controllable projecting pins (Braille transducers) or the tactile signal may be in the form of an imposed vibration, as in the stick shaker which is employed in aircraft cockpits to warn of the onset of stall conditions.

24.3   Information transfer: from man to machine

In order to control a machine or system, the operator must be capable of changing the settings on the machine to modify its operation as required. The range of information transfer associated with this task again ranges from the simple – an on/off switch – to the complex. In ergonomic terms, it is important that controls are designed taking account of the operator requirements, for example the introduction of some progressive resistance to the motion of a joystick to provide operator feel. Table 24.4 sets out the characteristics of some common mechanical control elements.

In systems where operation is via a computer terminal and/or graphics display there is an increasing movement away from the use of conventional keyboards for the entry of information and instructions. The introduction of menu driven software combined with the use of a mouse and pull-down windows has had a significant effect in increasing the ease with which the operator can communicate with the system.

In environments such as aircraft, there is a variety of ways in which man communicates his intentions. The term which is becoming commonly used to describe such MMI devices is 'inceptor'. This covers buttons, keypads, levers, handles, wheels, keys and pressure pads.

In many mechatronic systems the instructions are issued at a high level and the actual control system is transparent to the operator. For example, in a car the instructions are issued via the accelerator, brake and steering wheel and the on-board systems automatically monitor and adjust the engine settings to maximize the performance obtained. The introduction of microprocessor based controllers also enables the removal of cumbersome, and often unreliable, mechanical components and their replacement by solid state devices while still retaining a familiar layout for the operator control panel.

24.4   Safety

Safety is a concern in two areas. Firstly, and most important, there is the protection of the operator from any harm resulting from the operation of the system. Secondly, there is the protection of the system itself from harm as a result of external effects. As priority is given to operator safety these two requirements are not necessarily compatible, and the system may be sacrificed to improve operator safety. For example, deformable bumpers are intended to protect a vehicle from damage at low impact speeds such as those that result during parking, while the vehicle is designed with energy absorbing crumple zones to protect the occupants during high velocity collisions.

Table 24.4 Mechanical control elements

Control	Force transmitted	Response	Settings	Sensitivity	Visual feedback	Space required
Rotary switch	n/a	Fast	Many	Good	Yes	Medium
Thumbwheel	n/a	Fast	Many	Good	Yes	Low
Push button	n/a	Very fast	On/off	n/a	Yes	Low
Toggle switch	n/a	Very fast	On/off	n/a	Yes	Low
Knob	Small	Slow	Continuous	Good	Possible	Medium
Lever	High	Slow	Continuous	Medium	Possible	High
Pedal	High	Slow	Continuous	Low	Possible	High

24.4.1 OPERATOR SAFETY

Any item of machinery, even if working correctly, is capable of inflicting harm if misused or abused. The main areas of operator risk are:

Trapping   The body or limbs can be trapped by the action of a machine in closing or passing by. In the case of gears, belts and other rotating systems, trapping could occur as a result of the motion drawing the limb or body into the system.

Entanglement   Rotating machinery in particular can snag hair and loose items of clothing such as ties or shirt cuffs in the mechanism, drawing the operator into the system with the possibility of trapping as a result.

Contact   Contact with sharp or abrasive surfaces, hot surfaces and electrically live surfaces.

Impact   Impact of a stationary operator with a moving object or vice versa.

Ejection   Many operations, particularly in manufacturing, eject waste material in the form of swarf, chips, sparks or molten material. In addition, there is the possibility of the ejection of material as a result of a system failure, for example the bursting of a grinding wheel or the overspeed of a DC machine.

(c) ENVIRONMENT

Human beings are capable of operating effectively in a relatively limited range of environmental conditions, with the combination of temperature, humidity, air speed and work rate particularly important. Humans also have a limited visual and audible response range; high intensity sources, such as pneumatic drills and lasers, are possible sources of damage or injury. Tolerance of factors such as vibration and g forces is also relatively low and can lead to injury. Increased attention to operator comfort, for example through improved truck cab suspension systems, can reduce the risk from these sources. Environmental protection must also be introduced to safeguard operators from a wide range of pollutants, in particular those such as radiation, liquid and gaseous chemicals and dust whose effects are long term.

In an industrial environment, operator protection is generally provided in the form of guards and enclosures which separate the operator from the system and maintain a safe zone around and within the system during operation. Alternatively, operators may be provided with their own personal environment, of which the space suit is perhaps the most spectacular example, in order to enable them to function within a hazardous external environment. Design does, however, have a major role to play by creating systems which are intrinsically safe and by removing features, such as sharp corners, which are possible sources of injury.

24.4.2   SYSTEM SAFETY

Systems are at risk from a variety of sources and their design must be such as to minimize the effects of these hazards while retaining the system in a operational condition. Risks include:

Shock   All systems may be subject to shock forces from a variety of sources such as dropping or impact by other objects.

Vibration   Vibration induced stresses can lead to fatigue failure in materials. The choice of materials and the use of vibration mountings and dampers are therefore important elements of the design.

Corrosion   Systems may be required to operate in corrosive environments, demanding a proper choice of materials and/or surface coating to ensure their functioning.

Environment   Systems may be required to operate under a wide range of environmental conditions where they are subject to variations in parameters such as temperature, humidity and pressure.

Fire   Not only must systems be designed so that they themselves do not present a fire hazard; it may also be necessary to ensure the safe working of the system in the presence of fire, if only for a limited period.

Misuse   Operators have demonstrated a great ability to misuse a system or to develop previously unconsidered uses which lie outside the nominal operating regime. The possibility of vandalism must also be considered.

(a)   FAILSAFE DESIGN

Where possible, the design adopted should be such that failures will themselves not increase the risk to the operator. Such a failsafe approach involves undertaking a proper risk analysis of the system to identify the safety critical failure modes.

(b)   REDUNDANCY

Once the safety critical areas have been identified then steps can be taken to eliminate them. One approach that is often adopted is to introduce redundancy into the system. This may take the form of increased structural safety margins or the provision of parallel circuits, often of different types and routed differently, so that the failure of a part of the system does not result in a failure of the whole of the system.

In areas of particular concern, multiple systems incorporating a voting arrangement can be deployed. Here, each system is continually checking itself against the other equivalent systems to check for any deviation. Should any deviation be detected then that circuit identified as being at fault can be isolated, allowing operation to continue on the remaining systems.

(c)   SOFTWARE

With the increasing complexity of systems, software engineering has an important role to play in the development and control of all software but particularly that which is safety critical. Techniques such as the production of parallel but independent code running on different processors increase the overall system reliability, while the introduction of software management tools such as CORE and MASCOT will assist in the production of effective, efficient and reliable code.

Part Five

Case studies

Chapter 25

Introduction to case studies

The scope of mechatronic systems is exceedingly wide and often hierarchical, with the essential sensor/microcontroller/actuator relationship appearing at several levels. This is illustrated in the following case studies, which are included not so much with the purpose of giving detail of particular systems, but rather to show how mechatronic principles have been applied in commercial practice and to provide encouragement for further applications. The case studies illustrate the three broad areas or categories into which most mechatronic applications would be found to fit.

In the first we have a **stand-alone product** as exemplified by the Canon EOS autofocus camera and the Autohelm boat autopilot (Chapters 26 and 29). The boat autopilot is much the simpler of the two but gives the most readily apparent illustration of the mechatronic relationship. The early prototypes were implemented in analogue circuitry, but the contribution of digital electronics in the current models has been in the avoidance of drift, and the provision of a degree of reliability and refinement which could not otherwise have been achieved within such a compact package. The sophistication of the design is demonstrated by the way in which a fluxgate compass (entirely designed and manufactured in-house) has been incorporated within the same package as the controller board and drive motor while avoiding problems with electromagnetic interference. The Canon EOS camera is a more complex system containing several mechatronic control loops within a single enclosure. The necessity of minimizing weight and bulk has driven the company to some novel solutions, of which the leading example is the ultrasonic motor (USM) used in focusing. Both of these products emphasize functionality without taxing the users intellectually or physically; they can concentrate on the main objective, that is reaching a port or composing a picture. Both demonstrate the point that in a stringent commercial and marketing environment, if the projected manufacturing quantities are large enough, the benefits of mechatronic integration from the conceptual stage of design may often best be obtained by using bespoke rather than proprietary components, including sensors and CPUs.

The second category, illustrated by the fly-by-wire aircraft (Chapter 27), is of a larger system having the topology of a **distributed system** while being mobile and self-contained. Less sophisticated examples on road vehicles could also be cited, where the emerging controlled area networks (CANs) used for linking systems such as suspension control and braking modules have their equivalent in the MIL-STD digital data transmission systems employed on the aircraft. The BAe Experimental Aircraft Programme (EAP) introduces advanced control laws designed to allow the pilot to fly the aircraft with an unrestrained enthusiasm, as one might say, but with the assurance that the structural limits cannot be exceeded. In aircraft technology, system integrity is of the highest priority, with quadriplexed control computers and data lanes and force-summed actuator outputs. Use is made of software watchdogs and comparison techniques to check the integrity of each lane, one against another. Undoubtedly such approaches will find their way into less esoteric applications such as vehicles, and indeed must do so if the mechanical simplifications deliverable by steer-by-wire are ever to be acceptable. Aircraft and military equipment procurement also relies upon the very precise specification of components and systems for manufacture by approved suppliers and subcontractors. The accent is upon control of the specification and approval of the design, but the detail design proposals are usually the responsibility of the subcontractor.

In the third category we have a large static, or factory, system which is also a distributed system but which links a number of major subsystems which are themselves mechatronic systems in their own right (Chapter 28). We could describe this as a **mechatronic systems group.** All large flexible manufacturing systems (FMSs) are of this type. The principal hardware elements visible in such a system are machining centres, robots for parts handling or tool changing, automatic guided vehicles (AGVs), conveyors and automated inspection stations employing coordinate measuring machines (CMMs). Characteristically each of these classes of item will come from a different and specialist supplier offering substantially standard equipment to meet the needs of a range of customers. Sophisticated CNC machine tools with or without automatic tool change and in-process gauging have long been available as productive stand-alone items of plant with their own control systems and computers. Their attraction has led sometimes to piecemeal application and the creation of the so-called islands of automation. User initiated attempts to link these retrospectively to realize the benefits of computer integrated manufacture (CIM), perhaps through to scheduling, process planning and even design, have often been frustrated due to the inability of the various machine controllers to communicate. However, protocols and interface modules such as support the manufacturing automation protocol (MAP) and communications media such as Ethernet are now doing for manufacturing equipment what MIL-STD-1553 has done for aircraft systems. The large FMS described in the case study was specified and designed from the outset as an integrated system – but by an organization whose in-house expertise was of a level which not only assured the success of the system itself but enabled the company to market software products developed within the execution of the project.

Chapter 26

Canon EOS autofocus camera

Automatic, autofocus cameras such as the Canon EOS 620 and 650 of Fig. 26.1 provide the user with a powerful, simple and flexible system capable of operating under a wide range of conditions. Their design encompasses the whole of mechatronics, requiring as it does the integration of sensor and drive technologies with the internal distributed processing power and the user interface.

The operation of such a camera requires control over the following areas:

1. The automatic setting of film speed and number of exposures using the DX code on the film cassette;
2. The measurement of the received light through the lens;
3. In the fully automatic mode, the setting of both aperture and shutter speed;
4. In other exposure modes, the setting of the aperture or the shutter speed as appropriate;
5. The control of focus including focusing on off-centre objects, and the control of depth of field taking into account the type and focal length of the lens being used;
6. Automatic film advance including multiframe action sequences and multiple exposures;
7. The automatic rewind of film after the last exposure;
8. Integration of operation with flash guns.

The control system for the Canon EOS is shown in Fig. 26.2. Operation is under the overall control of the main microprocessor in the body of the camera. This communicates with the lens microprocessor and also with those in the flash unit and data back if these are being used. The basic system, camera plus lens, also incorporates a total of five actuators. In addition, the flash gun includes its own separate actuator to operate its zoom function.

Sensors monitor not only the main functions of exposure control and focusing but also the internal operation of the camera, including checking the correct attachment of the lens and the correct closure of the camera back. The user communicates with the camera by means of a series of buttons and a thumbwheel which together enable the user to select the operating mode to be

Figure 26.1   Canon EOS 620 and 650.

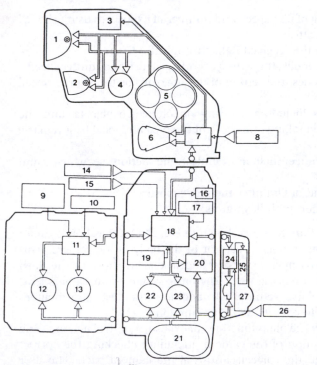

Flash
1 Marin flash element
2 Pre-flash
3 Display
4 Zoom M
5 Battery
6 Auxiliary light LED
7 Flash microprocessor
8 Input buttons

AF lens
9 M ring or electronic ring
10 Zooming ring
11 Lens microprocessor
12 AFD or USM
13 EMD

Body
14 Input buttons
15 Electronic input dial
16 Display
17 AE sensor
18 Main microprocessor
19 BASIS
20 Shutter magnet
21 Battery
22 Motor 1
23 Motor 2

Interchangeable back
24 Back microprocessor
25 Display
26 Input buttons
27 LED

Figure 26.2   Control system structure.

used. Tables 26.1 to 26.5 list various features of the cameras and summarize their functions.

26.1   Main microprocessor

The main microprocessor is a high speed, single chip, 8 bit device with a built-in EEPROM. The processor has a minimum processing time of 1 μs and uses 16 bit arithmetic for addition, subtraction and division and 8 bit arithmetic for multiplication. The main microprocessor is responsible for the control of all automatic exposure and autofocus functions and provides sequence control for

Table 26.1 Microprocessors used in the EOS system

Microprocessor	Function
Main	Control of automatic exposure and focus; sequence control
Lens	Data communication; control of focusing drive
Flash	Data communication; control of the flashing element
Data back	Data communication and storage; camera drive control

Table 26.2 Actuators used in the EOS system

Actuator	Function
Motor I	Film transport
Motor II	Shutter and mirror action; film rewind
Motor III	Focusing drive
Motor IV	Lens aperture drive
Motor V	Flash zoom drive
Shutter magnet	Exposure time control

Table 26.3 Sensors in the EOS system

Parameter sensed	Comments
Focus	Distance measurement
Focal length	Encoder in lens used to detect change in focal length of a zoom lens
Lens drive	Motion of lens is detected by a photoelectric sensor which generates a series of pulses as the lens rotates and focus changes
Lens mounting	Leaf switch
DX code	Pins detect a pattern on the outside of the film cassette
Back open/closed	Leaf switch
Film transport	Movement of the film generates a series of pulses

Table 26.4 Input and output functions

	Function
Input device	
Dial switch	Mode select, used in combination with other controls
Push buttons	Function settings, exposure and metering control
Slide switch	Selection of automatic or manual focusing
Rotary switch	On/off, ready and fully automatic settings
Output device	
Body panel	LCD provides information on exposure and focusing modes, film speed, frame count
Viewfinder	Displays aperture value and shutter speed, automatic/manual operation and metering mode; in-focus indication is provided by a flashing LED

Table 26.5 Power supplies

Power source	Function
Lithium battery	Main power supply for body and lens
Alkaline battery	Power supply for flash

the complete camera, including the film transport motors. A block diagram of the layout of the internal electronic system is given in Fig. 26.3.

26.2   Exposure control

The main metering mode used on the EOS is linked to the focusing operation, and makes use of the fact that the main subject is in the centre of the image when focus is achieved. Information on light levels from a combination of six sensors arranged in relation to the image as shown in Fig. 26.4a is first grouped into three areas involving sensors L_1 (main subject and central area of image), L_2 (main subject and adjacent background) and L_3, L_4, L_5 and L_6 (peripheral areas of image) respectively. This data is recorded and analysed to obtain information on the size of the subject and the type of scene (normal, backlit, dark background, contrast levels and so on) and hence to determine the required exposure value (Fig. 26.4b). As an alternative to the evaluative metering function, a partial metering function is available using only the central 6.5% of the image.

Once the exposure value has been found then, in automatic mode, this information is translated into appropriate speed and aperture values for the particular lens being used. The aperture value is transmitted to the lens microprocessor which will then operate the aperture drive motor when the

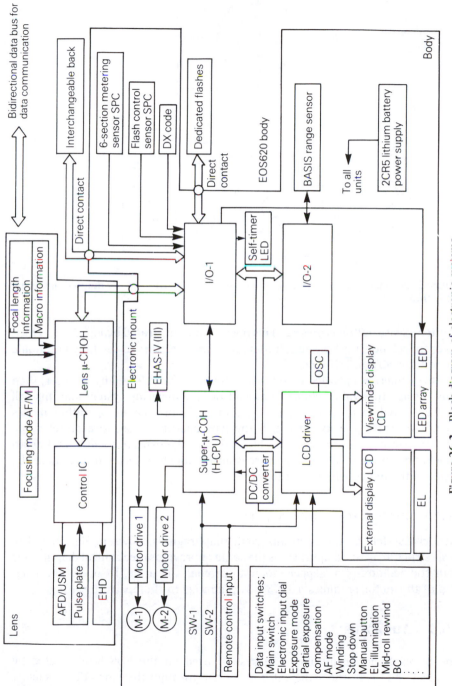

Figure 26.3 Block diagram of electronic systems.

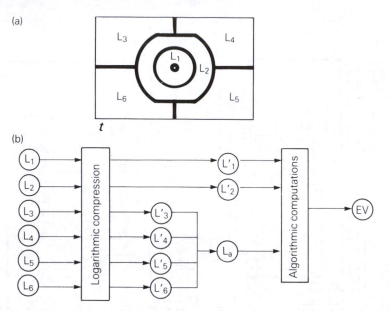

Figure 26.4 Metering: (a) sensor divisions (b) structure of evaluative metering algorithm.

shutter release button is pressed. This operation is illustrated by Fig. 26.5, which gives the relationship between shutter speed and aperture in the fully automatic mode for a 50 mm/f1.8 lens.

If the camera is operated with the shutter speed preselected by the user, then only the aperture value will be obtained from the metering system and this will be transferred to the lens microprocessor as before. Other operating modes include aperture priority, in which the lens aperture is set to a preset value and the required shutter speed is obtained from the exposure value, and depth of field priority, in which the aperture is automatically selected to maximize the depth of the image which is in focus.

26.2.1   DIAPHRAGM DRIVE SYSTEM

The drive system for the electromagnetic diaphragm is shown in Fig. 26.6. The basic structure of this diaphragm is the same as previous mechanical diaphragms with the addition of a stepper motor to provide the drive. In operation, each step of the motor produces a 12.5% step change in the aperture setting.

26.3   Autofocus

The principle of the autofocusing system used on the EOS is illustrated by Fig. 26.7, and is in the same category as the through the lens (TTL) focusing

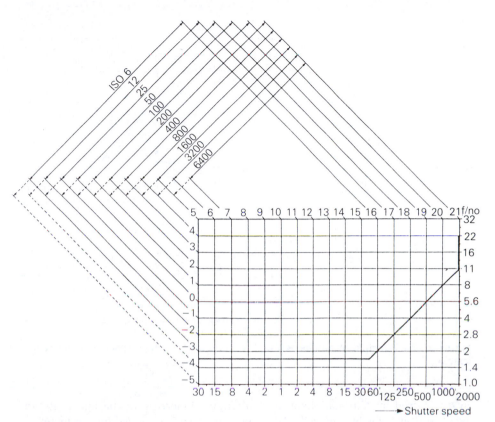

Figure 26.5 Nominal metering range against ISO film speed: solid lines indicate high/low temperature/humidity, dotted lines indicate normal temperature.

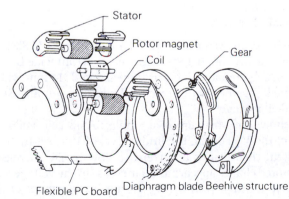

Figure 26.6 Outline of the electromagnetic diaphragm (EMD) mechanism.

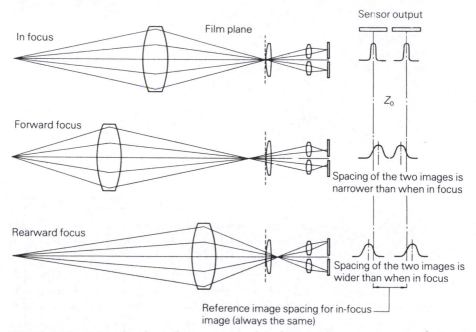

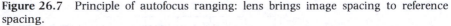

Figure 26.7   Principle of autofocus ranging: lens brings image spacing to reference spacing.

systems used by Minolta, Nikon and Olympus. Referring to the figure, it can be seen that focus is achieved when the two images produced by the focusing optical systems are positioned at the reference positions on the sensors. At other points, with these images either closer together or wider apart, the displacement of the images from the reference position varies in proportion to the amount by which the image is out of focus.

26.3.1   THE BASIS SENSOR

Focus is detected by means of the base stored image sensor (BASIS). This consists of two 48 bit linear arrays of photodetectors each $30\,\mu m$ wide by $150\,\mu m$ high. The total light sensing area for each array measures $1.44\,mm$ wide by $150\,\mu m$ high and is formed, together with associated circuitry, on an LSI chip housed in a clear moulded package.

Figure 26.8 shows the structure of a single BASIS cell. Following reset by the application of the V_{res} signal, the initial level of V_e is set by the V_{vrs} signal. The incoming light then causes a current to flow in TR, increasing the base voltage V_b; the stored charge appears as an increase in the voltage V_c. The charge levels at each cell are then read by the circuit arrangement of Fig. 26.9. As each cell has its own amplifier directly associated with it, this enables an

(a)

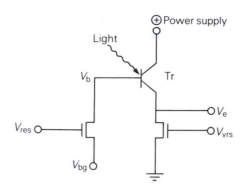

(b)

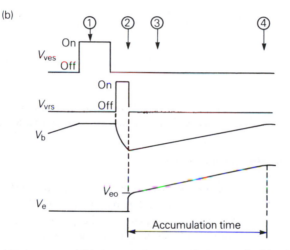

Figure 26.8 (a) Structure and (b) storage operation of a basic cell of the BASIS sensor.

improved signal-to-noise ratio to be achieved in comparison with conventional MOSFET readouts.

BASIS is sufficiently sensitive to enable operation in light levels down to exposure values of $+1$ (ISO 100 bright area luminance), while the incorporation of the peak output monitoring function at the individual cells aids in the optimization of the automatic gain control (AGC).

As has already been seen, focusing error is obtained in terms of the displacement of the images from the reference positions. Initially, the central 24 bits of each array of BASIS are checked to determine the error. If a result is obtained from these bits, the appropriate drive information is sent to the focusing drive motor. If a result is not possible using 24 bits then all 48 bits are used followed by a check using the central 24 bits.

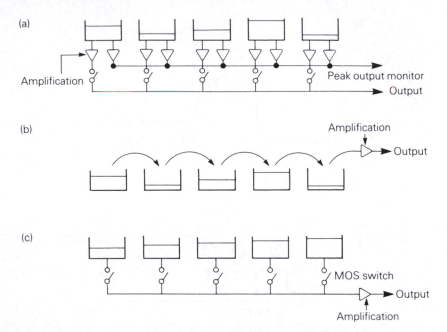

Figure 26.9 Comparison of signal readout from sensors: (a) BASIS, successive readout by switching; (b) CCD, successive readout by charge transfer; (c) MOS, successive readout by switching.

26.3.2   DEPTH OF FIELD SETTING

It is possible using the on-board processor to provide a depth of field control using the autofocus system. By first focusing on a near object and then on a distant object, the required depth of field can be preset. When the shutter is pressed the focus is automatically set to a value between the two preset values and the aperture is adjusted to give the maximum depth of field.

26.3.3   FOCUSING DRIVES

In order to avoid the structure of the autofocus lens becoming too large and bulky, compact high torque drives are required for the focusing mechanism. Two forms of drive are used on the lenses for the EOS to control the setting of the focus: the arc form drive and the ultrasonic motor. Each of these operates under the control of the lens microprocessor using information received from the main microprocessor.

(a)   ARC FORM DRIVE

The arrangement of the arc form drive (AFD) is shown in Fig. 26.10 and uses a brushless Hall motor as the actuator. This is basically a permanent magnet

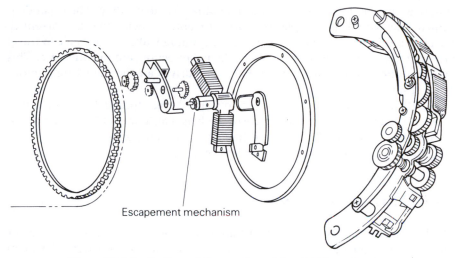

Figure 26.10   Outline of the arc form drive (AFD) mechanism.

stepper motor in which the position of the rotor is detected by means of a pair of Hall effect sensors. The drive from the motor is transmitted through the gears to the focusing mechanism. This uses a helicoid or cam to move the focusing lens along the optical axis, thus changing the focus.

The relationship between the rotor pitch and the movement needed to achieve focus is set according to the focusing accuracy of the individual lens, and the gear ratio is selected to ensure that the lens is in focus when within plus or minus one pitch rotation of the best position.

(b)   ULTRASONIC MOTOR

The structure of the ultrasonic motor (USM) is shown in Fig. 26.11. This uses a series of piezoelectric elements connected on to a backing plate to form a ring.

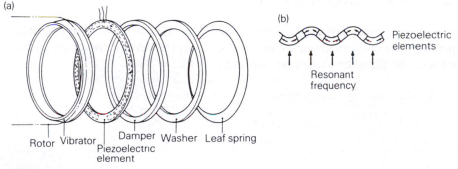

Figure 26.11   (a) Ultrasonic motor (USM) (b) bending wave produced by the piezoelectric elements.

When a current is supplied to the individual elements they will expand or contract according to the polarity of the current. By switching the current to the elements in the appropriate sequence, a bending wave can be made to travel in either the clockwise or the anti-clockwise direction around the ring. This motion is then transmitted to the rotor which drives a helicoid or cam to move the focusing element.

The USM provides a compact, silent, low speed, high torque drive whose ring shape is ideally suited to the structure of the lens. Efficiency is high as the motor is mounted alongside the drive mechanism, and optimization of the drive to suit the characteristic of a particular lens is possible.

26.4   Lens microprocessor

The lens microprocessor is a semi-custom 4 bit device and has responsibility for the control of all the lens functions and for communicating with the main microprocessor in the camera body. When the lens is connected to the body, information regarding zoom capability, macro (close focus) capability, maximum and minimum aperture values and the status of the focusing drive are all made available to the main microprocessor. This enables the latter to determine the type of lens being used and to set up the necessary internal references. Information transmitted to the lens will then include the aperture value required.

The operation of the lens microprocessor is supported by a control integrated circuit. This is a custom I^2L chip which has responsibility, on instruction from the lens microprocessor, for control of the focusing drive and the aperture setting.

The lens microprocessor, control IC and around 21 other circuit components are built into the barrel of the lens. This has necessitated the use of flexible printed circuit board located on the rear of the lens mount. Connection to the actuators is via flexible connector to allow for the movement of the lens.

26.5   Film transport system

The film transport system of the EOS is based on the use of two motors: a high performance coreless motor for film winding, and a separate motor for shutter return, charging the shutter mechanism and film rewind.

Chapter 27

Fly-by-wire

A demanding application of a microprocessor system is the flight control of an aircraft. The numerous control surfaces of the aircraft, such as its ailerons and rudders, must be controlled in response to the pilot's demand within the allowed flight envelope of the aircraft, which ensures stable flying conditions. The use of microprocessors and digital control techniques is not common among conventional aircraft owing to the high reliability requirements of the control system. However, older generations of microprocessor devices now yield reliability data which allows the necessary degrees of confidence to be met for systems in which they are used. Typically, the reliability level of the microprocessor system must be equal to or better than that of the mechanical system that it replaces, which is generally quoted as 10^{-7} per hour total system loss probability, or one chance in ten million.

Some of the benefits that a digital control system has over its mechanical equivalent are as follows:

1. Easier implementation of complex control laws;
2. Simpler revision of the control laws;
3. Exact and repeatable performance;
4. Relative insensitivity to the environment;
5. Insensitivity to power supply variations;
6. Partial self-test and diagnostic capability;
7. Size and cost reductions with time;
8. Absence of control input from structural deflections.

One disadvantage of the digital system is that it is easier to introduce stray inputs due to electrical noise or interference, which may result in a controller performance which is unpredictable.

Fly-by-wire is an electronic system in which the control signals from the pilot's stick to the final actuation devices are electrical and are conveyed by wires to the actuators. Analogue forms of fly-by-wire control systems have been in existence for some time; for example, Concorde has had such a system since 1969. More recently, digital fly-by-wire has been used in the Airbus A310 and subsequent models.

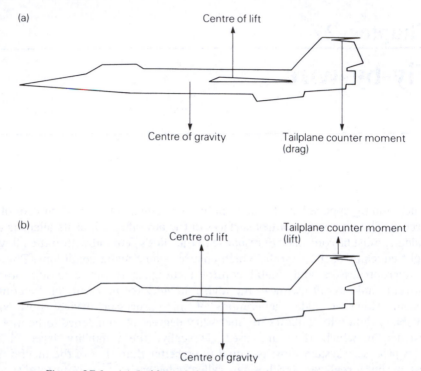

(a)

Centre of lift

Centre of gravity

Tailplane counter moment
(drag)

(b)

Tailplane counter moment
(lift)

Centre of lift

Centre of gravity

Figure 27.1   (a) Stable and (b) unstable aircraft configurations.

In the world of combat aircraft, fly-by-wire techniques have allowed new control strategies to be developed and successfully implemented. An aircraft is normally designed to be naturally stable in flight; this is achieved by arranging for the aerodynamic centre – the point at which act the aerodynamic lift forces generated by the wing – to be behind its normal centre of gravity, as shown in Fig. 27.1a. A change in the direction of air flow causes a corresponding change in the aerodynamic force at the centre of pressure which tends to rotate the aircraft into the new flow direction, and this balance effect provides the aircraft with natural stability. A major disadvantage of this, however, is that the turning moment caused by the centre of pressure being behind that of the centre of gravity must be counterbalanced by a downward force on the tailplane, which causes a reduction in the overall lift and increases the drag of the aircraft.

A solution to the problem is to reverse the relative positions of the centre of pressure and centre of gravity, such that the tailplane must now generate extra lift to neutralize the turning moment, as shown in Fig. 27.1b. The problem now is that the aircraft is no longer inherently stable, but the drag may be reduced by typically 15 to 20%, which in turn can be used to reduce the size and weight of the aircraft by as much as 15%. However, these improvements are achieved at the price of making the aircraft unstable, such that any disturbance in the oncoming air flow would cause the aircraft to pitch upwards in an uncontrollable motion. The aircraft may therefore only achieve stable flight with the assistance of a fast control system.

The need for such aircraft to have an active compensation controller has caused them to be known as control configured vehicles (CCVs); these have been the subject of research since the mid 1970s. In particular, British Aerospace have implemented this active control technology (ACT) in a modified Jaguar single seat attack aircraft as part of a demonstrator program; the technology developed has then been used in future aircraft designs. As a result; the Experimental Aircraft Programme (EAP) demonstrator, also developed by British Aerospace, uses a similar control system as that of the Jaguar, with the aim of proving the technology before it is transferred to the European Fighter Aircraft project.

27.1   The EAP systems architecture and flight control system

Figure 27.2 shows the overall systems architecture for the EAP aircraft, which consists of four major subsystems. These are the flight control system, the communications system, the cockpit control and pilot displays, and the systems management system. These subsystems are interconnected at various points by two dual-redundant communications systems based on the military standard MIL-STD-1553B which is described later. One of these communications systems is called the avionics bus and the other the utilities bus.

A more detailed diagram of the flight control system (FCS) shown in Fig. 27.3, from which it can be seen that many of the functions consist of four units operating in parallel, called a quadruplex architecture. This particular configuration was chosen as it was considered to represent the best compromise between the complexity and the level of fault tolerance for the system, which was designed such that it could suffer a single failure and continue to operate without loss of performance.

At the heart of the FCS are the flight control computers (FCCs), which interface the FCS to the other sybsystems via the avionics bus and utilities bus. The most important inputs to the FCCs come from quadruplex sensors with either digital or analogue outputs. Many of the sensors are complex devices having their own digital processing circuitry. For example, the aircraft motion sensor units (AMSUs) contain an orthogonal set of accelerometers and dual axis rate gyros which produce data which is processed within the AMSUs, and provide the FCCs with the aircraft body rate and accelerations with respect to the three orthogonal axes of the aircraft.

The four paths for the sensor input data are cross-linked between the FCCs by the interlane data links, which are optically isolated from one another so as to prevent electrical interference between the individual FCCs. The signals from the sensors are checked to ensure they fall within an acceptable range before

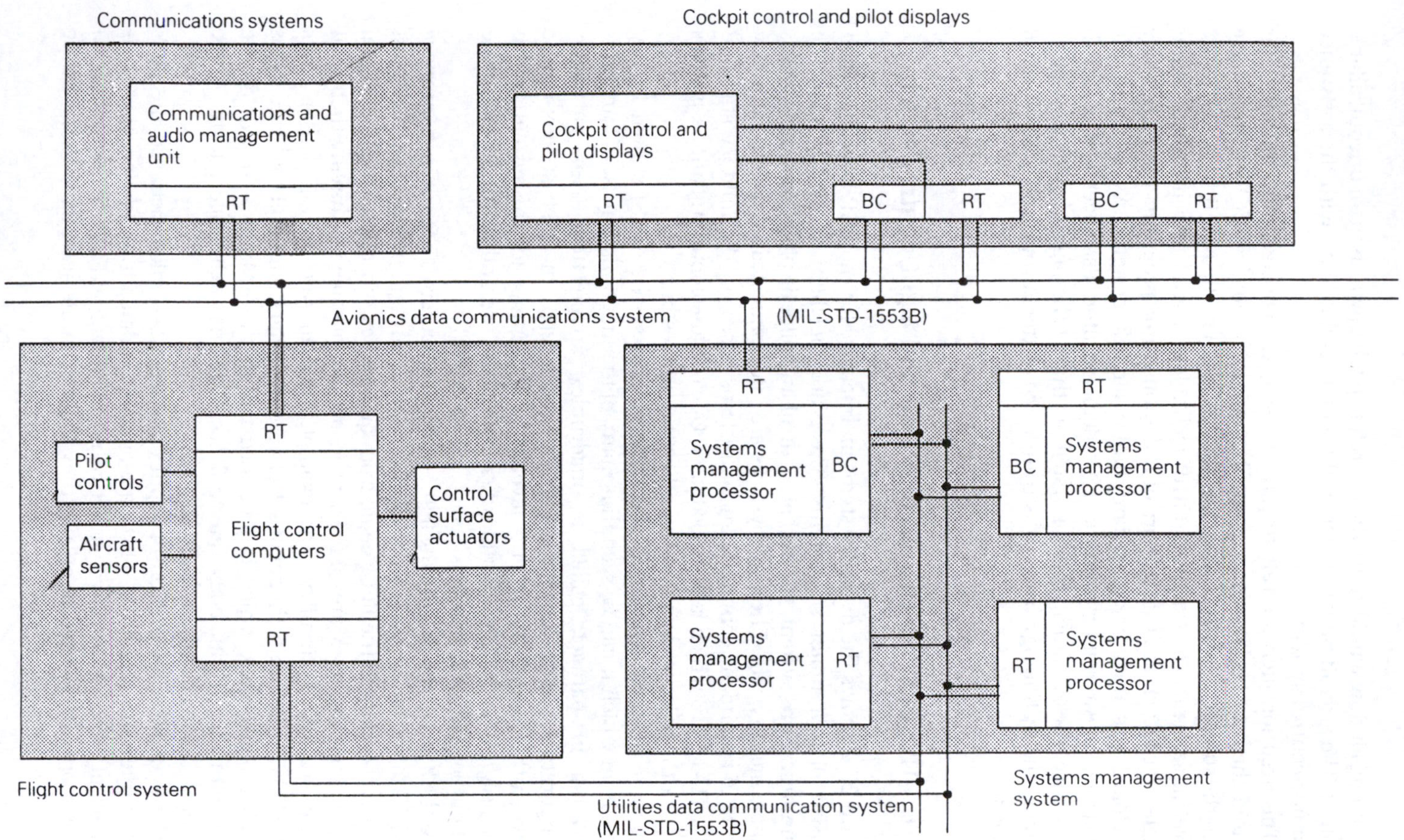

Figure 27.2 Simplified systems architecture of the EAP aircraft.

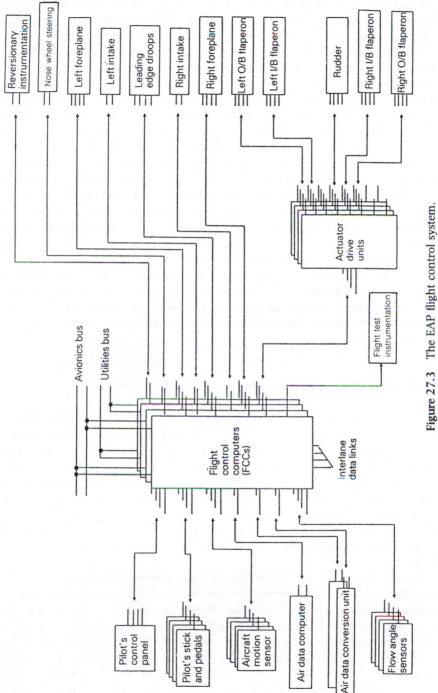

Figure 27.3 The EAP flight control system.

they are averaged to produce a single common input signal for all four computers. If any one of the input signals falls within a tolerance band then it is treated with suspicion by reducing its contribution to the average signal, thereby reducing the effect such a signal has on the averaged signal. Should an input signal fall outside the tolerance band then it is simply ignored.

From these averaged input signals, each of the four FCCs produces an identical output to drive the control surfaces of the aircraft. The foreplanes and leading edge droops are directly driven from the outputs of the FCCs, as are the engine intake cowl control and nose wheel steering actuation. The remaining control surfaces, such as the flaperons and rudder, are interfaced to the FCCs through the actuator drive units (ADUs), which are installed in the rear of the aircraft and connected to the FCCs by digital serial data links. These data links considerably reduce the weight of the aircraft in comparison with conventional wiring between the FCCs and the control surface actuators. Each of the four actuation output lanes has a first stage hydraulic actuator whose output is force summed together with the other three outputs. This signal then provides the input to a single power actuator connected to the control surface. The position of the first stage actuator is monitored with an LVDT (see section 3.3) which allows its operation to be checked by the FCC. Should the position of the actuator vary significantly from the other three, or from a position at which the FCC thinks the actuator should be, then a solenoid valve is actuated which bypasses the first stage actuator.

27.2   The flight control computers

Each of the four FCCs is based on microprocessor bit slice technology. Bit slice technology allows the designer to tailor the architecture and organization of the CPU, including its instruction set, to more closely meet his requirements, instead of settling for a general purpose device. In a bit slice design, each of the functional blocks of the CPU is available as a separate component. So, for example, the ALU (see section 11.2.1) may be available as a 4 bit ALU block which allows the designer to cascade several ALUs together to produce a CPU, with a data word of any width being a multiple of four. This is why the name 'bit slice' was coined, as the data paths of the CPU can be built up from 4 bit slices.

The functions of the instruction decode and control unit (see section 11.1.4) may also be available as separate components, and consist of microinstruction sequencers, registers, multiplexers and PROMs (see section 12.1.1) which store the instruction set of the processor. This may be defined by the designer according to those instructions he wishes to implement to control the hardware of the CPU.

The FCCs in the EAP contain a 16 bit processor made from four 4 bit bipolar slices (see section 13.3.3) together with bipolar fusible link PROMs giving 26 K words of program store. Bipolar technology is used because it is particularly robust in electrically noisy environments, which may destroy parts made by a MOS process. In addition, the FCCs must have an exceptionally high reliability under the extreme heat and vibration experienced during flight conditions. The reliability of the software and microcoded instructions of the FCCs must also be high, as any programming mistake is automatically duplicated in each FCC.

The program that runs in the FCCs implements sophisticated control functions to stabilize the aircraft and prevent the occurrence of dangerous flying conditions that may result in the aircraft spinning or stalling. Because the FCS is now the only link between the pilot and the control surfaces of the aircraft, the pilot has a very different feel of the aircraft. No force feedback from the actuators is given to the stick or pedal controls of the pilot. Only a simple spring feel system is implemented, and so the pilot must rely entirely on the FCS for the control of the aircraft.

27.3 MIL-STD-1553: A digital data transmission system for military applications

The four major sybsystems of the EAP are interconnected by an avionics data communications system and a utilities data communications system. Both of these are based on the military standard MIL-STD-1553B which defines the physical and data link layer of the communications system. Higher layers of the OSI model are not defined as they are not appropriate to the functional requirements of the system.

A 1553 data communications system is made up of a bus controller (BC) and up to 31 remote terminals (RTs) connected in serial bus configuration. Although Fig. 27.2 shows each communications system as having two BCs, only one is active at a time, with the other being available should the first one malfunction or be damaged. In addition, the communications system employs dual-redundant buses routed through different paths of the airframe so as to minimize the risk of both buses being affected by battle damage. The interface between the avionics system and the utilities system is by the systems management processor, which acts as an RT to the avionics communications system and as a BC to the utilities system.

At the physical level of 1553, binary data is transmitted in a Manchester biphase encoded format, at a data rate of 1 Mbps. Figure 27.4 shows this format. For each bit that is transmitted, there is at least one logic transition, which allows circuitry in the receivers to derive a clock signal which is then used to clock the received data into shift registers so that it may be converted into parallel data for processing by the microprocessor. The data and its inverse value are transmitted via twin wire twisted cable between the BC and the RTs. An important feature of the transceivers is the inclusion of a watchdog feature; if a device fails in such a way that it continuously transmits data, then the watchdog timer will inhibit the transceiver from transmitting data on to the bus.

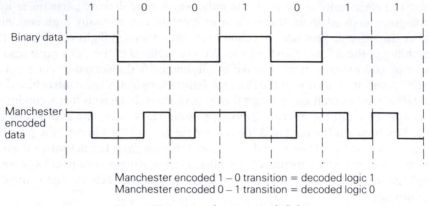

Manchester encoded 1 – 0 transition = decoded logic 1
Manchester encoded 0 – 1 transition = decoded logic 0

Figure 27.4   Manchester encoded data.

This prevents the failure of a single RT from causing a total system failure due to inoperability of the communications bus.

Data is transmitted as three different types of 20 bit words, as shown in Fig. 27.5. Each word has a synchronization pattern at its beginning, and a parity data bit at its end. A data word contains 16 data bits. The command and status words have several fields whose functions control, or indicate the status of, the operation of the BC and RTs.

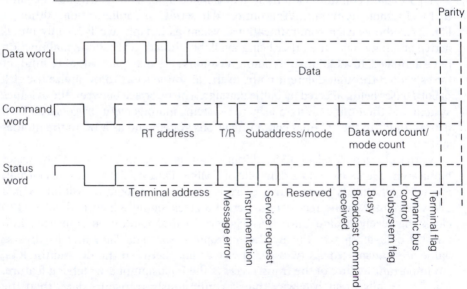

Figure 27.5   MIL-STD-1553B word formats.

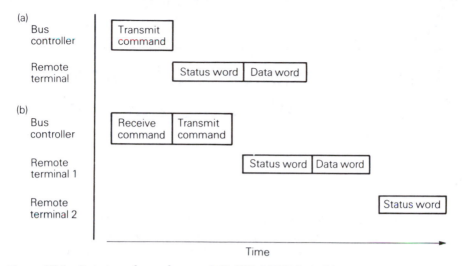

Figure 27.6   Data transfer modes over MIL-STD-1553B bus: (a) remote terminal to bus controller (b) remote terminal 1 to remote terminal 2.

Corresponding to the OSI data link level, data transfer between the BC and RTs is under the control of the BC in a master/slave configuration. In a simple transfer of data from an RT to a BC, the BC first issues a transmit command to the RT, which then responds with a status word followed by up to 32 data words. The address of the RT is contained in the command word, together with a subaddress field which allows the RT to have several ports, each of which may be addressed by the BC.

The 1553 also provides a data transfer mode directly between RTs. In this case, the BC first sends a receive command to the RT which is to receive the data, followed immediately by a transmit command to the RT which is to transmit the data. The RT which is to transmit responds by transmitting a status word followed by the data. When the data has been transmitted, the receiving RT transmits a status word to the BC to terminate the sequence. These sequences are shown in Fig. 27.6.

A basic 1553 interface typically consists of two chips: a data transceiver and a protocol sequencer. The transceiver implements the parallel/serial/parallel conversion of transmitted and received data, data encoding, clock recovery, RT address recognition, and parity and synchronization detection. The protocol sequencer is a state machine which implements all of the data link level functions, including error recovery. The protocol sequencer has a general purpose 16 bit data port which allows it to interface either to a microprocessor based system, or directly to devices with no intelligence, such as sensors and actuators.

Chapter 28

British Aerospace small parts flexible manufacturing system

This case study describes a large system composed of many mechatronic devices – machine tools, robots, automatic guided vehicles (AGVs), conveyors and cranes. A characteristic of all of these devices is their programmability, enabling the system to be run unmanned for part of the day.

This flexible manufacturing system (FMS) was an ambitious but evolutionary step for British Aerospace, and combines both off-the-shelf and specially designed devices. The whole configuration is then integrated by the control and scheduling software. In the design of such a system it is important to focus on the interfaces between devices to ensure compatibility and minimize commissioning problems.

The system comprises the following areas, as shown in Fig. 28.1:

billet preparation and store
aluminium workpiece handling and current kitting (Figs. 28.2 and 28.3)
aluminium machining (Fig. 28.4)
steel and titanium machining (Fig. 28.5)
inspection
cutter store (Fig. 28.6)
AGV system (Fig. 28.7).

28.1 Billet preparation

When an order is received, the billet is cut to the appropriate length and a bar code label is attached to it. Then, when the operator loads a particular billet to the FMS, he scans the bar code to enter the billet identification to the system. From then on, the FMS host software exerts complete control of billet identification and location.

First stop is the billet preparation cell. Here, two specially modified Automax machines equipped with NUM 760 controllers from GEC are fed with billets from storage racking by a robot. This runs up and down a track between the two lines of racking, one of which includes the machines. The racking has space for 48 hours of work on the FMS, which is about 400 billets. Typically, at any

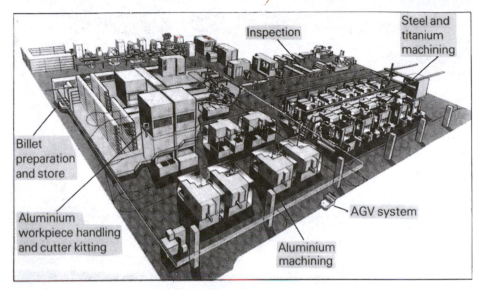

Figure 28.1   System layout.

Figure 28.2   Pallet preparation system.

Figure 28.3   Automax tool kitting.

Figure 28.4   Automax machines.

Figure 28.5   Mitsui machines with tool change robot.

one time it would hold equal numbers of raw billets and prepared ones. There are several hundred different sizes of billet and all can be handled by the robot.

When a billet is received at the billet preparation cell, the robot, under instruction from the host computer, will store it in the racking until required to be processed. When this time comes, the robot will retrieve the billet and deliver it to one of the Automax machines for drill, tap and facing. Each of the Automax machines in this cell is equipped with a specially designed fixture which allows automatic hydraulic clamping of the complete range of billets. When the robot drops the billet into the fixture, hydraulic connectors on the machine are set into action to drive the hydraulic clamps so that machining can begin. Both sides of the billet have to be machined; so, on completion of the first side, the robot is brought into use to remove the billet, turn it over, and drop it back into the fixture for machining on the second side.

In normal operation, one of the machines in the billet preparation cell will be used for aluminium and one for steel and titanium. However, if one breaks down, the other can take on all the work. Only about six or seven cutters are needed for machining each billet at this cell and, with 63 tools available at

Figure 28.6   Tool store with man rider cranes.

each machine, machining can continue without tool replacement for many hours. The machine tool controllers monitor the number of billets each tool has completed, and after a preprogrammed number will request a new set. One of the AGVs will then bring in a new crate of tools and automatically exchange it for the old one.

The AGV bringing the new tools enters the billet preparation cell through a gap in the storage racking and moves to the Automax machines directly along the path of the robot in the cell. Therefore, when the AGV is required to enter the cell, the robot is sent to a standing position at one end of the line of racking to allow the AGV to enter safely. Also, to enable the AGV to straighten up as it turns into the narrow aisle, a section of flooring moves across to cover up the robot track for the AGV wheels to traverse. Once it is straight, the AGV can continue along the line of the robot track as its wheels are situated on either side. In one journey the AGV can deliver a new crate of tools and remove the old one, returning it to the cutter preparation area for regrind or disposal. The robot in the billet preparation cell is unable to move until the AGV has completely left the cell.

The AGV system consists of four Digitron vehicles. They are not dedicated to specific tasks but will be allotted to tasks as required. They are guided via an

Figure 28.7 Automatic guided vehicle at base of paternoster.

inductive wire in the floor, under the control of a layout controller. Whenever idle, the vehicles move to a battery charger station.

The aluminium billets go through a separate production cycle from the titanium and steel ones. The aluminium is machined on a set of six Automax machines, while the titanium and steel are machined on ten Mitsui Seiki machines.

28.2 Steel and titanium parts

When the titanium and steel billets are required at the Mitsui Seiki machines, the robot retrieves them from the storage racks and delivers them to a load station at one end of the aisle, where it deposits them onto a simple tray-type pallet capable of taking four billets. From here an AGV will pick up the pallet and take it to one of the load stations adjacent to a Mitsui Seiki machine where it will unload the pallet.

The Mitsui Seiki machines are manually loaded and unloaded with workpieces. With a dual pallet table, the operator can load one fixture while the other is being machined. The fixture is of a cube type and, with four sides loaded, can

Figure 28.8   Nested aluminium components.

carry up to 12 hours of machining. The philosophy is to man the machines during the day shift and leave them with 16 hours work so that they can continue unmanned overnight. Not all the components machined on this line use the same subplate but the software aims to minimize changes. The machines feature adaptive control software to optimize the feed rate by monitoring the cutting force. They also have in-cycle gauging to detect broken tools and tool wear.

Machining the titanium and steel is very hard on cutting tools, and British Aerospace has recently installed an automatic tool change system to allow the unmanned periods of operation. It was supplied by SI Handling. The system includes a buffer store with seven positions for crates of tools at floor level, as seen in the lower right corner of Fig. 28.5. The AGV system transfers crates between here and the central cutter store. The method of replenishing tools in the machine magazines is by a Toshiba robot, which runs on rails at head height between the two rows of machines. New crates of tools have to be shuttled horizontally from the buffer store to a central position from where they are raised vertically to the robot level and then transferred horizontally to the robot trolley. The robot is able to remove finished tools and replace them with new ones at all ten Mitsui Seiki machines, thus making total automatic tool change possible. Owing to the high forces involved in removing the tools from the tool magazines, each machine tool has had to be fitted with a special actuator to push the tools out of their pockets.

The robot is a Toshiba SR200-6V with Toshiba control system. The transfer system is from SI Cartrac with GEM 80 controllers, and the Mitsui Seiki machines have Fanuc 7 controllers.

Automatic tool exchange is really most important in this cell. Although there is very little opportunity for reuse of cutters used on titanium and steel, the cost of the chucks is very high. Thus it is most important to minimize the number of chucks in the system, and this means returning worn tools to the tool store as frequently as possible for breaking down.

When parts have finished being machined on the Mitsui Seiki machines they are manually unloaded and transferred by means of an AGV to a finished parts buffer on the same trays as previously.

28.3   Aluminium parts

The aluminium billets take a different route to the same finished parts buffer. From the billet preparation cell, the robot transfers prepared billets to an overhead conveyor which moves the billets to a mezzanine floor, situated above the machining area containing the Automax machines. On this mezzanine level, there is a complicated system of conveyors where prepared billets and part-machined billets come to be palletized or turned over. Marwin was responsible for the supply of most of the equipment on this upper level. Some operator involvement is required.

The pallet for the aluminium billets is covered with a latticework of base pins. For each type of billet, a unique pattern of module caps has to be assembled on the base pins. Threaded pins are used for billets on their first operation, and dowels (giving more accurate locations) on the second. The billet can then be located on the pallet and the module caps tightened, and the assembly is ready for machining. The holes necessary for billet location have previously been drilled during the preparation operations.

On the mezzanine floor, a specially designed gantry robot is used to place the module caps in the required locations on the pallet. A view of the cell is shown in Fig. 28.2. The robot will attempt to insert each cap a number of times before giving up and rejecting it. The pallet moves by means of a conveyor system from the robot station to the billet load station, where an operator moves the raw billet from its entry conveyor on to the pallet, with the assistance of a hydraulic lift, and tightens up the locations. The pallet, now ready to go to a machine, is transferred by means of a conveyor and shuttle system into one of three paternoster stores where it waits until required at a machine.

Each paternoster contains 19 shelves, each with two positions. They serve as a bridge between the mezzanine level and the shopfloor level, so items can be stored at one level and retrieved at the other. Thus, when the schedule dictates, the prepared pallet will be retrieved at shopfloor level and taken by means of an AGV to the requisite machine tool. All transfers are totally automatic (see Fig. 28.7).

When the first machining operation has been performed on an Automax machine, the pallet will be returned to the paternoster store and thence to the mezzanine level, where an operator will take the semi-finished billet off the pallet, clean up the fixture, and relocate the billet on the fixture ready for the second machining operation. This pallet will be returned to the paternoster store again and delivered to shopfloor level for machining. The loading station for the paternoster and shopfloor level has been designed to decouple activity on the upper level from activity at the ground level.

Once both operations have been completed, the billets go for inspection. Having successfully passed inspection, they move to the finished parts buffer.

Automated tool replacement for the Automax machines has also been implemented. A huge number of tools are needed for the machining of these parts because of the very large variety of components, and because very few billets are required in a batch size greater than one. British Aerospace therefore decided to prepare a kit of tools for each aluminium billet to be processed. This kit is coordinated to arrive at the machine with its corresponding billet. When machining on that billet has been completed, the kit of tools is stripped and the tools are made available for other billets.

A local store of tools is held on the mezzanine level. The cell is shown in Fig. 28.3. It has capacity for about 4000 tools, and this would include several hundred different types. The tools are stored in two parallel racks each 21 rows high. Between the two racks is a track on which runs a small trolley supporting a Unimate Puma 700 robot and tool crate. The robot travels up and down the racks, picking up the required tools and placing them in the crate. When a complete kit has been assembled the crate is transferred through a gap in the racking on to a shuttle (the same shuttle used for palletted billets) which inserts the crate into one of the paternosters. The crate is then delivered to shopfloor level, removed from the paternoster and taken by AGV to a machine tool for cutting a billet.

A gantry robot on each machine, shown in Fig. 28.4, transfers the tools as required into the tool changer mechanism serving the machine spindles. The Automax machines can machine two billets at the same time and they have automatic tool length compensation.

When cutting is finished, the crate is returned to the paternoster and then to the robot which dekits it.

The host software is responsible for monitoring tool identification, location and condition. It will keep account of which tools have reached the end of their lives; when a reasonable number have done so, it will instruct the robot to load an empty crate with all the old tools and send it via a paternoster and AGV back to the central cutter preparation area for regrinding. In the same way, new tools from the central cutter store can be brought into use in the Automax machining cell.

Inspection on an LK Microvector coordinate measuring machine assists the host in monitoring tool performance. A minimum of two measurements on each completed billet will be taken near the end of cutter life. If this reveals cutter failure, the host will be able to stop the relevant cutter from being released for further work. It takes a maximum of seven minutes for the inspection process to be carried out. The AGV delivers the palletized billets directly on to the inspection table and then removes them and transfers them to the finished parts buffer.

A second measuring machine is also included in the system. This has a number of roles to perform. Firstly, it can be brought into use if the first breaks down. Secondly, should a billet fail the first inspection, it can be used to perform a more vigorous inspection procedure. Another task for this measuring machine is the proving of both cutting and inspection part programs.

28.4 Control

A VAX 11/750 controls the entire system through an Ethernet local area network. The control software is a recent development which is based on the British Aerospace control architecture (BCA). It was the subject of an 18 month collaborative venture with GEC. Designed to be modular with a high degree of standardization, the software is available commercially from BAe.

The VAX will download part programs to each machine as required. It sends transport orders to the AGV layout controller. Each AGV will carry out an optical 'handshake' with each machine during load transfer.

All other machines, robots, conveyors and devices have their own local controllers.

Chapter 29

Autohelm 800 boat autopilot

As the popularity of offshore cruising in sailing yachts has increased, with longer passages being made, so a market has developed for reliable and low cost automatic boat steering systems. As applied to boats and yachts up to perhaps 15 m waterline length, these are to be regarded as course correction devices; large rudder angles and rates are not normally to be expected. Even so, it is a major challenge to produce a compact and self-contained unit able to correct changes in heading caused by wind variations or wave action which requires no external service beyond a 12 volt DC supply and which sells for under £200 in its simplest form.

The Autohelm 800 tiller pilot is a particularly good example of an integrated mechatronic product. Full account has been taken of the need to simplify the user interface within an aesthetically pleasing design. At the same time it can withstand the rigours of its working environment from sunlight to icing, to submersion in green sea and other corrosive fluids, and to being snagged in ropes or dropped in moments of panic. Much attention to detail is evident in the design of items such as push buttons. A general view of the unit is shown in Fig. 29.1.

For the user, operation is simplicity itself. All that is needed is to set up the boat manually on a chosen course using a landmark or the boat's normal magnetic compass, and then push the AUTO button to lock on. Course changes and trimming adjustments of any size can then be made by using the +1, −1 degree and +10, −10 degree control buttons.

Functionally, the controller may be regarded as in the schematic Fig. 29.2. The direction of the earth's magnetic field is measured by a gimballed fluxgate compass linked to an 8049 single chip microprocessor. Output is to a 12 volt DC motor via the power IC board. The mechanical drive is then taken through a double reduction toothed belt drive on to a screw which propels the hollow ram using a low friction ball nut. The ram movement is calculated by integrating the motor emf and feedback loop is completed through the heading of the boat itself.

As in any mechatronic product, design must commence with the mechanical

Figure 29.1   Autohelm unit.

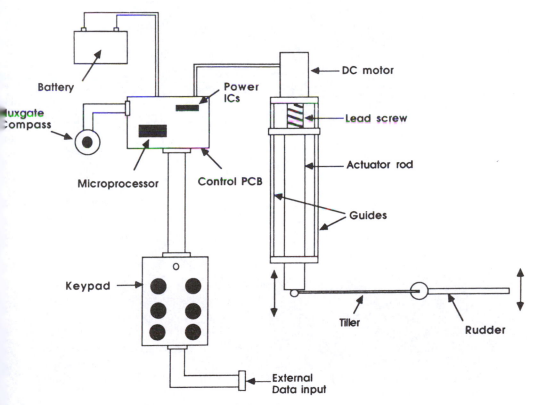

Figure 29.2   Autohelm schematic arrangement.

criteria to be met. Here these are the tiller forces to be generated at specified angles and rates of rotation, taking into account the hydrodynamic effects of particular envelopes of rudder configuration and forward speed. From this information the size of components can be determined and the required power levels of the driver and its associated power electronics established. The control strategy will be set out at an early stage in order to identify the interfaces between the drive circuits and the operational algorithms resident in the microprocessor.

Autohelm is available with programmable dodge and tacking features and accepts inputs from wind vanes. Units can even be interfaced to radio navigation receivers, with which it is possible to program waypoints and execute course changes automatically.

Such products could not have been brought to the marketplace without an integrated mechatronic design approach.

Appendix A

Definitions and terminology

Transducer
A transducer is a device which transfers information in the form of energy from one part of a system to another, including in some cases a change in the form of energy containing the information. The resulting output signal is directly related to the parameters of the signal in the first domain. Hence actuators may be considered as transducers.

A transducer may be an active device, in which case it acts as an energy converter, or a passive device, when it acts as an energy controller.

Sensor
A sensor is that part of a measurement system which responds directly to the physical parameter to be measured.

An analogue sensor provides a continuous analogue output, while a digital sensor provides a direct digital output.

Accuracy
This is a qualitative term used to characterize the ability of a measuring system or a component of that system to provide an indication of the true value of the measurand within certain specified probability limits.

The quantitative expression of accuracy is in terms of an uncertainty (inaccuracy), which is the accumulation of uncertainties contributed by factors such as non-linearity, hysteresis, drift, temperature effects, ageing and other physical conditions.

True value
This is the actual value of the measurand and is not obtainable in practice.

Conventional true value
This is a value used to represent the true value. It is usually used to refer to the value obtained for the measurand using the reference standard appropriate to the measurement being made. For example, in calibrating an electricity meter the conventional true value is taken to be the value indicated on a reference meter, and comparison is then made with this value.

Nominal value
This is a stated value which must be accompanied by an uncertainty or tolerance.

Postulated value
This is a value deduced from measurements giving the range within which the true value must lie.

Tolerance or uncertainty
This is that part of the expression of measurement which states the range of values within which the true value must lie. Defined as $\pm x$ units or $\pm x\%$.

Tolerance is best used when referring to a specification. Uncertainty is best used when referring to a measurement.

Error
The error is the algebraic difference between the result of the measurement and the true value of the quantity being measured. It may be expressed in units of measurement as an absolute error; as a percentage of the quantity measured as a relative error; or as a percentage of some specified value. In general the true value is replaced by the conventional true value.

Systematic errors
A systematic error is a consistent or fixed error which may be allowed for in the final result. Systematic error can arise from incorrect setting of a reference (zero offset), incorrect calibration, static or dynamic instrument errors, environmental effects, human error such as a consistent misreading, or arithmetic errors introduced by rounding. Systematic errors are cumulative in the final result.

Random errors
Random errors arise as a result of all forms of noise, variations in component tolerances and reading error. They cannot be eliminated but must be handled by statistical methods.

Possible random error
The difference between any given sample and the mean of the set of samples is referred to as the deviation of that sample. There is a 50% probability that the deviation of any single sample is greater than or less than the probable random error. For a Gaussian distribution the probable random error is $\pm 0.675\sigma$, where σ is the standard deviation.

The mean value of a group of N samples is

$$x_m = \frac{1}{N} \sum_{i=1}^{N} x_i$$

The population standard deviation of the group is then

$$\sigma = \sqrt{\left[\frac{1}{N} \sum_{i=1}^{N} (x_i - x_m)^2 \right]}$$

The square of the standard deviation is referred to as the variance.

For groups of samples for which $N \leqslant 20$, the sample standard deviation

$$\sigma = \sqrt{\left[\frac{1}{N-1} \sum_{i=1}^{N} (x_i - x_m)^2 \right]}$$

would be used.

For a Gaussian distribution the probability that any single sample will lie within $\pm \sigma$ of the mean value is 68.3%, rising to 95.4% within $\pm 2\sigma$ and 99.7% within $\pm 3\sigma$.

Quantization error
This is the error introduced in analogue to digital conversion by the representation of the analogue signal amplitude by a discrete level.

Resolution or discrimination
This is the smallest increment of the measurand that can be detected with certainty.

Sensitivity
This is the ratio in the change in magnitude of the output signal to the corresponding change in the measurand. It is sometimes referred to as the incremental gain or scale factor.

Linearity
In a linear system the sensitivity does not vary with the value of the measurand. Variation in the linearity is expressed as a percentage of the maximum (full scale) value. Figure A.1 illustrates these conditions.

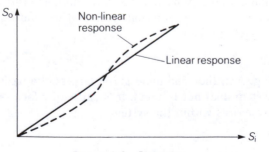

Figure A.1   Linearity.

Repeatability
This is the closeness of agreement between a group of measurements or repeated actions using the same methods and equipment and under the same conditions. It may also be used to refer to the ability of an item of equipment to function with a particular level of uncertainty when subject to a variation in operational parameters. An example is the ability of a robot arm to move objects of different masses over the same path.

Bandwidth
This defines the range of frequencies over which a measuring system is suitable for use. The most commonly used form of bandwidth is the 3 dB bandwidth, measured at the point at which the signal amplitude is 3 dB below the level in the pass band. The effective noise bandwidth is the bandwidth of the ideal filter of the same pass band amplitude that transmits the same power as the practical filter when supplied from a source having a constant power spectral density with frequency (white noise).

Form factor
The form factor is the ratio (RMS value)/(peak value) for a time varying signal.

Friction
A number of different types of friction may be present in a system:

> *Static friction or stiction*   This is the force or torque that is required to initiate motion from rest.
> *Coulomb or dynamic friction*   This is a friction force or torque that opposes motion. It is normally less than the stiction, and its magnitude is dependent upon the velocity component.
> *Viscous friction*   This is a function of velocity. Viscous friction opposes motion and introduces damping into and modifies the response of a system by introducing a lag into the motion.

Drift
This is a variation in output not caused by any change in the input. It is usually caused by factors such as internal temperature variations, component instabilities and ageing effects.

Dead band
Illustrated by Fig. A.2, the dead band is the largest change of the measurand to which the system does not respond. It is caused by factors such as friction, backlash and hysteresis within the system.

Hysteresis
For a given value of the measurand, the output from the system may differ depending on whether the measurand is increasing or decreasing, as shown in

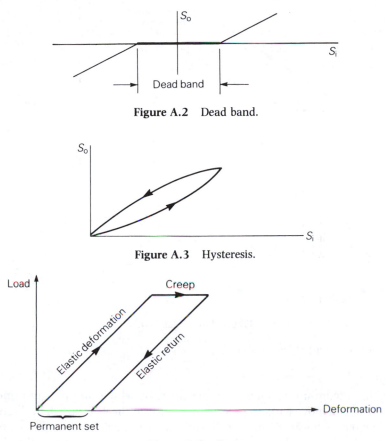

Figure A.2   Dead band.

Figure A.3   Hysteresis.

Figure A.4   Creep.

Fig. A.3. The level of hysteresis is expressed by the algebraic difference between the measurements when the measuring point is approached from opposite directions. The area of the hysteresis loop is related to the work done on the system while cycling around the loop.

Creep
As shown by Fig. A.4, creep is the variation in system conditions with time that occurs as a result of component deformation under load due to viscous flow in the material of the component.

Elastic after-effect
As can be seen from Fig. A.5, this is similar to creep in that a material under constant load exhibits deformation with time. However, unlike creep, on removing the load the material exhibits a time-dependent relaxation to its original position, resulting in little or no residual deformation of the material.

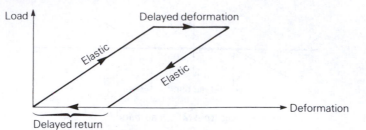

Figure A.5   Elastic after-effect.

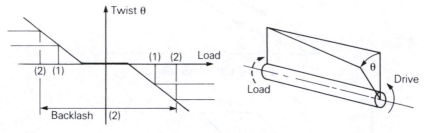

Figure A.6   Backlash.

Backlash
This is the maximum distance or angle through which any part of a mechanical system can move without producing motion in the next part of the system.

The degree of backlash may well vary with system loading. For example, consider a load driven by a shaft. For any driven load the twist in the shaft is proportional to the load. Referring to Fig. A.6, on reversing the direction of the drive, and once motion has ceased, the shaft must first unload (wind down), then pass through any dead band, and then take up load in the reverse direction (wind up).

Frequency domain
The operations of observation, measurement and analysis are carried out with frequency as the principal variable.

Time domain
The operations of observation, measurement and analysis are carried out with time as the principal variable.

Impulse response
As shown in Fig. A.7, this is the output of a system when a unit impulse is applied at its input.

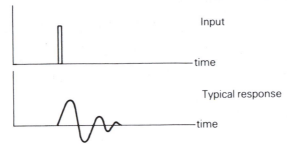

Figure A.7   Impulse response.

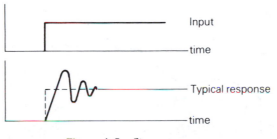

Figure A.8   Step response.

Step response
As shown by Fig. A.8, this is the output of a system when a step function is applied at its input.

Frequency response
This is the variation in the output of a system when a fixed amplitude, variable frequency signal is applied at its input.

Transient response
This is the variation in the output of a system when subject to a sudden variation in the input conditions and before steady state conditions have become established.

Rise time
This is the time required for a pulse to rise from 10% to 90% of its final value.

Stability
A stable system is one that will return to an equilibrium state following any disturbance. An unstable system is one which under similar conditions will enter a state of uncontrolled oscillation.

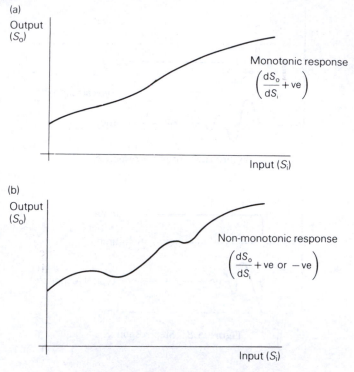

Figure A.9

Natural frequency
This is the frequency of free, undamped oscillations in the system.

Time constant
The time constant is taken as the time required for a 63.2% fractional variation in the amplitude of a signal varying exponentially with time.

Multiplexing
This is the sharing of a single channel by a number of devices by means of some form of controlled switching.

Monotonicity
Referring to Fig. A.9, a system is monotonic if a change in the input level always produces a change of the same sense, either positive or negative, in the output level.

Appendix B

Inertial loads

B.1 Tangentially driven loads

Tangentially driven loads include belts, pulleys, conveyors, rack and pinion drives and pinch wheels. For the arrangement of Fig. B.1, the torque required to accelerate the load at a constant angular acceleration is given by

$$T = J_l + J_m + J_p \tag{B.1}$$

in which

$$J_l = M_l r^2 \tag{B.2}$$

where M_l is the mass of the load (kg); r is the radius of the pulley (m); and J_m and J_p are the inertias of the motor and the pulley (kg m^2 or N m s^{-2}).

B.1.1 SYSTEMS WITH A RATIO CHANGE

Consider the arrangements of Fig. B.2. For the belt drive (Fig. B.2a), the effective inertia at the motor shaft is given by

$$J_e = J_1 + J_2(D_1/D_2)^2 \tag{B.3}$$

Similarly, for the geared drive (Fig. B.26),

$$J_e = J_1 + J_2(T_1/T_2)^2 \tag{B.4}$$

where J_1 is the inertia of pulley 1 or gear 1; J_2 is the inertia at the load shaft of the combination of the load, the load shaft and pulley 1 or gear 1; D_1 and D_2 are the effective diameters of pulleys and 2; and T_1 and T_2 are the number of teeth on gears 1 and 2.

B.2 Leadscrew driven loads

For the system of Fig. B.3, the torque required to accelerate the load at a constant linear acceleration a is given by

$$T_a = (J_l/e + J_s + J_m)2\pi a/p \tag{B.5}$$

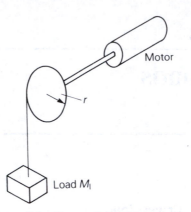

Figure B.1   Tangentially driven load.

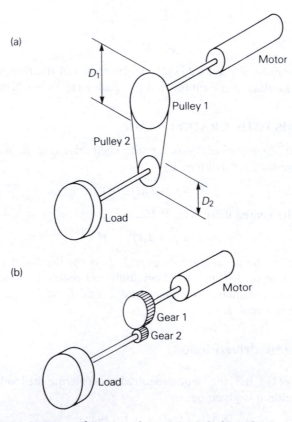

Figure B.2   System with a ratio change: (a) belt drive (b) geared drive.

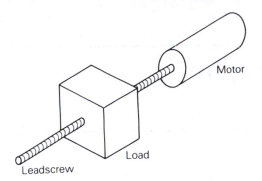

Figure B.3   Leadscrew driven load.

in which

$$J_l = M_l p^2 / (2\pi)^2 \qquad (B.6)$$

where M_l is the mass of the load (kg); p is the pitch of the leadscrew (m); e is the efficiency of the leadscrew; and J_s and J_m are the inertias of the leadscrew and the motor (kg m^2 or N m s^2).

Bibliography

Artificial intelligence and expert systems

BOOKS

Aleksander, I. (1984) *Designing Intelligent Systems*, Kogan Page.
Jackson, P. (1986) *Introduction to Expert Systems*, Addison-Wesley.
Pham, D.T. (1988) *Expert Systems in Engineering*, IFS.
Rich, E. (1983) *Artificial Intelligence*, McGraw-Hill.
Van Horn, M. (1986) *Understanding Expert Systems*, Bantam.

PAPERS

Glickstein, I. and Chen, M. (1986) AI/expert system processing of sensor information. Proceedings IEEE/AIAA 7th Digital Electronics Conference, Forth Worth, Texas, 382–8.
McClelland, S. (1987) Artificial intelligence: sensors need it. *Sensor Review*, 7 part 2, 133.

Communications

Hutchison, D. (1988) *Local Area Network Architectures*, Addison-Wesley.
Sutcliffe, A. (1988) *Human Computer Interface*, MacMillan.
Tolhurst, M.R. (1988) *Open Systems Interconnection*, MacMillan.

Operational amplifiers

Barna, A. and Porat, D.I. (1989) *Operational Amplifiers*, Wiley.
Berlin, H.M. (1978) *Design of Op-Amp circuits*, Sams.
Horrocks, D.H. (1990) *Feedback Circuits and Operational Amplifiers*, Van Nostrand Reinhold.

Vision systems

BOOKS

Ballard, D.H. and Brown, C.M. (1982) *Computer Vision*, Prentice-Hall.
Fairhurst, M.C. (1988) *Computer Vision for Robotic Systems*, Prentice-Hall.
Horn, B.K.P. (1986) *Robot Vision*, MIT Press (McGraw-Hill).

PAPER

Beiser, L. (1986) Imaging with laser scanners. *Optics News*, November, 10–16.

Optoelectronics and lasers

BOOKS

Watson, J. (1988) *Optoelectronics*, Van Nostrand Reinhold.
Wilson, J. and Hawkes, J.F.B. (1987) *Lasers: Principles and Applications*, Prentice-Hall.
Wilson, J. and Hawkes, J.F.B. (1988) *Optoelectronics: An Introduction*, Prentice-Hall.

PAPER

Main, R. (1986) Optical Technologies – an agenda for the future. *Sensor Review*, 7, part 1, 33–8.

Sensors, transducers and measurements

BOOKS

Banister, B.R. and Whitehead, D.C. (1986) *Transducers and Interfacing*, Van Nostrand Reinhold.
Barney, G.C. (1988) *Intelligent Instrumentation: Microprocessor Applications in Measurement and Control*, Prentice-Hall.
Broch, J.T. (1984) *Mechanical Vibration and Shock Measurements*, Bruel and Kjaer.
Cluley, J.C. (1983) *Transducers for Microprocessor Systems*, MacMillan.
Dobelein, E.O. (1983) *Measurement Systems: Application and Design*, McGraw-Hill.
Gregory, B.A. (1981) *An Introduction to Electrical Instrumentation and Measurement Systems*, MacMillan.
Jones, B.E. (1977) *Instrumentation, Measurement and Feedback*, McGraw-Hill.
Loxton, R. and Pope, P. (eds) (1986) *Instrumentation: A Reader*, Open University Press.
Morris, A.S. (1988) *Principles of Measurement and Instrumentation*, Prentice-Hall.
Neubert, H.K.P. (1975) *Instrument Transducers: An Introduction to their Performance and Design*, Oxford University Press.
Pugh, A. (ed.) (1986) *Robot Sensors. Volume 1: Vision*, IFS.
Pugh, A. (ed.) (1986) *Robot Sensors. Volume 2: Tactile and Non-Vision*, IFS.
Ruocco, S.R. (1987) *Robot Sensors and Transducers*, Open University Press.
Usher, M.J. (1989) *Sensors and Transducers*, MacMillan.
Van Putten, A.F.P. (1988) *Electronic Measurement Systems*, Prentice-Hall.
Vaughan, J. *Strain Measurements*, Bruel and Kjaer.

JOURNALS

IEEE Transactions on Instrumentation and Measurement
Journal of Physics E: Scientific Instrumentation
Sensors and Actuators
Sensor Review

PAPERS

GENERAL

Bergveld, P. (1986) Sensors for biomedical applications. *Sensors and Actuators*, **10** (3/4), 165–80.

Buchy, F. (1987) Silicon sensors lead pressure transmitter technology. *Industrial and Process Control*, **60**, part 2, February, 37–9.

Evans, J.M. (ed.) (1986) Measurement technology for automation in construction and large scale assembly. *Robotics*, **2**, part 2, June, 87–101.

Heginbotham, W.B. (1985) The future of sensors for intelligent machine control. *J. Phys. E: Scientific Instrumentation*, **18**, 766–9.

Goodenough, F. (1985) Sensor IC's processing materials open factory doors. *Electronic Design*, 18 April, 131–48.

Horner, G.R. and Lacey, R.J. (1987) Rotary velocity and position transducers under review. *Sensor Review*, **7**, part 2, 85–94.

Igarashi, Z. (1986) New technology of sensors for automotive applications. *Sensors and Actuators*, **10** (3/4) 181–94.

Jordan, G.R. (1987) Sensor technologies of the future. *GEC Review*, **3**, part 1, 23–32.

Kuroda, S., Jitsumori, A. and Inari, T. (1984) Ultrasonic imaging system for robots using an electronic scanning method. *Robotica*, **2**, 47–53.

McClelland, S. (1987) Sensor fabrication: micromachining marches on. *Sensor Review*, **7**, part 2, April, 83–4.

Middelhoek, S. and Hoogerwerf, A.C. (1986) Classifying solid state sensors: the 'sensor effect cube'. *Sensors and Actuators*, **10** (1/2), 1–8.

Middelhoek, S. and Noorlag, D.J.W. (1981) Silicon microtransducers. *J. Phys. E: Scientific Instrumentation*, **14**, 1343.

Morten, B., Prudenziati, M. and Taroni, A. (1983) Thick film technology and sensors. *Sensors and Actuators*, **4**, 237–45.

Petersen, K.E. (1982) Silicon as a mechanical material. *Proceedings IEEE*, **70** (5), May, 420–57.

Prudenziati, M. and Morten, B. (1986) Thick film sensors: an overview. *Sensors and Actuators*, **10** (1/2), 65–82.

Reeder, T. M. and Cullen, D.E. (1976) Surface acoustic wave pressure and temperature sensors. *Proceedings IEEE*, **64** (5), May, 754–6.

Regtien, P.P.L. (1986) Sensors for applications in robots. *Sensors and Actuators*, **10** (3/4), 195–218.

OPTICAL SYSTEMS

Collet, M.G.(1986) Solid state image sensors. *Sensors and Actuators*, **10** (3/4), 287–302.

Culshaw, B. (1982) Optical fibre transducers. *Radio and Electronic Engineer*, **52** (6), June, 283–90.

Culshaw, B. (1986) Photodetectors and photodetection. *Sensors and Actuators*, **10** (3/4), 263–86.

Du Chastel, M.H. (1987) Fibre optic sensors begin moving from laboratory to market place. *Laser Focus/Electro-Optics*, **23**, part 5, 110–17.

Fisher, S. (1986) Sensing with light. *Australian Electronics Engineering*, **19**, part 10, October, 84–6.

Fluitman, J. and Popma, Th. (1986) Optical waveguide sensors. *Sensors and Actuators*, **10** (1/2), 25–46.

Giallorenzi, T.G., Bucaro, J.A., Dandridge, A., Sigel, G.H., Cole, J.H., Rashleigh, S.C. and Priest, R.G. (1982) Optical fibre sensor technology. *IEEE Trans. Microwave Theory and Techniques*, **MIT-30** (4), April, 472–511.

Haruna, M. and Nishihara, H. (1986) Optical waveguide pressure sensors. *Integrated Optical Circuit Engineering* 3, Vol 651, April 276–9.
Idesawa, M. (1986) New type of miniaturised optical range sensing methods. *Journal of Robot Systems*, **3**, part 2, 165–81.
Jackson, D.A. and Jones, J.D. (1986) Fibre optic sensors. *Optica Acta*, **33**, part 12, 1469–503.
Lagakos, N., Cole, J.H. and Bucaro, J.A. (1987) Microbending fibre optic sensor. *Applied Optics*, part 11, June.
Roef, P. (1987) Attention focuses on optical fibre biosensors. *Sensor Review*, part 7, July, 127–32.

SMART SENSORS AND SIGNAL PROCESSING

Brignell, J.E. (1986) Sensors in distributed instrumentation systems. *Sensors and Actuators*, **10**, (3/4), 249–62.
Brignell, J.E. and Dorey, A.P. (1983) Sensors for microprocessor-based instrumentation. *J. Phys. E. Scientific Instrumentation*, **16**, 952–8.
Giachino, J.M. (1986) Smart sensors. *Sensors and Actuators*, **10** (3/4), 239–48.
Huijsing, J.H. (1986) Signal conditioning on the sensor chip. *Sensors and Actuators*, **10** (3/4), 219–38.
Middelhoek, S. and Noorlag, D.J.W. (1982) Signal conversion in solid state transducers. *Sensors and Actuators*, **2**, 211–28.

MAGNETIC MEASUREMENT

Billat, A. and Villermain-Lecolier, G. (1985) Some applications of a flat magnetic proximity sensor. *Mikrowellen Mag* (FRG), **13**, part 3, May, 330–2.
Cooper, A.R. and Brignell, J.E. (1985) An integrated circuit silicon sensor for magnetic fields. *J. Inst. Electronic and Radio Engineers*, **55**, 263–7.
Dibbern, U. (1986) Magnetic field sensors using hte magnetoresistive effect. *Sensors and Actuators*, **10** (1/2), 127–40.

Microprocessors

BOOKS

Cahill, S. (1987) *The Single Chip Microcomputer*, Prentice-Hall.
Clements, A. (1985) *The Principles of Computer Hardware*, Open University Press.
Downton, A. (1984) *Computers and Microprocessors*, Van Nostrand Reinhold.
Ferguson, J. (1985) *Microprocessor Systems Engineering*, Addison-Wesley.
Kernigan, B. and Ritchie, D. (1987) *The C Programming Language*, Prentice-Hall.
Leventhal, L. (1979) *Introduction to Microprocessors – Software, Hardware, Programming* Prentice-Hall.
Lipovsky, G. (1988) *Single and Multiple Chip Microcomputer Interfacing*, Prentice-Hall.
Loveday, G. (1984) *Practical Interface Circuits for Micros*, Pitman.
Sommerville, I. (1985) *Software Engineering*, Addison-Wesley.

JOURNALS

IEEE Micro, IEEE Computer Society.
Microprocessors and Microsystems, Butterworth.

PAPERS

Armitage, B., Dunlop, G., Hutchinson, D. and Shousan Yu (1988) Fieldbus: an emerging communications standard. *Microprocessors and Microsystems*, **12** (10) 555–62.

Weston, R., Sumpter C. and Gascoigne, J. (1986) Industrial computer networks and the role of MAP – part 1. *Microprocessors and Microsystems*, **10** (7), 363–70.

Weston, R., Sumpter C. and Gascoigne J. (1987) Industrial computer networks and the role of MAP – part 2. *Microprocessors and Microsystems*, **11** (1), 21–34.

Robotics and manufacturing technology

Appleton, E. and Williams, D.J. (1987) *Industrial Robot Applications*, Open University Press.

Asfahl, C.R. (1985) *Robots and Manufacturing Automation*, Wiley.

Blacker, F. and Osbourne, D. (eds) (1987) *Information Technology and People: Designing for the Future*, British Psychological Society.

Craig, J.J. (1987) *Introduction to Robotics*, Addison-Wesley.

Greenwood, N.R. (1988) *Implementing FMS*, MacMillan.

Groover, M.P., Weiss, M., Nagel, R.N. and Odrey, N.G. (1986) *Industrial Robotics: Technology, Programming and Applications*, McGraw-Hill.

Groover, M.P. and Zimmers, E.W. (1984) *Computer Aided Design and Manufacturing*, Prentice-Hall.

Hordeski, M. (1988) *Computer Integrated Manufacturing*, TAB.

Lhote, F., Kauffmann, J.-M., André, P. and Taillard, J.-P. (1984) *Robot Components and Systems*, Kogan Page.

Redford, A. and Lo, E. (1986) *Robots in Assembly*, Open University Press.

Rhodes, E. and Wield, D. (eds) (1985) *Implementing New Technologies*, Basil Blackwell.

Rushton, D.F.H., Hodkinson, G.D., Broughton, T. and Winstone, J. (1986) *Guide to Manufacturing Strategy*, Institution of Production Engineers.

Senker, P. (1986) *Towards the Automated Factory? The Need for Training*, IFS.

Synder, W.E. (1985) *Industrial Robots: Computer Interfacing and Control*, Prentice-Hall.

Taguchi, G. (1986) *Introduction to Quality Engineering*, American Supplier Institute.

Thring, M.W. (1983) *Robots and Telechirs*, Ellis Horwood.

Todd, D.J. (1985) *Walking Machines – An Introduction to Legged Robots*, Kogan Page.

Warnock, I.G. (1988) *Programmable Controllers*, Prentice-Hall.

Drives

BOOKS

Bradley, D.A. (1987) *Power Electronics*, Van Nostrand Reinhold.

Dorf, R.C. (1989) *Modern Control Systems*, Addison-Wesley.

Joucomatic (1986) *Industrial Pneumatics*, Joucomatic/Techno-Nathan, La Nouvelle Libraire, Paris.

Kenjo, T. (1984) *Stepping Motors and their Microprocessor Controls*, Oxford Scientific.

Sauer, H. (1986) *Modern Replay Technology*, Alfred Huethig.

Shepherd, W. and Hulley, L.N. (1987) *Power Electronics and Motor Control*, Cambridge University Press.

Tonyan, M.J. (1985) *Electronically Controlled Proportional Valves – Selection and Application*, Marcel Dekker.

PAPERS

Chapple, P.J. and Dorey, R.E. (1986) The performance comparison of hydrostatic piston motors – factors affecting their performance and use. Proceedings of the 7th international symposium on fluid power, Bath, BHRA, 1–7.

Editorial (1988) Fast drives position loads accurately. *Eureka*, Innopress, September, 90–2.

Maskrey, R.H. and Thayer, W.J. (1978) A brief history of electrohydraulic servo-mechanisms. Moog Inc. Controls Division. Originally in *ASME Journal of Dynamic Systems Measurement and Control*, June 1978.

Park, E.D. and Gat, E. (1986) Fibre optic commutation system for brushless DC motors. Proceedings of the 9th International motor conference, Boston, 53–9.

Vaughan, N.D. and Whiting, I.M. (1986) Microprocessor control applied to a non-linear electrohydraulic position servo system. Proceedings of the 7th international symposium on fluid power, Bath, BHRA, 187–97.

Signal processing

Brook, D. and Wynn, R.J. (1988) *Signal Processing: Principles and Applications*, Edward Arnold.

Connor, F.R. (1982) *Noise*, Edward Arnold.

Lam, H.Y.-F. (1979) *Analogue and Digital Filters: Design and Realisation*, Prentice-Hall.

Lynn, P.A. (1982) *An Introduction to the Analysis and Processing of Signals*, MacMillan.

Meade, M.L. and Dixon, C.R. (1986) *Signals and Systems: Models and Behaviour*, Van Nostrand Reinhold.

Terrell, T. J. (1980) *Introduction to Digital Filters*, MacMillan.

Engineering design

Andreasen, M. and Hein, L. (1985) *Integrated Product Development*, IFS Publications.

Dieter, G.E. (1986) *Engineering Design: A Materials and Processing Approach*, McGraw-Hill.

Flurscheim, C. H. (ed.) (1983) *Industrial Design in Engineering: A Marrige of Techniques*, Design Council.

French, M.J. (1985) *Conceptual Design for Engineers*, Design Council.

Irons, B. and Ahmad, S. (1980) *Techniques of Finite Element Analysis*, Ellis Horwood.

Kalpakjian, S. (1984) *Manufacturing Processes for Engineering Materials*, Addison-Wesley.

Matousek, R. (1973) *Engineering Design: A Systematic Approach*, Blackie.

Ryder, G.H. (1979) *Strength of Materials*, MacMillan.

Tse, F.S., Morse, I.E. and Hinkle, R.T. (1978) *Mechanical Vibrations: Theory and Practice*, Allyn and Bacon.

Index